INTEGRATING HUMAN ASPECTS IN PRODUCTION MANAGEMENT

IFIP – The International Conference for Information Processing

IFIP is a non-governmental, non-profit umbrella organization for national societies working in the field of information processing. It was established in 1960 under the auspices of UNESCO as an aftermath of the first World Computer Congress held in Paris in 1959. Today, IFIP has several types of Members and maintains friendly connections to specialized agencies of the UN system and non-governmental organizations. Technical work, which is the heart of IFIP's activity, is managed by a series of Technical Committees.

IFIP's mission is to be the leading, truly international, apolitical organization which encourages and assists in the development, exploitation and application of Information Technology for the benefit of all people. Its principal elements include:

1. To stimulate, encourage and participate in research, development and application of Information Technology (IT) and to foster international co-operation in these activities.
2. To provide a meeting place where national IT Societies can discuss and plan courses of action on issues in our field which are of international significance and thereby to forge increasingly strong links between them and with IFIP.
3. To promote international co-operation directly and through national IT Societies in a free environment between individuals, national and international governmental bodies and kindred scientific and professional organizations.
4. To pay special attention to the needs of developing countries and to assist them in appropriate ways to secure the optimum benefit from the application of IT.
5. To promote professionalism, incorporating high standards of ethics and conduct, among all IT practitioners.
6. To provide a forum for assessing the social consequences of IT applications; to campaign for the safe and beneficial development and use of IT and the protection of people from abuse through its improper application.
7. To foster and facilitate co-operation between academics, the IT industry and governmental bodies and to seek to represent the interest of users.
8. To provide a vehicle for work on the international aspects of IT development and application including the necessary preparatory work for the generation of international standards.
9. To contribute to the formulation of the education and training needed by IT practitioners, users and the public at large.

IFIP's principal aims were and are to foster international cooperation, to stimulate research, development and applications and to encourage education and the dissemination and exchange of information on all aspects of computing and communication. IFIP's creation was well timed. In the 1960s there began a veritable explosion in the growth of the computer industry and in the application of its products. Within the life-span of IFIP information technology (as it is widely known today) has become a potent instrument affecting people in everything from their education and work to their leisure and in their homes. It is a powerful tool in science and engineering, in commerce and industry, in education and administration and in entertainment.

Further information on IFIP can be found at the following URL: http://www.ifip.or.at/

INTEGRATING HUMAN ASPECTS IN PRODUCTION MANAGEMENT

IFIP TC5 / WG5.7 Proceedings of the International Conference on Human Aspects in Production Management
5-9 October 2003, Karlsruhe, Germany.

edited by

Gert Zülch
University of Karlsruhe
Germany

Harinder S. Jagdev
University of Manchester Institute of Science and Technology (UMIST)
Manchester, United Kingdom

Patricia Stock
University of Karlsruhe
Germany

Gert Zülch
Institute für
Arbeitswissenschaft und
Betriebsorganisation
Universität Karlsruhe (TH)
Postfach 6980
76128 Karlsruhe
Germany

Harinder S. Jagdev
UMIST
P.O. Box 88
Manchester M60 1QD
United Kingdom

Email:
Hjagdev@umist.ac.uk

Patricia Stock
Ifab-Institute of Human
and Industrial Engineering
University of Karlsruhe
Kaiserstr. 12
76128 Karlsruhe
Germany

Tel: +49-721-608-4839
Fax: +49-721-608-7935
EMAIL:
Patricia.stock@ifab.uni-karlsruhe.de

A C.I.P. Catalogue record for this book is available from the Library of Congress.

INTEGRATING HUMAN ASPECTS IN PRODUCTION MANAGEMENT
Edited by Gert Zülch, Harinder S. Jagdev, and Patricia Stock
p.cm. (The International Federation for Information Processing)

ISBN: 1-4899-9740-7 / (eBOOK) 0-387-23078-5 Printed on acid-free paper.

9 8 7 6 5 4 3 2 1 SPIN 11319832 (HC) / 11320890 (eBook)
springeronline.com

CONTENTS

PART THREE – Human Aspects in Production Planning & Control

PART FOUR – Knowledge Management

PART FIVE – Management of Distributed Work

PART SIX – Service Engineering

Preface

In recent years the situation of production enterprises has been aggravated by the change from a vendors' market to a buyers' market, the globalisation of competition, a severe market segmentation and rapid progress in product and process technologies. Beside cost and quality, time has taken on an increasingly important role, forcing enterprises to become ever more dynamic and versatile. Therefore, in all areas of production management, novel, effective concepts, procedures and tools have been developed in order to meet these new requirements.

But beyond these more technical, organisational and information technology related aspects there is certainly another one which has to be considered more closely than ever before, namely that of human resources. Is not group technology also related to group work? Do partners in a global network only operate according to predefined process schemes with no personal contact? Are the mental process models of the programmers of ERP-systems the same as those of the users? What is the impact of human behaviour and what consequences are to be expected if organisational and individual objectives are separated? And finally, how do necessary technological changes affect the workforce and the individual needs and wishes of the employees.

As a consequence, production management should consider human aspects in greater detail for a better understanding of its double role within production management: Humans are not only regarded as managed resources, as they are looked upon from a traditional perspective. They are also managing resources, not only on the executive level of an enterprise, but in many cases also on the shop floor level, as demonstrated by many examples of continuous improvement teams in industry.

This book brings together the opinions of a number of leading experts, analysts, academics, researchers, vendors and industrial practitioners from around the world who have been engaged extensively in integrating human aspects in production management. Through individual chapters in this book, authors put forward their views, approaches and new tools. Still, other authors present a glimpse of the nature of solutions that may be developed in the near future.

This book is loosely structured to allow chapters which address common themes to be grouped together. In these chapters, the reader will learn key issues which are currently being addressed in production management research and practice throughout the world. In short, this book presents some of the latest thinking and solutions for integrating human aspects into industrial practice. The book is composed of six parts, each focusing on a specific theme:

- Human Resource Planning,
- Human Aspects in the Digital Factory,
- Human Aspects in Production Planning & Control,
- Knowledge Management,

- Management of Distributed Work, and
- Service Engineering.

The oral versions of the included papers were presented at the International Working Conference *"Human Aspects in Production Management"*, held in Karlsruhe, Germany, on 5th through 9th October 2003. Following this conference, the papers have been extended by the authors and passed a peer review process. The conference was supported by the International Federation of Information Processing (IFIP) and was organised by its Working Group 5.7 "Integration in Production Management". The conference was hosted by the ifab-Institute of Human and Industrial Engineering of the University of Karlsruhe.

May 2004

Gert Zülch (*University of Karlsruhe, Germany*)
Harinder Singh Jagdev (*UMIST, Manchester, UK*)
Patricia Stock (*University of Karlsruhe, Germany*)

PART ONE

Human Resource Planning

Modelling Human Systems in Support of Process Engineering

Joseph O. Ajaefobi and Richard H. Weston
Loughborough University, MSI Research Institute, Leics LE 11 3TU, Loughborough, United Kingdom.
Email: {j.o.ajaefobi, r.h.weston}@lboro.ac.uk

Abstract: There is a general requirement to resource the business processes of manufacturing enterprises with suitably structured human and technical systems. Further custom and practice is based upon specifying, building, utilizing and developing multiple human and technical systems so that they must be capable of interoperating in a customized way in order to concurrently fulfill the goals of multiple business processes. Various Enterprise Modelling techniques have been developed in recent decades which offer support for enterprise design and thereby help specify system requirements and solutions. However those techniques are generally known to be deficient in the support they provide for human systems engineering. This paper proposes means of characterizing the competencies and capacities of human resources, with reference to strategic, tactical and operational aspects of business processes. Also it explains how these 'models' of human resource can be deployed within the wider context of Enterprise Modelling to match the abilities and behaviors of stereotypical human systems to specific business process requirements.

Key words: Enterprise modelling, Business processes, Human systems engineering, Competencies, Capacities

1. INTRODUCTION

According to ISO 14258 (British Standards Institute 1998), three classes of activity are associated with plan/build, use/operate, and recycle/dispose life-phases of a Manufacturing Enterprise (ME). These classes of activity

1. decide *what* the ME should do (so called *W* activities),

2. decide *how* the ME should operate (so called *H* activities) and
3. *do* those activities needed to realize products and services for customers (so called *D* activities), in conformance with *W* and *H* activity outcomes.

It follows that the business processes (BPs) of manufacturing enterprises (Mes) should comprise needed groupings of *W*, *H* and *D* enterprise activities.

It is also noted that invariably MEs realize their BPs by deploying a number of human and technical systems, the purpose of which is to structure and support the use of human and technical resources. Typically in any given ME, the systems deployed need to interoperate in order to concurrently fulfill multiple BP goals in a timely and cost effective manner (WESTON 1999). The human systems (HSs) deployed commonly comprise structured groupings of people that may be supported by technology, and by so doing form ad hoc groups, teams, business units, departments or the complete ME.

Thus a general requirement to design a complex set of human and technical systems arises, so that in the short term MEs operate innovatively and competitively, and in the longer term MEs can reform and recompose themselves as product markets and environmental conditions change. A consequent need arises to match the abilities, capacities and availabilities of human and technical resources to functional and behavioral requirements of enterprise activities, and to do this sufficient knowledge of wider BP requirements in a given ME is needed so that valuable resources are appropriately and efficiently assigned and utilized.

Public domain enterprise Modelling EM concepts, architectures and methodologies such as *CIMOSA* (AMICE 1993), *GRAI* (CHEN, DOUMEINGTS 1996) *IEM* (VERNADAT 1996), *PERA* (WILLIAMS 1996) are known to provide means of decomposing, specifying, communicating, analyzing, visualizing and integrating complex enterprise requirements and thereby provide concepts to support the design of complex, interoperating resource systems. However, both theoretical and practical limitations have been observed in respect of the current solution provisions of EM concepts, particularly with respect to characterizing human systems and their potential roles in an enterprise (KOSANKE 2003; WESTON et al. 2003). Therefore this paper proposes and illustrates means of characterizing human resources, such that the potential of EM may be enhanced.

2. AN OVERVIEW OF THE MEANING OF COMPETENCY

ME processes (and their *W*, *H* and *D* elemental activities) interoperate to deliver products or services to specified customers within or outside the ME. These products or services are the observable measurable outcomes that can be judged in terms of their degree of excellence, timeliness and quantity. HSs are frequently deployed in association with other resources to realize process instances. An individual or a team is said to be competent with respect to any activity or process (or instance[1] of these) if they satisfy prescribed requirements in a timely and cost effective manner. A competent individual (or team) can be trusted and relied upon to own an activity or process (i.e. by assigned responsibility) and to deliver the required products or services, other things being equal (AJAEFOBI 2004).

To understand the term competency, the following independent variables need to be kept in view: the *activity* (or process) to be executed competently; the *required output* (or activity performance) to be delivered at the *desired level of competency*; the *HSs* (and their elements) that come to the work with their *available competencies*. Bearing in mind the need to interrelate these variables within any ME context, the authors define the term "competency" as follows:

> That property possessed by a HS that is willingly brought to bear on work, resulting in the effective and efficient delivery of the prescribed work outcomes.

Such a property possessed may encompass: *natural traits* (underlying traits), *acquired competency* (skills, experiences, education, trainings etc), *adaptation competency* (ability to adapt to new challenges and changes in job), and *performing competency* (i.e. ability to achieve an observable output). This view of competency builds upon observations of GARETH (1997) and thereby views competency from perspectives of: *being* (underlying traits), *knowing* (acquired skills, knowledge, experience) and *doing* (action or behavior that authenticates the other two).

The above definition also unifies two widely published but complementary views of competencies: namely

1. the input approach subscribed to by many US authors (BOYATZIS 1982) and
2. the output approach favored more in Europe (BROPHY, KIELY 2002).

Such a unification is required in MEs where it is necessary to define: *Required Competencies* (competencies necessary to achieve a purpose, mission, goal, objective, process, task or an activity) and *Available Competencies* that can be brought to bear to accomplish the required purpose, mission, etc (HARZALLAH, VERNADAT 2002). HSs especially teams can be classified with respect to the type of activities they execute as

[1] Activity process execution has a defined "start", "body" and "end". Activities and processes may need to execute many times. Each time execution occurs this is referred to as an 'instance'

strategic, tactical and operational teams (BYER, WESTON 2001). It follows that associated sets of competencies at strategic, tactical and operational levels of abstraction are required to match people to *W*, *H*, *D* activities (and their parent processes) in MEs.

3. MODELLING FRAMEWORK TO MATCH CAPABILITIES OF HUMAN SYSTEMS TO REQUIREMENTS OF ENGINEERING PROCESSES

The MSI Research Institute, which operates out of the Wolfson School at Loughborough University (UK) has developed and tested an EM approach known as Multiple Process Modelling (*MPM*) method (CHATHA et al. 2003). Essentially, *MPM* lends structure to the use of state-of-art public domain EM approaches by organizing:

1. the ongoing capture of a coherent and semantically rich picture of dependent processes, in such a way that key dependencies can explicitly be represented during the life time of dependent processes and
2. the reuse of the multiple process models and their modeled dependencies for different enterprise engineering purposes.

Enterprise engineering projects can be supported by *MPM* through a number of stages of model development and deployment. The different types of model created, developed and deployed is illustrated conceptually in Figure 1. Enterprise modelers and process and system engineers can move iteratively through modelling stages, developing and reusing model elements coherently as needed. *MPM* builds upon previously proven modelling concepts; this enables the capture of static, multi-perspective requirement models of a specified enterprise as well as means of creating dynamic requirement models, so that behaviors of candidate systems can be matched to defined requirements.

The first four modelling stages depicted by Figure 1 correspond to static model development. In *MPM*, *CIMOSA* concepts (AMICE 1993) have provided a backbone of modelling constructs onto which other new and previously established modelling concepts and constructs have (and can progressively be) connected. The backbone also allows model fragments (be they related to enterprise processes or enterprise systems) to be interpreted within a well-defined context. This content dependency was established because *CIMOSA* domain process, business process, enterprise activity and functional operation decomposition principles have been adopted and as appropriate that decomposition is reinforced by new *W*, *H* and *D* decomposition principles.

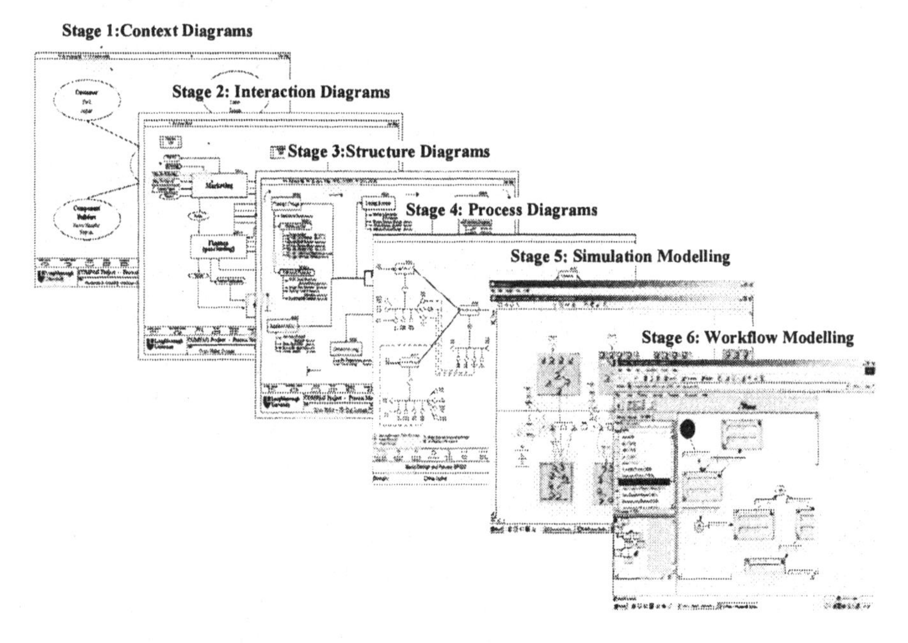

Figure 1. MPM modelling stages and formalisms

In the study reported in this paper the authors have exploited the eclectic capabilities of *MPM* by extending its framework to include means of specifying, analysing and attributing competency constructs to ME processes, and by so doing have enabled the required competencies of elemental activities (that constitute a given process) to be formally described.

Further, it was proposed that all *W*, *H*, and *D* activity types deployed by MEs can be represented in an abstract manner via the use of four semi-generic activity types (that can be particularized to suit specifics of any W, H, and D context), so that groupings of activity can be resourced at suitable levels of modelling abstraction by a limited number of different classes of human system competency. The competency constructs conceived for such a purpose were designated CCL1 (Competency Class 1), CCL2, CCL3, and CCL4 and corresponding respectively to "simple operation", "skilled operation", "tactical operation" and "strategic operation" aspects. Table 1 illustrates examples of activities and their competency requirements as proposed. It was observed that HSs can posses these competency types as they interact with one another and competency possessed by other enterprise resource systems namely: technological (e.g. CNC machines and software applications), structural (e.g. operating procedures and shared objectives) and transitional elements (e.g. information entities and process states) used to accomplish enterprise processes that collectively achieve defined goals.

Table 1. Semi-generic competency classes and matching activity types

Competency Class	Enterprise activity types
CCL 1	Competency to execute defined set of general operations based on specified methods, procedures and order. Here activities are essentially routine and results predictable
CCL 2	Competency to understand, interpret and implement concepts, designs, and operation plans linked to specific product realization and to apply them in solving practical problems e.g. system installation, operation and maintenance.
CCL 3	Competency needed to translate abstract concepts into shared realities in the form of product designs, process specifications, operation procedures, budgeting and resource specifications
CCL 4	Competency needed to formulating high level business goals, mission, policies, strategies and innovative ideas

A systematic approach to process resourcing was conceived in which available human competencies are matched to enterprise activity requirements at needed abstraction levels. This approach attributes modelling constructs (used to define competencies required from human resource systems) to models of enterprise activities and their associated functional operations[2], in the manner shown conceptually by Figure 2.

By such means semantically enriched models of processes can be captured, so as to more completely define process requirements. The addition of the competency modelling capability, and associated systemic approaches to matching required and available resource systems has been termed *E-MPM* (Enhanced-Multi Process Modelling method) because, essentially at modelling stage 4 of Figure 1, it builds upon and extends the modelling constructs incorporated into its predecessor *MPM*.

[2] Activities are carried out by human and technical resources. Resources possess functional capabilities (often termed competencies in the case of human resources). Sometimes activities are described in terms of more elemental functional operations, the needed capabilities of which can be matched to functional capabilities possessed by resources.

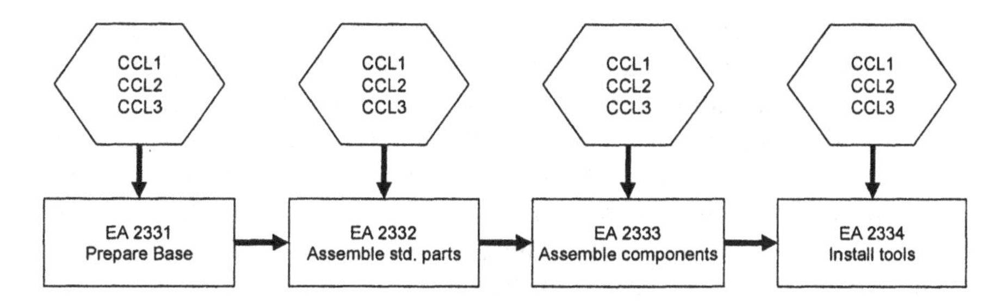

Figure 2. Conceptual illustration of the "attachment" of required competencies to enterprise activities[3]

E-MPM therefore integrates the use of *MPM, CIMOSA* and new HS concepts, constructs and modelling templates in such a way that complex engineering processes can naturally be grouped into organized groupings of enterprise activities (EAs) and functional operations (FOs), for which competency requirements can be formally specified. Also as an integral part of *E-MPM* a Hierarchical Task Analysis (HTA) technique (SHEPHERD 2000) can be applied to further decompose functional operations into elemental operations (things to do) and plans (how to do them). By such means needed human competencies can be encoded at four levels of abstraction corresponding to generic competencies required to satisfy general operational, skilled operational, tactical, and strategic activity needs, also illustrated by Figure 2 (see footnote). This naturally leads to an enriched graphical representation of business process (BP) models that supports a first stage matching of available HS competencies to BP competency requirements. The enterprise activities illustrated by Figure 2 are a small subset of activities that belong to a multi-process model of a case study company encoded using the first four modelling stages of MPM, as described in greater detail in AJAEFOBI (2004). This particular case study company is a vendor of production machines and systems. Primarily they build these machines and systems in conformance with requirements specified by customer contracts (or orders).

This first stage of resource matching is illustrated conceptually by Figure 3 where specified classes of CCL requirement (attributed to processes, activities and functional operations[4]) are matched to CCL abilities possessed

[3] The nature of this particular process is such that in this case needed strategic competencies (normally attributed to CCL4) can be covered by aspects of tactical and operational competencies. In Figure 2 the abbreviation EA is used to denote an Enterprise Activity.

[4] The reader should note that processes comprise more elemental sub-processes and activities, while activities comprise more elemental functional operations. These levels of process decomposition facilitate matching of processes to needed resources at multiple levels of abstraction.

by candidate human resource systems, thereby enabling selection from amongst candidate systems to determine those that are viable candidates and those that are not. During this first matching stage therefore center of attention is on deciding whether HSs have the ability to do the jobs at hand. However at this stage no consideration is given to secondary concerns about the capacity of suitable resources to achieve activities in finite time frames. Nor at this stage is attention paid to the possibility that activity requirements might be concurrent or be subject to change.

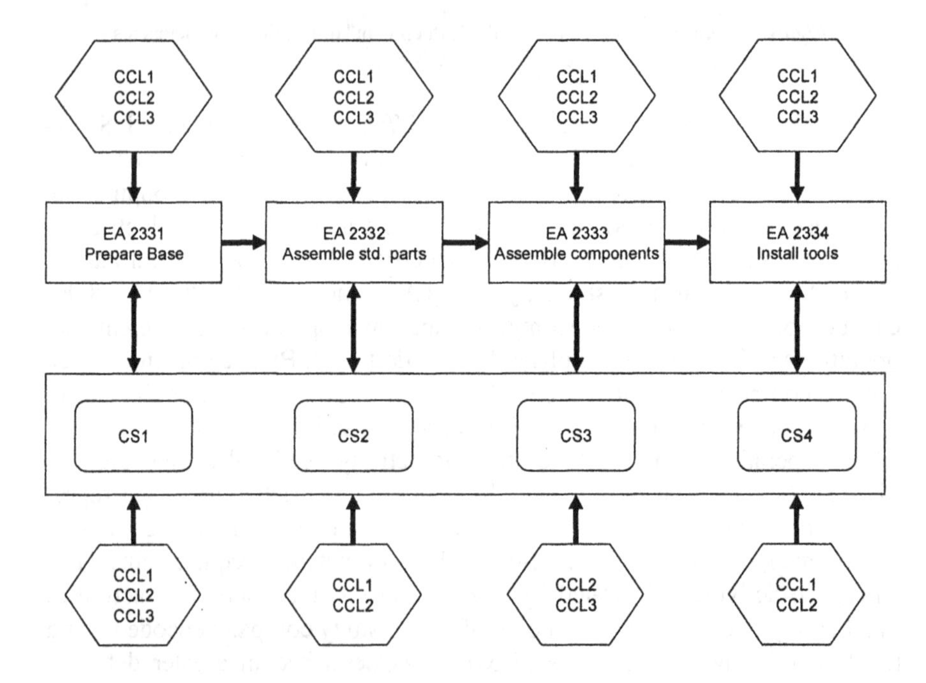

Figure 3. Conceptual illustration of stage 1 "Competency-Based Selection of Candidate Systems" (CS) – see also footnote 3

In general, it is envisaged that changes in job requirements will occur in two possible ways, namely

1. planned and anticipated changes and
2. unplanned and unanticipated changes.

It was observed that changes in job requirements (e.g. process specifications) can impact significantly on the adequacy of available human competencies. Therefore a second stage of matching was observed as being needed, during which competency behaviors could be simulated dynamically with a view to addressing issues like:

a) How can viable candidate systems react to changes in job competency requirements?
b) Can competencies of the viable candidates handle effectively current and predicted requirements?

Hence the simulation modelling capabilities of *MPM* were also enhanced at its modelling stages, so as to create *E-MPM* which supports a second stage of human resource matching based upon dynamic (process and system) modelling.

During this second stage of human resource system selection, computer executable models of activity requirements and candidate human resources are recoded, using modelling constructs provided by the *ithink* dynamic systems simulation tool (ITHINK 2002). To encode "capacity" information in a generic and reusable manner, HS workload modelling concepts were conceived.

4. WORKLOAD CONCEPT IN MATCHING TASKS TO HUMAN SYSTEM

It was determined that workload (WL) modelling concepts and constructs needed to be deployed to enable dynamic assessment to be made about the competencies and capacities of human resources, with respect to their assigned roles under different conditions of load. In the example case study company referred to in the previous section, it was necessary to consider workload (WL) constraints arising from concurrent instances of the design and build processes used to make manufacturing machines for various customers. A given HS may be competent to undertake one or more operations but the imposition of WL constraints can impair the effectiveness and timeliness of needed activity and process outputs. Literature on WL was reviewed with focus on a consideration of mental WL concerns, and particularly upon total attentional demands on HSs of an assigned activity or operation (ALDRICH et al. 1989; NORTH and RILEY 1989; WICKENS 1984).

Based on the literature reviewed, a set of WL constructs was defined to characterize enterprise activities in the following respects: *operational criticality, OC* (the significance the operation to entire process completion), *operational demands, OD* (sensory modalities and conflicts that may occur during execution), *operational uncertainty, OU* (availability and accessibility of needed information), *operational precision, OP* and *temporal demand, Tr/Ta*. Subjective ratings of 1.0 to 7.0 were attached to these WL constructs, with 7.0 denoting very high pressure and a likelihood that performance break down might occur. Humans are capable of handling operations requiring concurrent responses in accordance with a "multiple resource" theory

(WICKENS 1984) but greater risk of performance breakdown occurs when the use of common sensory modalities arise (NORTH, RILEY 1989) such as where there is simultaneous demand in respect to visual operations. Knowledge of WL issues was found to support the design of and selection between alternative candidate HS, on grounds of their capacity to do different classes of enterprise activity.

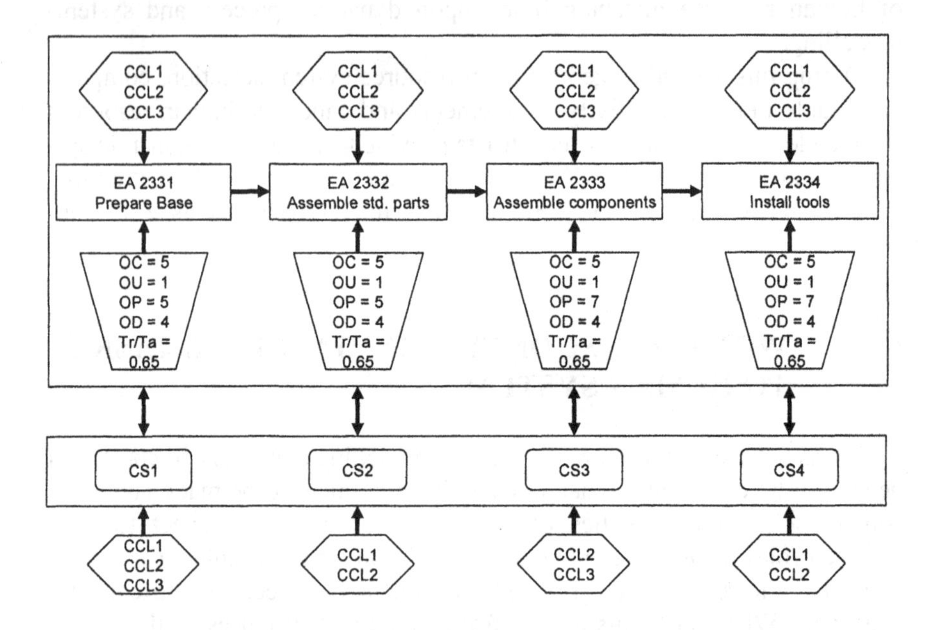

Figure 4. Conceptual illustration of the attachment of workload modelling constructs to enterprise activities[5]
(Source: AJAEFOBI, WESTON 49)

Figure 4 illustrates conceptually how WL modelling constructs were attached to graphical process models. As mentioned previously at the fifth modelling stage these enriched ME process models are recoded using the *ithink* tool. Figure 5 illustrates such an *ithink* model, which is a simulation model, corresponding to earlier models shown in Figures 2 and 3.

Whereas Figure 6 shows one of many forms of predicted performance information which can be generated during selection stage 2, so as to enable the dynamic performance of viable human system candidates (selected out during selection stage 1) to be contrasted and compared, primarily on grounds of their timeliness and/or cost effectiveness.

[5] The additional graphical modeling constructs attached to enterprise activities illustrates the ratings (on a scale of 1 to 7) attributed to workload (WL) constructs.

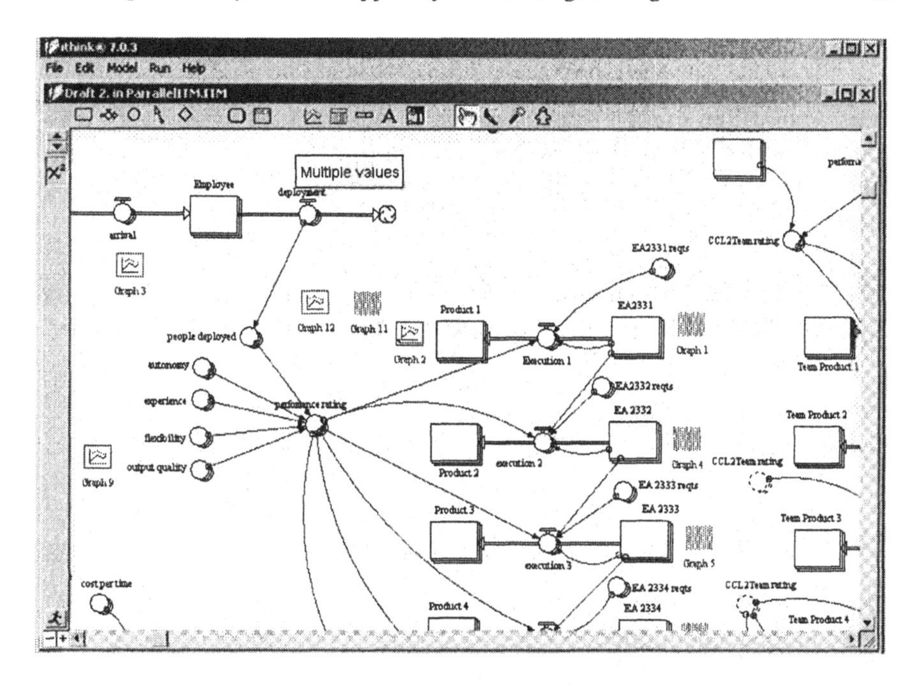

Figure 5. Illustrative extract from a semi-generic ithink model created to support second stage HS selection on grounds of dynamic performance measures

ithink was selected as the simulation tool because its underlying approach to modelling system dynamics characterizes complex systems as a set of interdependent differential equations. These equations are created and parameterized in a fairly user friendly way, via graphical modelling constructs. The equations are solved when simulations are run using a Numerical Integration technique and a set of input/output devices.

Of course other types of simulation tool could have been selected and deployed to fulfill requirements of the fifth stage of *MPM/E-MPM* modelling. But bearing in mind the need to quantify process lead times and costs when deploying alternative candidate resource systems, use of the *ithink* tool was found to provide appropriate support for decision making.

5. CONCLUSION

The paper describes in outline how a multiple process modelling method was enriched via the provisions of HS modelling concepts. It also explains how the enriched modelling method facilitates a two stage matching of HSs in terms of static competency modelling and dynamic capacity modelling. The modelling facility so developed has been shown to be capable of supporting business process engineering, by comparing alternative ways of resourcing activities and operations from both lead time and cost perspectives. Application of the competency and capacity modelling concepts, constructs and systemic approaches has been outlined when resourcing complex engineering activities in an example build-to-order (BTO) manufacturing business.

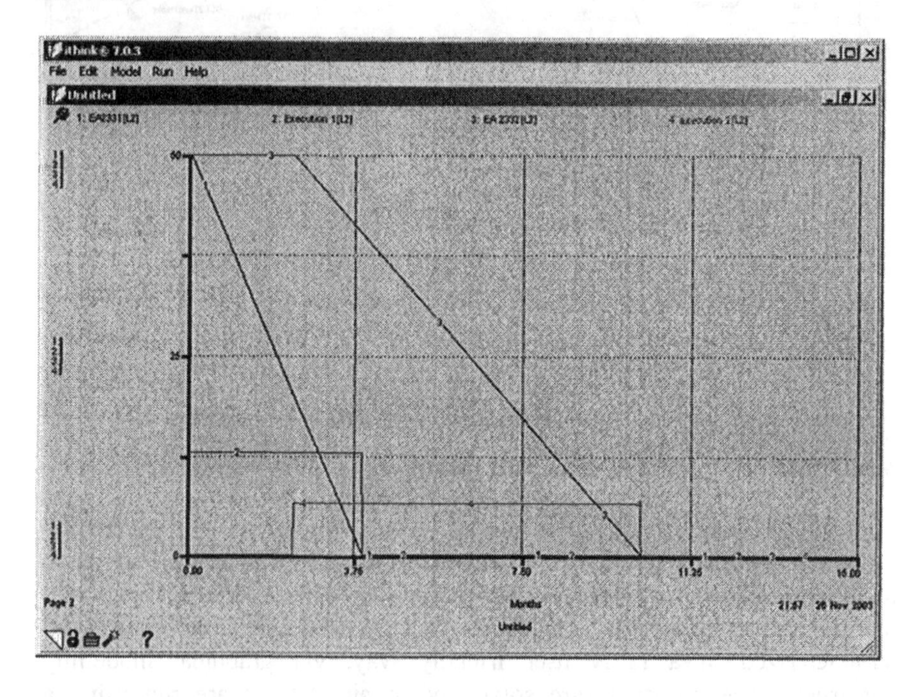

Figure 6. Illustrative dynamic performance information, predicted by running the semi-generic ithink model with specific HS competency and capacity model attributes assigned

AJAEFOBI (2004) describes in greater detail how HS modelling concepts reported in this paper have been used to enhance *MPM* and thereby enable improved matching of specified groupings of engineering activities to organized systems of available human resources. It has been observed that two stage selection can be made between candidate individuals, teams and workgroups, on joint grounds of competency and workload capacity to realize requirements within defined cost and time constraints. Further, at stage six of *MPM/E-MPM* application, off-the-shelf workflow technology can be deployed to enact computer executable models of the process flows created at earlier modelling stages. Thereby specification and management of BTO (and other) workflows can be enabled, so that programmable organizational structure is overlaid onto the collective working of structured units of engineering personnel.

REFERENCES

AJAEFOBI, J.O.:
Human Systems Modeling in Support of Enhanced Process Realisation.
Loughborough, University PhD thesis, 2004.

AJAEFOBI, Joseph O.; WESTON, Richard H.:
An Approach to Modelling and Matching Human System Competencies to Engineering Processes.
In: Human Aspects in Production Management.
Eds.: ZÜLCH, Gert; STOWASSER, Sascha; JAGDEV, Harinder S.
Aachen: Shaker Verlag, 2003, pp. 46-52.
(esim – European Series in Industrial Management, Volume 5)

ALDRICH, T. B.; SZABO, S. M.; BIERBAUM, C. R.:
The Development and Application of Models to Predict Operator Workload.
In: Applications of Human Performance Models to System Design.
Eds.: McMILLAN, G. R. et al.
London: Plenum Press 1989, pp. 65-80.

BOYATZIS, R. E.:
The Competent Manager: a model for effective performance.
New York, NY: Willey, 1982.

BROPHY, M.; KIELY, T.:
Competencies: a new sector.
In: Journal of European Industrial Training,
Bradford, 26(2002)2, pp. 165-176.

CHATHA, K. A.; WESTON, R. H.; MONFARED, R. P.:
An Approach to Modelling Dependencies Linking Engineering Processes.
In: Journal of Engineering Manufacture,
London, 217(2003)B5, pp. 669-687.

CHEN, D; DOUMEINGTS, G.:
The GRAI-GIM reference model, archtecture and methodology.
In: Architectures for Enterprise Integration.
Eds.: BERNUS, P. et al .
London et al : Chapman & Hall, 1996 pp.103-126.

CIMOSA:
Open System Architecture for CIM..
Ed.: AMICE-ESPRIT Consortium.
Berlin: Springer Verlag, 2nd edition, Volume 1, 1993.

GARETH, R.:
Recruitment And Selection A Competency Approach.
London: Chartered Institute of Personnel and Development (CIPD), 1997.

HARZALLAH, M.; VERNADAT, F.:
IT-based competency modelling and management: from theory to practice in enterprise engineering and operations.
In: Computers in Industry,
Amsterdam, 48(2002)2, pp. 157-179.

ISO 14258:
Industrial automation systems - Concepts and rules for enterprise models.
September 1998.

ITHINK: ithink® Version 7.
Lebanon: High Performance Systems Inc., 2002.

KOSANKE, K.:
Standardisation in Enterprise inter and intra Organisational Integration.
In: Proceedings of 10th Int. Soc. for Productivity Enhancement international conference on Concurrent Engineering: The Vision for the Future Generation in Research and Applications.
Eds.: JARDIM-GONCALVES, R. et al.
Lisse: Swets & Zeitlinger, 2003, pp. 873-878.
NORTH, R. A.; RILEY, V. A.:
W/INDEX: A Predictive Model of Operator Workload.
In: Applications of Human Performance Models to System Design.
Eds.: McMILLAN, G. R. et al.
London: Plenum Press 1989, pp. 81-89.
SHEPHERD, A.:
Hierarchical Task Analysis.
London: Taylor & Francis, 2000.
VERNADAT, F.:
Enterprise modeling and integration: principles and applications.
London et al.: Chapman & Hall, 1996.
WICKENS, C. D.:
Engineering psychology and human performance.
Columbus, OH: Bell and Howell Company, 1984.
WILLIAM, T. J.:
An Overview of PERA and the Purdue Methodology.
In: Architectures for Enterprise Integration.
Eds.: BERNUS, P. et al.
London et al.: Chapman & Hall, 1996, pp. 129-161.
WESTON, R. H.; BYER, N.; AJAEFOBI, J.O.:
EM in support of team system engineering.
In: Proceedings of 10th Int. Soc. for Productivity Enhancement international conference on Concurrent Engineering: The Vision for the Future Generation in Research and Applications.
Eds.: JARDIM-GONCALVES, R. et al.
Lisse: Swets & Zeitlinger, 2003, pp. 865-872.
WESTON, R. H.:
A Model-Driven, Component-Based Approach to Reconfiguring Manufacturing Software Systems.
In: International Journal of Operations and Production Management,
Bradford, 19(1999)8, pp. 834-855.

Human Aspects of IT-Supported Performance Measurement System

Sai S. Nudurupati and Umit S. Bititci
University of Strathclyde, Centre for Strategic Manufacturing, 75 Montrose Street, Glasgow G1 1XJ, United Kingdom.
Email: sai.nudurupati@strath.ac.uk

Abstract: In the field of performance measurement, a lot of literature is emerging, which largely focuses on designing performance measurement systems, with few studies on issues in implementation and use of performance measurement systems. The objective of this paper is to study the people related issues surrounding the implementation and use of IT-supported performance measurement system (IT-PMS). Action research was chosen as the main methodology and IT-PMS was implemented in three companies. The study has identified different human factors, such as senior management commitment, drive, overcoming resistance, training, etc., that would enable IT-PMS implementation and bring positive impact on management behaviour and business results.

Key words: Performance measurement, Human factors, IT, Management and business implications

1. INTRODUCTION

The performance measurement revolution started in the late 1970s and early 1980s with the dissatisfaction of traditional backward looking accounting systems. Since then, there has been constant development in this field. Most of the focus has been on designing performance measurement systems, with few studies illustrating the issues in implementing and using performance measurement systems (BOURNE 2000; NUDURUPATI, BITITCI 2003a; KENNERLY 2003). According to BITITCI (2002) and HOLLOWAY (2001), there is little solid research evidence that illustrates the impact of performance measurement systems. The objective of this paper is to identify the human aspects (factors) of an IT-supported Performance Measurement System (IT-PMS) and understand how it impacts management behaviour and business results.

2. BACKGROUND

There has been constant development in the field of performance measurement. There are a number of frameworks and models developed for designing performance measures from strategy. However, while implementing these performance measures, usually indicators are poorly defined (SCHNEIDERMAN 1999), which can lead to misunderstanding. Hence, the measures and indicators should be clearly defined (BOURNE et al. 1998, NEELY et al. 1996), understood and communicated. According to BOURNE et al. (2000), MARR et al. (2002) and NUDURUPATI et al. (2000), for implementing each measure, the following tasks are required:

- *Data Creation:* The policies, procedures and systems required to create the data required.
- *Data Collection:* The collection of data at regular intervals.
- *Data Analysis:* The conversion of the collected data into useful information, in the form of trend charts, comparison charts, summary reports, statistical analysis, etc.
- *Information Distribution:* The communication of this information to the right people in the business for assisting decision-making.

According to BITITCI et al. (2000), BOURNE et al. (2000), HUDSON et al. (2001), NEELY (1999), BIERBUSSE et al. (1998), IT plays a major role in performance measurement implementation. However to demonstrate that IT-supported performance measurement system (IT-PMS) will have a positive or negative impact on management and business, it is necessary to design, implement and monitor IT-PMS in organisations. For doing this, in addition to performance measurement, it is also necessary to review the literature on understanding and managing change in the organisations.

According to LEWIN's (1947; 1951) theory of force field-analysis, there are two sets of forces in operation within a social system - one driving the force to operate for a change and the other trying to increase the resisting forces. In order to maintain a successful change, the implementation team either should increase the driving forces or decrease the resisting forces. BURNS and STALKER's (1961) approach highlights the importance of an organisation's ability to adapt to the turbulent environment and includes two contrasting management systems: mechanistic system and organic system.

GALBRAITH (1973) identified that within an organisation, research and development departments have organic systems and production departments have mechanistic systems.

In the late 1980's DUNPHY and STACE (1990) proposed a two dimensional framework based on the scale of change, such as fine tuning, incremental adjustment, modular transformation or corporate transformation, as well as the style of leadership, such as collaborative, consultative, directive or coercive. Two independent researches by PETTIGREW (1990) and CHILD et al. (1987) identified that a contextualist approach is required, which includes three elements:

- the *content* for the change programme,
- the *process* of change programme, and
- the *context* in which changes occur.

ORLIKOWSKI (1996) and MINTZBERG (1987) criticised most of the change approaches for neglecting a distinction between deliberate and emergent change. They argue that emergent change can only be realised in "action".

However, the success of IT-supported performance measurement implementations lie in availability of performance information, as well as how people use and behave with the available performance information (PRAHALAD et al. 2002; ORLIKOWSKI 2000; MARCHAND et al. 2000; DAVENPORT 1997; ECCLES 1991). MEEKINGS (1995) argues that making people use measurements properly not only delivers performance improvement but also becomes a vehicle for a cultural change, which helps liberating the power of the organisation.

Hence, when a performance measurement system is implemented in an organisation, there are several factors (both technical and human factors) that will affect its design, implementation and use. Some of these factors include:

- lack of time and effort required (BOURNE et al 2000; HUDSON et al. 2001),
- lack of IT support (BITITCI et al. 2002; BOURNE et al. 2000; HUDSON et al. 2001),
- lack of accuracy and timeliness of information (HUDSON et al. 2001; NEELY 1999),
- lack of senior management drive and commitment (BITITCI et al. 2002; BOURNE et al. 2000; HUDSON et al. 2001), and
- resistance (BITITCI et al. 2002; BOURNE et al. 2000).

3. METHODOLOGY

However, human factors (extraneous variables) can only be realised in *action* and cannot be anticipated or planned. Hence, the researcher should be *part of the organisation* and understand how and why their actions with IT-PMS in the organisation will bring about change in working practices as well as behaviours of people, groups and organisation (COGHLAN et al. 2001). In order to achieve this and identify the human factors, *action research* was chosen as the main methodology.

IT-PMS was implemented at three companies, ADL (bottling company), HSL (mineral water company) and SLC (label company). Throughout the research, the authors played *a dual role* as facilitators and/or personal observers. At ADL and SLC the authors acted as facilitators and implemented IT-PMS. However, at HSL an internal team within the company had implemented IT-PMS, whereas the authors acted as participant observers. A number of *interviews* were also held with the senior management and middle management teams, which led the authors to understand the *before* and *after* scenarios and validate the results with their personal observations at these three companies.

4. ACTION CASES: IMPLEMENTING AND USING IT-PMS

Prior to implementing IT-PMS in all three companies, data was distributed and scattered throughout the organisation. There were many difficulties and obstacles for implementing and using performance measurement. Since data was available from different sources, a significant amount of time and effort was required to collect, analyse and communicate the information to different people. Together with out of date business priorities and performance measurement (often lagging indicators), this raised questions on accuracy and timeliness of information, as shown in Figure 1. This made the management loose their confidence in the information available. Together with fear of exposure as people were pulled into light, this had initiated resistance from people in implementing performance measurement.

As a result of these obstacles and difficulties for successful implementation and use of performance measurement system, the management at these three companies decided to implement IT-PMS. In all of the three companies, the data collection, analysis and communication were automated to the maximum extent possible. Information (password protected) was made available and accessible through the intranet in near real-time.

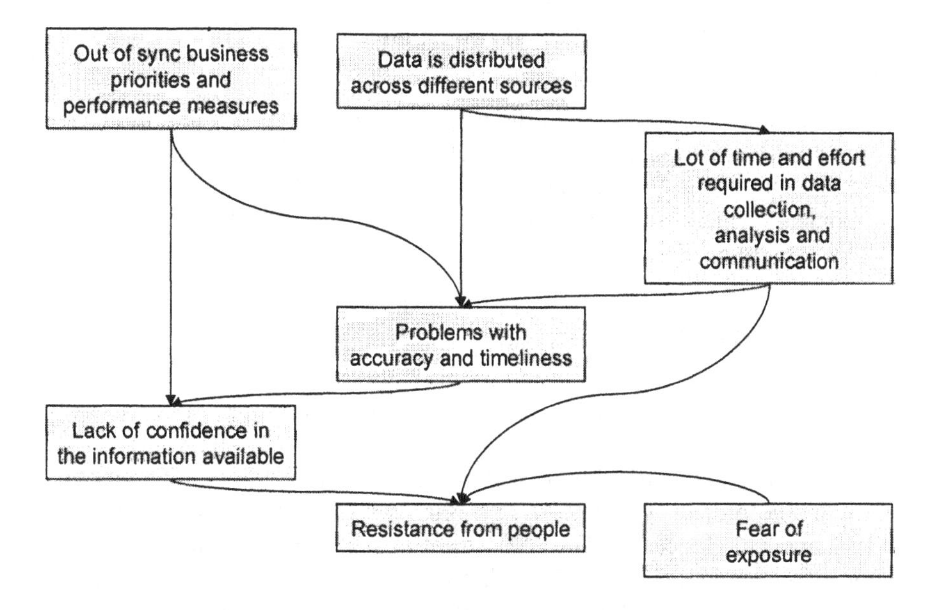

Figure 1. Factors obstructing the IT-PMS implementation (Source: NUDURUPATI, BITITCI 2003b, p. 63)

4.1 Action Case: ADL

At ADL, IT-PMS was only implemented for operational activities of the companies, using in-house technology together with a Statistical Process Control (SPC) charting tool (a software tool used to produce control charts, attribute charts, Pareto analysis, etc.). However, there were significant improvements on quickly identifying the weaknesses of business, facilitating continuous improvement of production lines, as well as facilitating pro-active decision-making. Also, it made only little improvement on encouraging positive behaviour of the people. The significant improvements were due to the following reasons:

- The senior management were committed to the project.
- Management gained confidence in performance information generated by the system in near-real-time.
- The performance results were communicated to everyone in the organisation, especially all team leaders and people on the shop floor.
- People were encouraged for identifying trends and, hence, identify root causes of the problem using continuous improvement techniques.

However, the reasons why the system did not make significant impact on someone are:

- The people on the shop floor did not start using the system because they did not have the facilities (i.e. computers, displays, etc.) to display information on the production lines.
- Some managers lacked vision in terms of what they wanted to gain from this implementation.

4.2 Action Case: HSL

At HSL, IT-PMS was implemented for the whole organisation, with inhouse technology using Cognos reporting tool (NUDURUPATI et al. 2000). Immediately after implementing IT-PMS, there were no improvements as the project was no longer a priority and more important projects were launched. Hence, the IT-PMS was not used. However, the implementation of new production facilities led to productivity problems, requiring senior management's attention to productivity-related performance issues and measures. This led to a senior management drive for proper use of the IT-PMS, resulting in significant improvements by quickly identifying and focusing on the weaknesses of the business, facilitating continuous improvement, facilitating pro-active decision-making and encouraging positive behaviour. The reasons behind this improvement were:

- Senior management was committed to the system and its use. They used it every day to ask questions relating to the performance.
- The management team gained confidence in performance information generated by the system because the data was accurate and timely.
- The performance results were communicated to everyone in the organisation, especially all team leaders and people on the shop floor.
- Senior management was taking the drive in making people use the system as their routine everyday business, making decisions based on facts.
- Senior management encouraged a cross-functional, flexible, team-based working culture.
- Senior management encouraged people to use the system through non-threatening management style and communicated the potential benefits of IT-PMS.
- People were encouraged to identify performance trends and root causes of the problem using continuous improvement techniques.

4.3 Action Case: SLC

At SLC, IT-PMS was only implemented for the operational activities of the company, using in-house technology together with an SPC charting tool. However, there were no improvements in this company because of the following reasons:

- The IT-PMS was implemented only for the operational activities of the company.
- There were technical limitations in the data collected on the shop floor (the automated data capturing system cannot be changed).
- Senior management did not develop confidence in the system due to the limitations in the data capturing and, hence, did not drive the company-wide use of the system.
- Consequently the performance results were not communicated to in the organisation, especially team leaders and people on the shop floor.

5. HUMAN FACTORS OF IT-PMS AND THEIR IMPACT

The human factors identified while implementing and using IT-PMS at these three action case companies are tabulated in Table 1. These factors are also classified, based on the level of significance of their existence in each company.

Table 1. Summary of human factors identified at action cases

Human Factor	ADL	HSL	SLC
1. Senior management commitment	Significant	Significant	Moderate
2. Drive from senior management	Significant	Significant	None
3. Communicate the benefits to people	Significant	Significant	Moderate
4. Usage of the IT-PMS	Significant	Significant	None
5. Proactive decision-making	Significant	Significant	None
6. Overcome resistance from people	Moderate	Significant	None

From these three action cases it was observed that these human factors of IT-PMS would impact management behaviour and business results, as shown in the Table 2. The people and technical factors identified from the individual cases and cross-case analysis would act as enablers between IT-PMS implementation and the management and business implications, as demonstrated in Figure 2.

Table 2. Summary of management and business implications at action cases

Management and Business Implication	ADL	HSL	SLC
1. Identifying weaknesses of the business	Significant	Significant	None
2. Facilitating continuous improvement	Significant	Significant	None
3. Facilitating pro-active decision-making	Significant	Significant	None
4. Improving positive behaviour of the people	Moderate	Significant	None

IT-PMS implementation should ensure that the positive impact on the following technical factors were realized:

- Up-to-date information in real-time is presented
- Data accuracy is ensured
- Information is communicated

If the information is available up-to-date in real-time and is ensured for accuracy, it would create confidence in people. These factors would lead to senior management commitment, and hence they would take the drive into their hands and ensure the following people factors are in place:

- Provide necessary training for people to overcome resistance and make them use IT-PMS
- Communicate perceived benefits to make people realize the importance of IT-PMS
- Overcome resistance from people
- Initiate team culture
- Empower people in making decisions based on the information available from IT-PMS

All these factors lead people to use IT-PMS as a routine part of the business in decision-making. These people factors would result in management and business implications. As the management is confident about the information and use it as a routine part of business, it would lead to a *proactive management style,* which would also lead them in *identifying weak areas of the business.* This together with team culture would result in *continuous improvement.* As the business experiences most of these people factors it would build *positive behaviour* of the people.

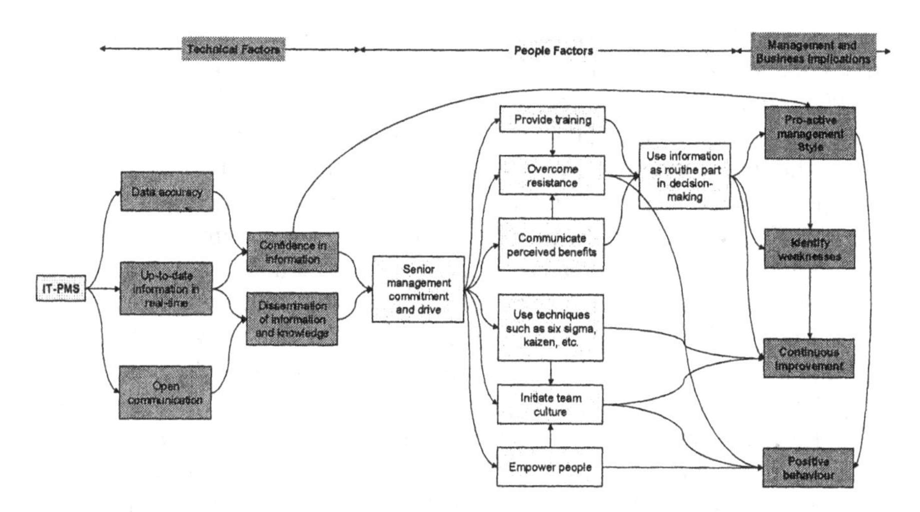

Figure 2. Technical and human factors affecting management and business

6. DISCUSSION

As demonstrated in Tables 1 and 2, at ADL and HSL senior management commitment and active participation (i.e. drive) played a significant role to make the implementation a success. In both these cases the initiative was taken to ensure that the performance information presented was accurate and consistent. Since the information available was accurate, reliable and consistent, management used IT-PMS as part of their daily routine business, identifying trends and making pro-active decisions. With an open and non-threatened management style, they insisted that everyone used the system.

Initially, in both these companies there was some resistance at lower levels (including some management levels) to using IT-PMS. However, to overcome resistance in these companies, the senior management explained and communicated the implications and benefits of using IT-PMS in their management briefings. They provided necessary training on IT-PMS for people who required it. They also insisted that all performance related discussion should be focused around the information presented on the system and not on information available elsewhere. This made the management at lower levels realise the benefits of the system and so they started using IT-PMS in their decision-making.

Most of the managers at ADL are bureaucratic and based their decisions on experience rather than on factual information. Prior to IT-PMS, some of the managers were unfocused and not clear about their objectives. However, the implementation of IT-PMS enabled the managers to augment their experience with accurate and timely performance information based on facts.

Even though the management style at HSL was bureaucratic, when IT-PMS was implemented, the company soon launched another project to change the management structure to a more flat structure. This enabled the self-managed autonomous teams to use the performance information available from IT-PMS in decision-making and drive continuous improvement.

Even though the senior management at SLC were initially committed to the project, they lost confidence in the accuracy of information generated by IT-PMS due to the limitations of data capturing. Hence, they were not committed to the rest of the project. As a result they did not drive the people into using the system. Thus people at lower levels did not bother using the IT-PMS. Therefore, in contrast to ADL and HSL, there was no change in the management style at SLC, as the management did not use IT-PMS in decision-making. The IT-PMS was suspended after six months of its implementation.

7. CONCLUSION

The action research, through these three cases, demonstrated that appropriately designed IT-PMS (by selecting the right set of performance measures), appropriately implemented in a company, would ensure the following technical factors:

- data accuracy,
- up-to-date information presented in real-time, and
- open communication.

Appropriately used (with drive and commitment from senior management) IT-PMS would ensure the following people factors:

- communicating perceived benefits of IT-PMS,
- providing training where necessary,
- overcoming resistance (the above two factors help for this factor),
- making people use IT-PMS in decision-making for the business,
- initiating team culture, and
- empowering people.

These technical and people factors of IT-PMS, in turn, would impact the management behavior and business results as follows:

- increases pro-active management style,
- increases positive behaviour such as focusing on facts, communication, empowerment and teamwork,

- identifies weak areas of the business, and
- promotes continuous improvement.

REFERENCES

BIERBUSSE, P.; SIESFELD, T.:
Measures that Matter.
In: Journal of Strategic Performance Measurement,
Boston, MA, 1(1998)2, pp. 6-11.
BITITCI, U. S; TURNER, T.; BEGEMANN, C.:
Dynamics of Performance Measurement Systems.
In: International Journal of Operations and Production Management,
Bradford, 20(2000)6, pp. 692-704.
BITITCI, U. S.; NUDURUPATI, S. S.; TURNER, T. J.; CREIGHTON, S.:
Web enabled Performance Measurement System: Management Implications.
In: International Journal of Operations and Production Management,
Bradford, 22(2002)11, pp. 1273-1287.
BOURNE, M.; WILCOX, M.:
Translating strategy into action.
In: Manufacturing Engineer,
London 77(1998)3, pp. 109-112.
BOURNE, M.; NEELY, A.:
Why performance measurement interventions succeed and fail.
In: Proceedings of the 2nd International Conference on Performance Measurement.
Ed.: NEELY, A.
Cranfield: Centre for Business Performance, Cranfield University, 2000, pp. 165-173.
BURNS, T.; STALKER, R.:
The Management of Innovation.
London: Tavistock Publications, 1961.
CHILD, T.; SMITH, C.:
The Context and Process of Organisational Transformation.
In: Journal of Management Studies,
Oxford, 24(1987)6, pp. 565-593.
COGHLAN, D.; BRANNICK, T.:
Doing Action Research in Your Own Organisation.
London: Sage Publications, 2001.
DAVENPORT, T. H.:
Information Ecology.
Oxford: Oxford University Press, 1997.
DUNPHY, D.; STACE, D.:
Under New Management: Australian Organizations in Transition.
Sydney: McGraw-Hill, 1990.

ECCLES, R. G.:
The Performance Measurement Manifesto.
In: Harvard Business Review,
Boston, MA, 69(1991)1, pp. 131-137.
GALBRAITH, J. R.:
Designing Complex Organizations.
Reading, MA: Addison-Wesley Publishing, 1973.
HOLLOWAY, J.:
Investigating the impact of performance measurement.
In: International Journal of Business Performance Management,
Leicester, 3(2001)2/3/4, pp. 167-180.
HUDSON M.; SMART A.; BOURNE M.:
Theory and practice in SME performance measurement systems
In: International Journal of Operations and Production Management,
Bradford, 21(2001)8, pp. 1096-1115.
KENNERLY, M.; NEELY, A.:
Measuring performance in changing business environment.
In: International Journal of Operations and Production Management,
Bradford, 23(2003)2, pp. 213-229.
LEWIN, K.:
Frontiers in Group Dynamics.
In: Human Relations, London, 1(1947)1, pp. 5-41.
LEWIN, K.:
Field Theory in Social Science.
London: Harper and Row, 1951.
MARCHAND, D.; KETTINGER, W.; ROLLINS, J.:
Company Performance and IM: the view from the top.
In: Mastering Information Management.
Eds.: MARCHAND, D.; DAVENPORT, T.; DICKSON, T.
London: Financial Times Prentice Hall, 2000, pp. 10-16.
MARR, B.; NEELY, A.:
Balanced Scorecard Software Report.
Stamford, CT: InfoEdge, Two Stamford Landing, 2002.
MEEKINGS. A.:
Unlocking the Potential of Performance Measurement: A Practical Implementation Guide.
In: Public Money & Management,
Oxford, 7(1995)4, pp. 5-12.
MINTZBERG, H.:
Crafting Strategy.
In: Harvard Business Review,
Boston, MA, 65(1987) July-August, pp. 66-75.
NEELY, A.; MILLS, J.; GREGORY, M.; RICHARDS, H.; PLATTS, K.; BOURNE, M.:
Getting the measure of your business.
Cambridge: University of Cambridge, Manufacturing Engineering Group, 1996.
NEELY, A.:
The performance measurement revolution: why now and what next?
In: International Journal of Operations and Production Management,
Bradford, 19(1999)2, pp. 205-228.

NUDURUPATI, S. S.; BITITCI, U. S.:
Review of Performance Management Information Systems (PerforMIS).
Internal Report.
Strathclyde: University of Strathclyde, Centre for Strategic Manufacturing, DMEM, 2000.
NUDURUPATI, S. S.; BITITCI, U. S.:
Implementation and Impact of IT enabled Performance Measurement.
IFIP 3rd International Workshop.
Bergamo, Italy, 19-20 June 2003.
(=2003a)
NUDURUPATI, Sai S.; BITITCI, Umit S.:
Human Aspects of IT-Supported Performance Measurement System.
In: Human Aspects in Production Management.
Eds.: ZÜLCH, Gert; STOWASSER, Sascha; JAGDEV, Harinder S.
Aachen: Shaker Verlag, 2003, pp. 60-67.
(esim – European Series in Industrial Management, Volume 5)
(=2003b)
ORLIKOWSKI, W. J.:
Improvising Organisational Transformation Over Time: A Situated Change Perspective.
In: Information Systems Research,
Linthicum, MD, 7(1996)1, pp. 63-92.
PETTIGREW, A.:
Longitudinal Field Research on Change: Theory and Practice.
In: Organization Science,
Linthicum, MD, 1(1990)3, pp. 267-292.
PRAHALAD, C. K.; KRISHNAN, M. S.:
The Dynamic Synchronisation of Strategy and Information Technology.
In: MIT Sloan Management Review,
Cambridge, MA, 2(2002)Summer, pp. 24-33.
SCHNEIDERMAN, A. M.:
Why Balanced Scorecards Fail.
In: Journal of Strategic Performance Measurement,
Boston, MA, 2(1999), pp. 6-11.

Competence and Preference-based Workplace Assignment

An Application of a Weighted, Non-pre-emptive and Pre-emptive Goal Programming Model

Malte L. Peters and Stephan Zelewski
University of Duisburg-Essen, Institute of Production and Industrial Information Management, Universitaetsstrasse 10, D-45141 Essen, Germany.
Email: {malte.peters ,stephan.zelewski}@pim.uni-essen.de

Abstract: Employees should be assigned to workplaces according to their competences and preferences to ensure motivated employees carrying out tasks effectively and efficiently. This contribution presents a goal programming model for workplace assignment, which takes into account both employee competences and preferences and workplace competence requirements and attributes.

Key words: Goal programming, Workplace and employee assignment, Competences

1. INTRODUCTION: PROBLEM IDENTIFICATION

In practice, employee workplace assignment can lead to ineffective and inefficient job performance for several reasons. On the one hand, employees probably might not satisfy the competence requirements associated with the workplaces to which they have been assigned and, as a result, they cannot cope with workplace strains. Employees should therefore be assigned to workplaces according to their competences to ensure effective and efficient job performance. On the other hand, employees may have been assigned to their workplaces against their preferences regarding the competences and the general conditions of the workplace.

Non-fulfilment of competence requirements as well as assignment contrary to employee preference may easily lead to employee demotivation. Firstly, workplace assignment based upon individual competences enables the employees to select the appropriate activities to perform the tasks. As a result, they are able to complete their tasks more easily. Secondly, the consideration of competence preferences leads to higher motivation since employees are normally more motivated to complete tasks related to their interests and abilities. Moreover, this consideration affords employees the opportunity to rank their interests and abilities. An employee might prefer to speak Spanish rather than French, for example. Such an employee would prefer a workplace with a higher relative importance of the competence "Spanish language knowledge" in comparison to the competence "French language knowledge". Thirdly, the consideration of preferences regarding general workplace conditions also leads to higher motivation. A mother raising children on her own may prefer a workplace with a reduced number of work hours per day or reduced work days in a week.

2. PRELIMINARIES

2.1 Terminological Preliminaries

2.1.1 Competences, Competence Hierarchies, Competence Profiles, and Competence Levels

A competence is defined as the ability of an employee to use his or her knowledge to achieve a predefined goal, such as an effective and efficient execution of a task. An example of a competence is an employee's ability to use his or her language knowledge to negotiate with foreign business partners. Competences can be structured in a so-called competence hierarchy. In such a hierarchy, similar competences are grouped under a node. The competences "French language knowledge" and "Spanish language knowledge", for example, may be grouped under the node "Foreign language knowledge". Moreover, competence profiles are suitable for competence documentation. On the one hand, employee competence profiles are required to document the competences of each employee. An employee's competence profile consists of a certain set of competences and an assessment of how well an employee meets each competence. The competences require assessment because the level of proficiency in each competence (employee competence level) may vary from employee to employee. One employee might reach a higher level of proficiency in the competence "French language knowledge" than another, for example. How well an employee meets each competence is assessed using an ordinal scale with levels ranging from 1 (dilettante) to 5 (expert). Other scales may be more appropriate (e.g., scales with 9 levels), depending on the desired degree of detail. ERP Systems (e.g., SAP R/3) and project management systems (e.g., Primavera P3e) provide appropriate functions. On the other hand, workplace requirement profiles are needed to document the minimum or specific fixed competences required for each workplace. A requirement profile has to be created for each workplace using the same ordinal scale. Each workplace requirement profile encloses a minimum or specific fixed level of proficiency for each competence. Moreover, a workplace requirement profile contains an assessment of the importance of a competence to a workplace.

2.1.2 Workplace Attributes, Attribute Hierarchies, and Attribute Profiles

Workplace attributes describe the general conditions associated with a workplace: "hours of work per day", "number of work days in a week", "operational safety", "building security", and "office space" serve as examples of such attributes. Analogous to competences, workplace attributes can be structured in an attribute hierarchy. The workplace attributes "building security", and "office space" may be grouped under the node "location attributes". All workplace attributes with the accompanying attribute values describing a certain workplace can be combined into a workplace attribute profile. The preferences of each employee for specific attribute values can be combined into an employee attribute profile in the same way.

2.2 Methodological Preliminaries

2.2.1 Analytical Hierarchy Process

The goal programming model presented here requires several assessments such as competence importance to the specific workplaces. These assessments can be undertaken using the analytical hierarchy process (AHP; e.g., SAATY 2001). The standard AHP requires paired comparison judgments concerning the dominance of one element (e.g. to assess the importance of a competence to a workplace) over another for each of a set of elements using a 1-9 scale, in order to obtain an importance weight for each element. The paired comparison judgments are entered in a square matrix A. If an element i is judged to be moderately important by comparison with another element j, for example, a 3 is entered as the value for the paired comparison judgment a_{ij} in the matrix A while the reciprocal value is entered for the paired comparison judgment a_{ji}. The importance weights are derived by computing a normalised eigenvector of the matrix A. Making pairwise comparison judgments in a model that has a large number of elements (e.g. competences, preferences) can be very time consuming.

Accordingly, if a large number of elements is considered, it could be practical to create a hierarchy of the elements. While a hierarchy of the elements can be used to reduce the number of pairwise comparisons, it can also be helpful to have well-structured elements. Further refinements of the AHP are addressed in the specialised literature (e.g., SAATY 2001).

2.2.2 Goal Programming

Goal programming is a mathematical programming technique designed to handle multiple conflicting objectives. CHARNES and COOPER (1961, p. 215) introduced goal programming. The utilisation of its models has spread in many, diversified fields of interest, such as site selection (HOFFMAN, SCHNIEDERJANS 1994), and project selection (MUKHERJEE, BERA 1995). In goal programming, each objective is viewed as a goal. The technique enables a decision maker to consider one-sided goals and two-sided goals. If the objective is to reach or exceed a one-sided goal, it is called an upper one-sided goal; otherwise, if the objective is to reach or fall below a one-sided goal, it is called a lower one-sided goal. If the objective is to meet a goal, it is called a two-sided goal (HILLIER, LIEBERMAN 2001, p. 332).

The aim of the application of goal programming is to minimise the deviations of the goals considered. So-called deviational variables measure the amount by which the values delivered by the solution of the goal programming model deviate from the respective goal. If an upper (a lower) one-sided goal is considered, the objective function will contain a non-negative underachievement (overachievement) variable which measures the amount of failing to reach or exceed (or fall below) the desired goal. If a two-sided goal is considered, the objective function will contain both an underachievement and an overachievement variable. Further, the basic goal programming model can be enhanced by considering differences in the relative importance of goals. This enhanced approach is named weighted goal programming and assigns importance to the underachievement or overachievement variables according to their relative importance.

Moreover, two types of goal programming models are differentiated: the pre-emptive case and the non-pre-emptive case (e.g., HILLIER, LIEBERMAN 2001, pp. 333-339). In pre-emptive goal programming models different objectives are prioritised as first-priority-goals, second-priority-goals and so on. In an initial step a first goal programming model is solved which only incorporates the first-priority-goals. If the execution of the initial step leads to more than one optimal solution of the first model, a second model incorporating the second-priority-goals is solved keeping the optimal achievement-level of first-priority-goals constant. As long as the execution of a step leads to more than one optimal solution of the respective model and as long as there are goals of lower priority defined, a further model is applied in an additional step. Lower-priority-goals are not considered unless the higher-priority-goals are optimally satisfied and this optimal solution is many-valued. By contrast, in non-pre-emptive goal programming models, different objectives are considered simultaneously as goals in an aggregated objective function. In the non-pre-emptive case, different importance can also be assigned both to the different objectives and to individual goals.

3. MODEL FORMULATION

3.1 Input Data Preparation

The goal programming model presented here makes it possible to consider three different objectives:

- assignment of employees to workplaces according to the fit of their actual and the required levels of competence, respectively;
- assignment of the employees to workplaces according to the fit of their preferences regarding the competences and the relative importance of the competences to the workplaces;
- assignment of the employees to workplaces according to the fit of their preferences regarding the general conditions of the workplace and the actual values of workplace attributes.

The first step is to assess the relative importance of the objectives employing the AHP. In the pre-emptive case presented below, the first objective is considered as single first-priority-goal, while the second and the third objectives are considered as second-priority-goals. The pre-emptive case requires the assessment of the relative importance v_2 and v_3 of the second and the third objective. In the non-pre-emptive case all three objectives are considered simultaneously as first-priority-goals. Thus it requires the assessment of the relative importance v_1, v_2, and v_3 of all three objectives. If an objective is not to be considered, it has to be assigned a value of 0 to the respective importance v_1, v_2, or v_3. Only the pre-emptive case enforces the consideration of the first objective. If the first objective is not to be considered, the non-pre-emptive case has to be utilised. As a rule of practical relevance, however, the highest importance value should be assigned to v_1, since the employees have to be enabled by their competence levels to select the appropriate activities to perform the tasks at the respective workplaces. The second step is to decide which workplace attributes should be covered. In the third step, it is necessary to assess the competence levels for each competence under consideration and each employee. Combining the competences and these competence levels yields the employee competence profiles. Moreover, the employees have to assess their preferences regarding the competences and the values of workplace attributes. The employees have to assess how much they prefer using one competence over another employing the AHP. In the fourth step, which can take place at the same time, the workplace requirement profiles have to be set up. Apart from the determination of the minimum required competence levels, the importance w_{ij} of a competence i to a workplace j has to be assessed. The AHP is also employed for this assessment. Furthermore, the actual values of workplace attributes have to be assessed (e.g. the number of work days needed per week to complete the work specific to a workplace).

In the following table, all inputs needed to solve the goal programming model are listed:

Table 1. Input for the goal programming model

I	number of competences
J	number of workplaces
N	number of employees
K	number of workplace attributes
a_{in}	actual level of competence i for employee n
b_{kj}	actual value of workplace attribute k for workplace j
g_{ij}	required level of competence i for workplace j
p_{in}	preference of employee n regarding competence i
h_{kn}	preferred value of workplace attribute k for employee n
w_{ij}	relative importance of competence i to workplace j
q_{kn}	importance of workplace attribute k to employee n
v_1	importance of the fulfilment of the required competence levels
v_2	importance of the fulfilment of the employee's preferences regarding the competences
v_3	importance of the fulfilment of the employee's preferred values of workplace attributes

3.2 The Goal Programming Model

3.2.1 The Pre-emptive Case

The first goal programming model presented below is utilised to establish the set of optimal assignments of employees to workplaces (shortly "optimal assignment sets" or "optimal solutions") which ensure that the respective workplace's competence level requirements are maximally fulfilled by the selected assignment sets. Each admissible assignment is represented by the decision variable x_{jn} which is set to the value 1 (0) if employee n is (not) assigned to workplace j. If more than one assignment set maximally fulfils the competence level requirements, the preferences regarding both the competence levels and the values of (general) workplace attributes are additionally considered in a second goal programming model. The two types of employee preferences are weighted with their relative importance $v_2 > 0$ and $v_3 > 0$, respectively. The objective function (1) covers the objective of optimally assigning employees to workplaces according to their actual and required competence levels, respectively. This objective is viewed as the first-priority-goal, since the employees have to be able to select the appropriate activities to perform the tasks at the respective workplaces. The objective function only contains the underachievement variables d_{ij}^- as (individually weighted) first-priority-goal variables and not the overachievement variables d_{ij}^+ of the required competence levels, since, in the example considered here, the objective is only to reach or to exceed the required competence levels (upper one-sided goals). Further, the objective function incorporates the AHP importance weights w_{ij}. Expressions such as those in equation (1) prevent a compensation of underachievements and overachievements of competence level requirements.

Objective function (first priority):

$$MIN \quad Z_1 = \sum_{i=1}^{I} \sum_{j=1}^{J} w_{ij} * d_{ij}^- \tag{1}$$

subject to the constraints:

$$\sum_{n=1}^{N} a_{in} * x_{jn} + d_{ij}^- - d_{ij}^+ = g_{ij} \quad for\ \forall i = 1,...,I \quad \forall j = 1,...,J \tag{2}$$

$$d_{ij}^{+}, d_{ij}^{-} \geq 0 \quad for\ \forall i = 1,...,I \quad \forall j = 1,...,J \tag{3}$$

$$N = J \vee N > J \rightarrow \sum_{n=1}^{N} x_{jn} = 1 \quad for\ \forall j = 1,...,J \tag{4}$$

$$N < J \rightarrow \qquad \sum_{n=1}^{N} x_{jn} \leq 1 \quad for\ \forall j = 1,...,J$$

$$N = J \vee N < J \rightarrow \sum_{j=1}^{J} x_{jn} = 1 \quad for\ \forall n = 1,...,N \tag{5}$$

$$N > J \rightarrow \qquad \sum_{j=1}^{J} x_{jn} \leq 1 \quad for\ \forall n = 1,...,N$$

$$x_{jn} = \{0;1\} \quad for\ \forall j = 1,...,J \quad \forall n = 1,...,N \tag{6}$$

The underachievement variables d_{ij}^{-} and the overachievement variables d_{ij}^{+} guarantee that constraint (2) can always be fulfilled. From another point of view, constraint (2) ensures in connection with constraint (3) that the values of the deviational variables d_{ij}^{-} and d_{ij}^{+} are implicitly (model endogenously) determined. The "technical" non-negativity constraint (3) prevents the underachievement variables and the overachievement variables from becoming negative. Constraint (4) ensures that each workplace receives exactly one employee while constraint (5) ensures that each employee is assigned to exactly one workplace. The binary constraint (6) for all decision variables x_{jn} completes the set of constraints in this model. If the (first) goal programming model above leads to more than one optimal solution, the (second) model below considering additionally the second-priority-goals has to be solved. The expression in the first row of the objective function (7) considers the deviations of the relative importance of the competences (for each assigned workplace j) from the employee preferences regarding the competences p_{in}. This expression contains the underachievement variables d_{in}^{-} as well as the overachievement variables d_{in}^{+}, since the objective is to meet exactly the preferences (two-sided goals). In the second row of the objective function (7), the deviations of actual values b_{kj} (for each assigned workplace j) from preferred values h_{kn} of each workplace attribute k are considered for each employee n. Since the preferred values should be met exactly (two-sided goals), such as hours of work per day, and number of work days in a week, underachievement variables d_{kn}^- and overachievement variables d_{kn}^+ are used once again. Other workplace attributes may be considered as one-sided goals. Some workplace attributes may require that a certain level is reached or exceeded. Examples of these workplace attributes are operational safety and office space. Other workplace attributes may need to be achieved equal to or below a certain level. Examples of such workplace attributes are air and noise pollution. To simplify matters, the objective function (7) does not take into account attributes viewed as one-sided goals, but it can easily be adjusted to do so.

The constraints (4) to (6) of the first model dedicated only to first-priority-goals also apply for the second model considering additionally the second-priority-goals. The constraints (8) and (9) ensure in connection with constraints (10) and (11), respectively, that the values of the deviational variables d_{in}^+, d_{in}^-, d_{kn}^+, and d_{kn}^- are implicitly (model endogenously) determined. The non-negativity constraints (10), and (11) prevent the deviational variables d_{in}^+, d_{in}^-, d_{kn}^+, and d_{kn}^- from becoming negative. Constraint (12) ensures that the maximal level Z_1^* of weighted fulfilment of all first-priority-goals which is realised by all optimal solutions of the first model also holds in the second model. Therefore constraint (12) plays the role of an integrity condition for consistently linking the second to the first model of the same real assignment problem.

Objective function (second priority):

$$MIN \quad \left[\begin{array}{l} v_2 * \left(\sum_{i=1}^{I} \sum_{n=1}^{N} d_{in}^- + d_{in}^+ \right) \\ + v_3 * \left(\sum_{k=1}^{K} \sum_{n=1}^{N} q_{kn} * (d_{kn}^- + d_{kn}^+) \right) \end{array} \right] \tag{7}$$

subject to the constraints:

$$\sum_{j=1}^{J} w_{ij} * x_{jn} + d_{in}^- - d_{in}^+ = p_{in} \quad for\ \forall i = 1,\ldots,I \quad \forall n = 1,\ldots,N \tag{8}$$

$$\sum_{j=1}^{J} b_{kj} * x_{jn} + d_{kn}^{-} - d_{kn}^{+} = h_{kn} \quad for \ \forall k = 1,...,K \quad \forall n = 1,...,N \quad (9)$$

$$d_{in}^{+}, d_{in}^{-} \geq 0 \quad for \ \forall i = 1,...,I \quad \forall n = n,...,N \quad (10)$$

$$d_{kn}^{+}, d_{kn}^{-} \geq 0 \quad for \ \forall k = 1,...,K \quad \forall n = 1,...,N \quad (11)$$

$$Z_1^{*} = \sum_{i=1}^{I} \sum_{j=1}^{J} w_{ij} * d_{ij}^{-} \quad (12)$$

3.2.2 The Non-pre-emptive Case

The non-pre-emptive goal programming model presented below is capable of considering all three objectives simultaneously. By contrast to the pre-emptive case, the following non-pre-emptive case allows compensation effects between the three objectives.

The objective function (11) seeks the optimal solutions of the assignment model each of which maximally fulfills the sum of weighted fits between two complementary aspects. Each row of the objective function (11) represents one of the three objectives. As in the pre-emptive case, the first objective refers to upper-one-sided goals, while the second and the third objectives refer to two-sided goals. The first row covers the assignment of the employees to workplaces according to the weighted (v_1) fit between weighted (w_{ij}) actual and required levels of competence for employees versus workplaces, respectively. The expression in the second row of the objective function (11) considers the weighted (v_2) fit of the employee preferences regarding the competences and the relative importance of the competences to the workplaces, while the expression in the third row covers the weighted (v_3) fit between weighted (q_{kn}) preferred versus actual values of workplace attributes. It is possible that there are several different assignment sets meeting the competence level requirements and preferences in the same optimal way, i.e., with the same maximal value of the objective function (11). Especially compensation effects between the three objectives can lead to several optimal solutions of the assignment model.

$$MIN \quad \left[\begin{array}{l} v_1 * \left(\sum_{i=1}^{I} \sum_{j=1}^{J} w_{ij} * d_{ij}^- \right) \\ + v_2 * \left(\sum_{i=1}^{I} \sum_{n=1}^{N} d_{in}^- + d_{in}^+ \right) \\ + v_3 * \left(\sum_{k=1}^{K} \sum_{n=1}^{N} q_{kn} * (d_{kn}^- + d_{kn}^+) \right) \end{array} \right] \qquad (11)$$

The constraints (2) to (6) and (8) to (11) from the pre-emptive case also apply in the non-pre-emptive case.

4. DIFFICULTIES OF THE MODEL

One obvious problem could be that the input data is not updated regularly. Hence, for example, the documentation of the competence profiles and the preferences of the employees might be out of date or incomplete. This may cause misleading assignment sets. Another problem may occur if the employee to be assigned has irreconcilable differences with another employee assigned to a workplace in the vicinity of the first employee's workplace, so that the enforced cooperation demotivates both employees and thus outweighs the efficiency gains of the workplace assignment delivered by the goal programming model. Moreover, the employee to be assigned may be content with her or his present workplace and may thus dislike an assignment to another workplace. This might be the case when the competence levels required for her or his present workplace deviate greatly from her or his actual competence levels or when his or her preferred competence levels are neglected or assessed to have low relative importance. One way to deal with this problem is to ask the employees, before utilising the goal programming model, whether they agree with an assignment to another workplace. In that case, only the employees agreeing with the assignment to another workplace may be considered in the goal programming model. If the employee's actual competence levels do not fulfil his present workplace's competence level requirements, but his or her preferred competence levels match the required competence levels of the workplace, another solution to this problem may be to offer the employee training opportunities with respect to the required competence levels.

5. CONCLUDING REMARKS

This paper has presented a formal model for the real problem of assignment of employees to workplaces considering workplace competence level requirements, importance of competences to workplaces, and actual workplace attribute values on the one hand and the employee's actual and preferred competences and preferred workplace attribute values on the other hand. The optimal solution of this model delivers at least one assignment set minimising the sum of the weighted discrepancies firstly caused by the weighted deviations between competence levels required for a workplace and the actual competence levels of the employees, secondly caused by the deviations between the importance of competences to a workplace and the employees' preferences regarding the competences, as well as thirdly caused by the weighted deviations between the actual workplace attributes and the employees' preferred values of the workplace attributes. The minimisation of the sum of the weighted discrepancies mentioned before equals to the maximal weighted fulfilment of workplaces' competence level and (general) attribute value requirements by each optimal assignment of employees to workplaces.

ACKNOWLEDGEMENTS

This work was partly funded by the Federal Ministry of Education and Research (BMBF) and coordinated both by the German Aerospace Centre (DLR) within the project "MOTIWIDI" and the Research Centre Karlsruhe (PFT) within the project "KOWIEN".

REFERENCES

CHARNES, Abraham; COOPER, William W.:
Management Models and Industrial Applications of Linear Programming, Volume I.
New York, London, Sydney: John Wiley & Sons, 1961.
HILLIER, F. S.; LIEBERMAN, G. J.:
Introduction to Operations Research.
Boston, MA: McGraw-Hill, 7th ed., 2001.
HOFFMAN, J. J.; SCHNIEDERJANS, M. J.:
Two-stage Model for Structuring Global Facility Site Selection Decisions.
In: International Journal of Operations & Production Management,
Bradford, 14(1994)1, pp. 79-96.

MUKHERJEE, K.; BERA, A.:
Application of goal programming in project selection decision – A case study from the Indian Coal mining industry.
In: European Journal of Operational Research,
Amsterdam, 82(1995)1, pp. 18-25.
SAATY, Thomas L.:
Decision Making for Leaders.
Pittsburgh, PA: RWS Publications, 3rd Edition, 2001.

Personnel Development and Assignment Based upon the Technology Calendar Concept

Sven Rottinger and Gert Zülch
University of Karlsruhe, ifab-Institute of Human and Industrial Engineering, Kaiserstrasse 12, D-76131 Karlsruhe, Germany.
Email: {sven.rottinger, gert.zuelch}@ifab.uni-karlsruhe.de.

Abstract: The optimal utilisation of human resources is one of the most important success factors contributing to long-term competitiveness. An essential requirement for the optimal utilisation is a goal-oriented planning of personnel development. The principal intention of personnel development and assignment planning is to cope with technology and product changes. These changes not only demand a quick and flexible adaptation of machinery resources but also that of an existing workforce. In addition to the productions logistical and financial objectives, which are generally in the foreground during the implementation of new technologies and products, the workers' personal concerns should be taken into account. In order to be able to offer a methodology for the continuous adaptation of a workforce to the process of change, the idea of the technology calendar concept is picked up and expanded here. Within this concept, the ifab-Institute of Human and Industrial Engineering at the University of Karlsruhe exploits the possibility of using simulation as a planning tool in order to support a goal-oriented planning of personnel development.

Key words: Manufacturing planning, Human-oriented simulation, Personnel development, Technology calendar

1. IMPORTANCE OF PERSONNEL DEVELOPMENT

New products and process technologies and a changing manufacturing programme demand a quick and flexible adaptation of the existing workforce. This adaptation is essential for the optimal utilisation of human resources which are in many manufacturing systems the most expensive but also the most flexible factor.

However, various enterprises use a traditional form of personnel development in order to adapt their personnel structures to the changes in manufacturing systems. This means that already existing qualification deficiencies are the trigger for development measures. Consequently the elimination of personnel bottlenecks occurs reactively, usually with single, improvised measures.

In contrast to the traditional form, the approach of a synchronised personnel development can be used. According to this concept, development measures are coordinated with the planned product and technology changes from the very beginning. The aim is for all affected workers to possess the necessary qualifications early enough to avoid qualification bottlenecks. For this purpose the technology calendar concept is used to adjust continuous product and technology changes with the personnel development and assignment.

In general, the planning process should encompass several technological phases. Therefore, it is necessary to define timeframes, which allow for a rough planning of the needed number and abilities of a workforce. These timeframes determine when product or technology changes should be expected. Which worker is suitable for further education and how the respective abilities should be achieved is a subject matter related to the determination of qualitative personnel requirements, whereas the number of needed persons is a quantitative issue. If the required qualifications are to be available at the right point in time, it is indispensable that the necessary qualification measures are defined and scheduled appropriately. The qualification concept which has been developed here describes the qualification measures, sorted into target groups, content, qualification costs, and time.

Dependent upon the type of personnel assignment problem to be solved, several instruments have been developed at the ifab-Institute of Human and Industrial Engineering of the University of Karlsruhe, e.g. the personnel-oriented simulation tool *ESPE* (in German: "Engpassorientierte Simulation von Personalstrukturen"; cf. HEITZ 1994; ZÜLCH 1996; ZÜLCH, VOLLSTEDT 2000). In order to support the described planning process, the possibility of using this simulation tool is exploited. For this purpose, a novel simulation tool has been adapted to the requirements of long-term personnel development and assignment planning.

2. FACTORS INFLUENCING PERSONNEL DEVELOPMENT PLANNING

In order to evaluate the development of a workforce, various criteria can be taken into account. In addition to the productions logistical and financial objectives, which are generally in the foreground during the implementation of new manufacturing technologies and product developments, the workers' personal concerns should also be observed.

2.1 Continuous Process Technology Changes

The development of the various product and process technologies is characterised by a further increase in the degree of automation. Furthermore, it can be seen that decentralised manufacturing systems are often used in order to improve flexibility and reactivity. This refers to an increased integration of indirect functions, such as maintenance and quality control, into the manufacturing process.

The technological developments usually provide important impulses for personnel development: Specialists who are qualified only for few work tasks are no longer the focus of interest. Rather, workers who can also take on up-stream or down-stream work tasks are desired to a far greater extent. This does not only imply an adoption of more, similarly structured work tasks, as in job enlargement, rather in particular the adoption of more responsibility, as in a vertical job enrichment (ULICH 1998, p. 159).

Another important impulse for personnel development planning is the increased likelihood of machinery failures during the start-up of new equipment. In order to avoid downtime, the personnel responsible for maintenance tasks must be present in adequate numbers and with the necessary qualifications.

2.2 Continuous Product Changes

In order to remain competitive in the market, a production enterprise must further develop its palette of available products and increase the number of variants according to market demands. The continuous product development has numerous effects upon future personnel developments:

On the one hand, questions regarding the quantitative personnel demands must be clarified when considering the development of product figures. On the other hand, the introduction of new product variants, the further development of existing products, and even new process technologies lead to more complex work requirements, which cannot be managed without a targeted, qualitative personnel development.

2.3 Considering Personnel Aspects

In order to take the preferences and restrictions of the existing workforce into account during development planning, the simulation procedure *ESPE* offers an appropriate approach (cf. HEEL 1999, p. 96). This simulation package includes a preference system, which permits the modelling of personal preferences and restrictions for each required qualification. This preference system has been supplemented with worker development potentials for future work tasks in order to support personnel development planning, and was implemented in *ESPE-PE* (*ESPE* for Personnel Development). The simulation procedure *ESPE-PE* can thus make a clear statement as to which person is best suited for the needed qualifications of an upcoming work task. The development potential can thereby be determined from worker surveys or from the guess of a superior.

Another personnel aspect is the sequential completeness of work functions (cf. HACKER 1987, p. 44). Sequential completeness refers to a work task which includes, in addition to the execution function, also upstream and downstream functions, for example set-up and maintenance functions as well as control functions. In particular persons without job alternation will obtain in the procedure a preferred enlargement of their abilities, on condition that these persons are suitable for further qualification.

3. COORDINATION OF MEASURES USING THE TECHNOLOGY CALENDAR CONCEPT

3.1 Synchronised Personnel Development

Numerous concepts for the integration of product and technology development into organisations development can be found in the literature (HAYES, WHEELRIGHT 1984, p. 197; SAVÉN, OLHAGER 2002, pp. 375). In particular various technology calendar concepts have been presented. These concepts permit the synchronisation of technology and product changes. The approaches differ with respect to their application areas and objectives (BURGSTAHLER 1997, p. 69).

However, none of these concepts consider the integration of personnel development planning in the process of change. In order to be able to offer a continuous methodology for the aforementioned problem with respect to the integration of personnel development in the process of change, the idea of the technology calendar concept was picked up and expanded. The concept has thereby been supplemented in order to coordinate personnel development measures, in the sense of a personnel development, which is synchronised with product and technology developments.

The technology calendar concept, which is implemented in *ESPE-PE*, recognises four sectors (cf. Figure 1): In the product programme sector the new products to be planned are listed along with the date of their planned production start. In the product and process technology sectors the new technologies are listed with the latest date for their application maturity (cf. BURGSTAHLER, 1997, p. 109). Due to the fundamental need for the integration of the personnel resources into this planning process, the personnel development has been incorporated into the fourth sector. In order to coordinate personnel development measures, the production system can be simulated for each modelled date of a product and technology development.

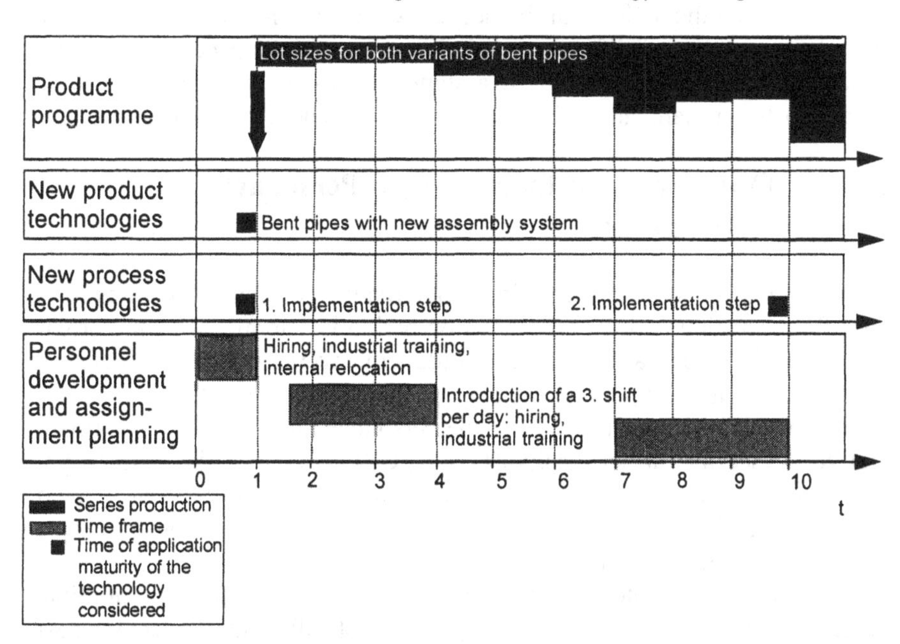

Figure 1. Technology calendar with synchronised timeframes for personnel development (Source: ROTTINGER, ZÜLCH 2003, p. 56)

3.2 Personal-oriented Modelling of Manufacturing Systems

In order to be able to evaluate the effects of personnel development measures on the manufacturing system using the simulation procedure *ESPE-PE*, an appropriate modelling of the technology and product change as well as a modelling of the personnel structure is necessary. The function-equipment-matrix of a manufacturing system (cf. Figure 2), used for the modelling of personnel structures, is a significant input mask. This matrix contains all functional elements appearing in the manufacturing programme for a specific time period. A functional element refers to the requirement of a specific function at a certain workplace or machine, e.g. the set-up function of a milling machine. Thus, the ability of a worker is defined by the set of all functional elements he can execute. This high degree of detail for modelling of the workers' abilities allows for personnel related bottlenecks, occurring after a technology or product change, to be identified in a simulation study and the according personnel development measures to be derived.

The problem of developing and assigning workers to functions and workplaces is characterised, on the one hand, by a large number of possibilities for varying the abilities of the needed workforce and, on the other hand, by a large number of different possibilities for allocating persons to work functions. Of course, the chosen assignment has an important effect on labour costs per manufactured unit and the achievement of logistical goals.

3.3 Objectives and Evaluation of Personnel Development

ESPE-PE supports an evaluation of alternative personnel structures for a specific evaluation period. An evaluation period is determined by the proximate point in times t and $t+1$ and considers the average lead time degree of orders, the system output and the average workload as logistical goals, and the calculated labour cost as well as the personnel adjustment costs as financial goals (see for further details ZÜLCH, GROBEL, JONSSON 1995, pp. 311). In addition to the logistical and financial goals, personnel-oriented aspects can also be evaluated, e.g. concerning personnel development potential or the completeness of work functions.

In order to evaluate a personnel development concept over multiple stages of technology and product changes it is not sufficient to select the best planning solution for each evaluation period separately. Beyond this it is necessary to take for each evaluation period several alternative planning solutions into account and to derive alternative personnel development paths for the entire sequence of changes. This approach ensures the generation of a personnel development concept, which contains the best solution over all development stages.

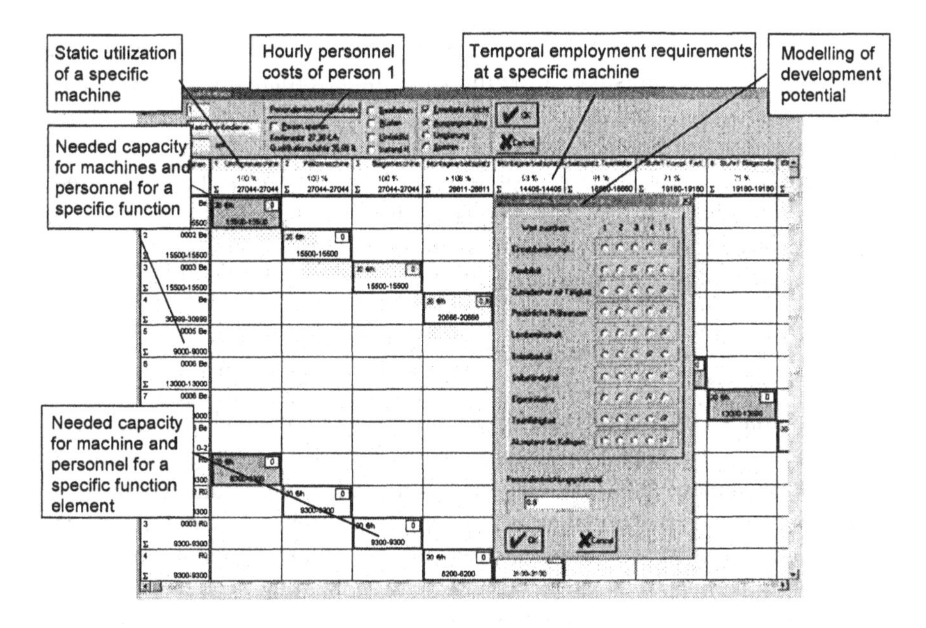

Figure 2. Function-equipment-matrix

4. TECHNOLOGY AND PRODUCT CHANGES FOR A COMPONENT SUPPLIER AS AN EXAMPLE

In the following case study the technology calendar concept is used for a component supplier from the automotive supply industry. In this project a personnel development concept which considers multiple stages of technology and product changes should be generated.

4.1 Problem Description

The considered production enterprise was faced with broad modifications to a specific manufacturing system. In the initial situation numerous component variants were manufactured in small lot sizes. In addition to this initial manufacturing programme, an order with large lot sizes and only two variants was envisaged by a certain customer from the automotive industry. This would possibly justify the application of a new manufacturing technology allowing for an assembly process without complex welding procedures. This new technology, resulting in an automated workplace, could be implemented in two stages.

In addition to this technology change, a quantitative development of the regarded product variants had to be modelled in accordance with the quantities demanded from the automotive company. Aside from a continuous quantitative production increase over time, one significant rise in output which would occur after the end of summer holidays had to be considered. The second significant rise was envisaged in a specific month in 2004, at which time the second implementation stage had to be concluded so that the required quantity could be produced.

4.2 Personnel Development Concept

In accordance with these multiple stages of technological changes, various scenarios were built. Their main aspects are the following: In order to obtain the personnel qualifications needed for the new manufacturing technology on time, a personnel development concept has to be set up. All qualitative and quantitative aspects of the personnel development, dependent upon the implementation stages of the new technology and the monthly enlargement of the production volume have to be defined in this concept.

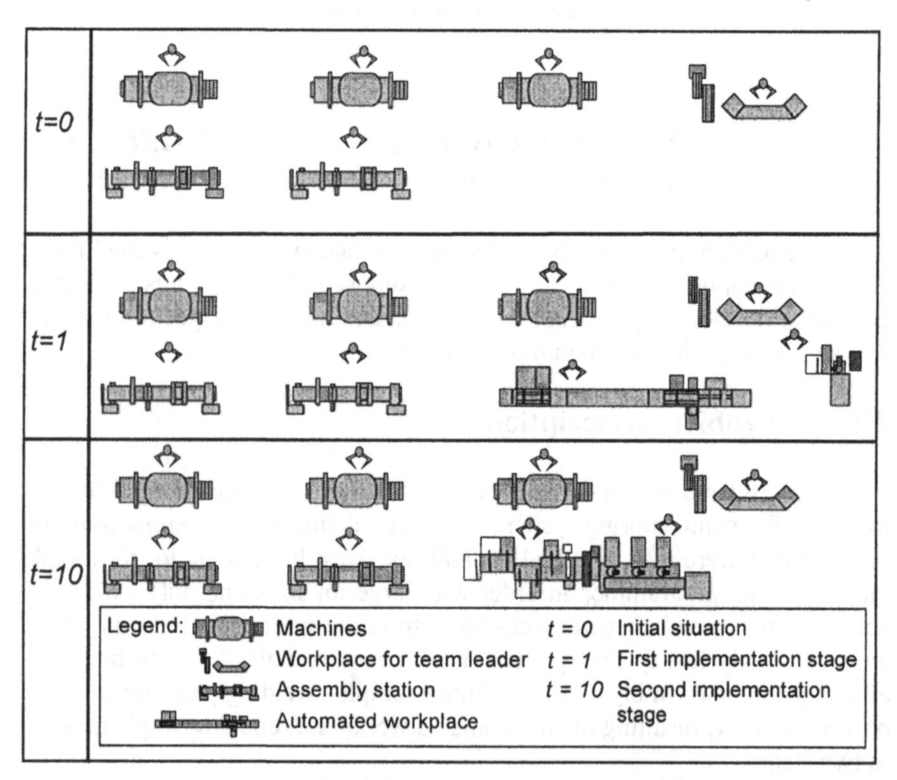

Figure 3. Development of the manufacturing system

Furthermore, an increased probability of machine failure after each implementation stages of the automated workplace has to be considered. In order to reduce machine disturbances, especially during the early introduction stages of the new technology, an adequate number of maintenance personnel have to be planned for the new tasks

The initial situation as well as the scenarios for the upcoming technology and product changes have been modelled and simulated in *ESPE-PE*. For each point in time of a technology or product change the scenarios have been evaluated and, where applicable, an adaptation of the existing personnel structure has been regarded.

The results contain a personnel development concept with the latest point in time for the necessary measures as well as the target group and content of all necessary measures for personnel development. Beyond continuous training of the existing staff, one measure includes the acquisition of skilled workers from the external employment market. The reason for the acquisition of additional workers is the high qualification requirements for the set-up and maintenance functions, created by the new technology, which cannot be fulfilled by the existing workforce while also taking their qualification preferences and restrictions into account. Furthermore, several restrictions for an extension of qualification of the present staff become evident, resulting from an assessment of the qualification development potential. The main reason for the qualification restrictions of the existing workforce is the high contingent of workers originating from external trades. Due to their lack of technical qualifications, such workers are only suitable for simple operating tasks.

	Measures	**Type of Personnel**	**Machine/Function Element**
t=1	External personnel hiring and industrial training	1 Equipment manager	AW1, AW2: Set-up, adjustment, control, minor disturbance elimination, maintenance
		2 Equipment operators	AW1, AW2: Material preparation, set-up, machine rigging
	Internal relocation and industrial training	1 Team leader	AW1, AW2: Set-up, minor disturbance removal, maintenance
	Instruction at the equipment	1 Machine operator 2 Equipment operators	AW1, AW2: Equipment fitting
t=4	External personnel hiring and industrial training	1 Equipment manager	AW1, AW2: Set-up, adjustment, control, minor disturbance elimination, maintenance

Legend: AW1, AW2 Automated Workplace

Figure 4. Personnel development concept

The derived measures contained an important input for those responsible for planning. Especially the determination of the time frame for the acquisition from the external employment market and the consequent realisation of this measure were decisive for the successful implementation of the automated workplace.

5. SUMMARY AND FURTHER RESEARCH

In order to be able to react quickly to product and technology changes, the flexibility potential of the available personnel structure must be registered, and exploited during the continuous planning process. Practical industrial experience shows that personnel development measures are often reactive or are carried out in an ad hoc or improvised manner. A simulation supported planning tool, with which measures for personnel assignment and development planning can be commenced in a targeted and timely manner, has been compiled here in order to counteract reactive personnel development.

The fundamental idea behind the technology calendar concept, which was originally conceived for the coordination of long-term change processes, can also be transferred onto problems arising from a shortened planning horizon in order to cope with demands of a market-driven production with increasing product diversity, ever shorter delivery times and higher delivery reliability. In order to be able to react quickly to the short-term changes to the sales market, many production enterprises turn to decentralised organisational structures.

However, as a result of a weekly or even daily varying order programme, these decentralised organisational structures are faced with a fluctuating workload. A future research requirement can be seen in the assignment of workers from weakly loaded organisational structures to organisational structures with higher personnel utilisation as a compensation for a daily varying order programme. Thus, the modelling of weekly or even daily variations in the quantity and type of the order programme, and the support of associated personnel assignment problems must be undertaken.

Building upon this, a systematic procedure, which supports the planning of short-term personnel assignment in decentralised organisations structures, should be developed. An objective of priority for this procedure is to plan group structures which, on account of their qualification, are able to exchange individual workers between the individual, decentralised production units in order to achieve high personnel utilisation, a complete processing of the manufacturing programme as well as low lead times.

ACKNOWLEDGEMENTS

The long-term generation of the personnel-oriented simulation tool *ESPE* was possible under the umbrella of the Special Research Area 346 "Computer-integrated design and manufacturing of parts". This research area was sponsored by the German Research Association (Deutsche Forschungsgemeinschaft - DFG) from 1990 – 2002. With this support the base for the presented possibilities for personnel development and assignment was created. With the objective of transferring the results of the fundamental research into industrial practice the Transfer Unit 48 emerged from the Special Research Area was established. The results of this article were gathered within the context of this transfer unit.

REFERENCES

BURGSTAHLER, Bernd:
Synchronisation von Produkt- und Produktionsentwicklung mit Hilfe eines Technologiekalenders.
Braunschweig, Univ. Diss., 1997.

HACKER, Winfried:
Software-Ergonomie: Gestalten rechnergestützter Arbeit?
In: Software Ergonomie 1987.
Edts.: SCHÖNPFLUG, Wolfgang.
Stuttgart: B.G. Teubner Verlag, 1987, pp. 31-54.

HAYES, Robert H.; WHEELRIGHT, Steven C.:
Restoring our Competitive Edge, Competing through Manufacturing.
New York NY: John Wiley and Sons, 1984.

HEEL, Jochen:
Reorganisation des Personaleinsatzes mit Hilfe der personalorientierten Simulation.
Aachen: Shaker Verlag, 1999.
(ifab-Forschungsberichte aus dem Institut für Arbeitswissenschaft und Betriebsorganisation der Universität Karlsruhe, Vol. 18)

HEITZ, Max-Jürgen:
Ein engpaßorientierter Ansatz zur simulationsunterstützten Planung von Personalstrukturen.
Karlsruhe Univ., Diss. 1994.
(ifab-Forschungsberichte aus dem Institut für Arbeitswissenschaft und Betriebsorganisation der Universität Karlsruhe, Vol. 7)

ROTTINGER, Sven; ZÜLCH, Gert:
Personnel Development and Assignment Based upon the Technology Calendar Concept.
In: Human Aspects in Production Management.
Eds.: ZÜLCH, Gert; STOWASSER, Sascha; JAGDEV, Harinder S.
Aachen: Shaker Verlag, 2003, pp. 53-59.
(esim – European Series in Industrial Management, Volume 5)

SAVÉN, Ruth Sara; OLHAGER, Jan:
Integration of Product, Process and Functional Orientations: Principles and a Case Study.
IFIP WG5.7 International Conference on Advanced Production Management Systems, APMS 2002, Pre-prints.
Eds.: JAGDEV, Hari u.a.
Eindhoven: Technische Universiteit, 2002, pp. 375-389.

ULICH, Eberhard:
Arbeitspsychologie.
Stuttgart: Schäffer-Poeschel Verlag, 4. edition 1998.

ZÜLCH, Gert:
A Heuristic Solution to the Personnel Structure Problem.
In: ORBEL10, Proceedings of the Tenth Conference on Quantitative Methods for Decision Making.
Ed.: JANSSENS, Gerrit K.
Brussels: Belgian Operations Research Society, 1996.

ZÜLCH, Gert; VOLLSTEDT, Thorsten:
Personnel-integrated and Personnel-orientated Simulation - A New Guideline of the German Association of Engineers.
In: Information and Communication Technology (ICT) in Logistics and Production Management.
Eds.: STRANDHAGEN, Jan Ola; ALFNES, Erlend.
Trondheim: Norwegian University of Science and Technology, Department of Production and Quality Engineering, 2000, pp. 185-192.

ZÜLCH, Gert; GROBEL, Thomas; JONSSON, Uwe:
Indicators for the Evaluation of Organizational Performance.
In: Benchmarking - Theory and Practice.
Ed.: ROLSTADÅS, Asbjørn.
London et al.: Chapman & Hall, 1995, pp. 311-321.

Reorganising the Working Time System of a Call-Centre with Personnel-oriented Simulation

Patricia Stock and Gert Zülch
University of Karlsruhe, ifab-Institute of Human and Industrial Engineering, Kaiserstrasse 12, D-76131 Karlsruhe (Germany).
Email: {patricia.stock, gert.zuelch}@ifab.uni-karlsruhe.de

Abstract: The configuration of a working time model is a very complex task due to the fact that different restrictions concerning work demands and employee preferences have to be taken into account. The use of a simulation tool is a promising approach as it allows for an objective, quantitative, efficient and prospective assessment of alternative working time models during the design phase. This paper presents a simulation study within a call-centre of manufacturer of electric devices.

Key words: Personnel-oriented simulation, Working time configuration, Service sector

1. REQUIREMENTS OF THE WORKING TIME CONFIGURATION

In recent years flexible working time systems have been established with the front office of many service departments due to the fact that they enable the adjustment of the assignment of personnel to the fluctuating appearance of customers. The configuration of a flexible working time model has turned out to be a very complex task since, in addition to the interests of the service department and those of the employees, further parameters, e.g. legal regulations or ergonomic recommendations, must be considered (cf. in detail HORNBERGER, KNAUTH 2000, pp. 25). Moreover, the following deficits in working time configuration aggravate the design of an appropriate working time model (cf. BOGUS 2002, pp. 24):

- Complexity arises from the numerous alternatively or complementary applicable working time models (cf. ACKERMANN 1990, p. 185). Approximately 10,000 different working time models are in use worldwide, clearly illustrating that "the" working time model does not exist (cf. KNAUTH 1995, p. 210). The simple transfer of a general working time model to a specific enterprise case rarely leads to a favourable outcome (cf. KRAMER et al. 1998, p. 17).
- No data regarding the prevalence of or concrete regulations for flexible working time models are available (cf. BUNDESMANN-JANSEN, GROß, MUNZ 2000, p. 23).
- In particular the configuration of flexible working time models with respect to economic and personnel related objectives has been investigated only limitedly. Furthermore, only very few consolidated findings regarding consequences for employee load and strain are available (cf. FERREIRA, LANDAU 2001, p. 245).
- The level of awareness of the involved planers, with respect to configuration possibilities and the consequences of working time models, is very low (cf. FERREIRA, LANDAU 2001, p. 245).
- Conventional analysis methods such as check lists or benefit analyses are only conditionally suitable for the support of the configuration process since an objective, quantitative, efficient and prospective assessment of working time models is only possible to a limited degree (cf. SCHREIER 1995, p. 21).

2. WORKING TIME CONFIGURATION WITH PERSONAL-ORIENTED SIMULATION

Therefore, the use of a personnel-oriented simulation tool is a very promising approach since, with its help, the effects of different working time models can be analysed with respect to their implications and advantages prospectively, during the planning phase (BOGUS 2002, pp. 42). Such a simulation procedure was developed at the ifab-Institute of Human and Industrial Engineering of the University of Karlsruhe. In order to implement the simulation-based approach into the configuration of working times in service departments, several modelling concepts had to be designed:

- A modelling concept for service departments was developed, with which various service departments could be characterised and represented in a simulation procedure. Resources (personnel types and work places), customer types and arrival times (with different arrival time distributions) as well as the corresponding activities (with different operation time distributions) can be modelled according to the business facility being considered.

- A concept was developed, with which working time models can be modelled. The configuration elements duration, placement and reference time frame of the working times are considered. Furthermore, due to the necessity for precise employee employment for the assessment of working time models using simulation, a heuristic has been developed for the definition of flexible working times. Thus, not only single working time models can be assessed, rather also complex working time systems (meaning the entirety of different working time models used in the business facility).
- An assessment concept was developed, containing not only enterprise organisational and financial goals, rather also specifically, in simulation determinable, employee-oriented goal criteria. The possibility to comparatively assess various working time models with respect to their expected workload and the effects thereof, using the example of time stress and aggregated physical load, was thereby created.

The modelling concepts described above were integrated into the object-oriented simulation procedure *OSim* (*O*bject *Sim*ulator), developed earlier at the ifab-Institute (see ZÜLCH, FISCHER, JONSSON 2000). The simulation procedure *OSim-GAM* (*OSim* for Working Time Configuration), which was thereby created, can assess working time models efficiently and in a quantitative manner. A detailed representation of the simulation procedure *OSim-GAM* can be found in BOGUS (2002) or BOGUS and STOCK (2002).

3. WORKING TIME CONFIGURATION OF A CALL-CENTRE

In a simulation study the working time system of a call-centre in the semi-industrial service sector was modelled and simulated. In the following the call-centre, the approaches and the results are sketched (see also ZÜLCH, STEIH, STOCK et al. 2004).

3.1 The Call-Centre

The manufacturer of electric devices has set up a call-centre to take care of the "after sales service" for his large array of products. The call-centre can be reached through three different telephone numbers, which are structured according to the type of inquiry:

- the order processing, in which, in addition to the registration of orders, information regarding delivery, prices and discounts, as well as general information is provided and in which complaints are processed,
- the spare parts ascertainment, in which the specification of spare part is carried out, and
- the workshop information desk, in which questions regarding guarantees, goodwill, backorders etc. are answered and information regarding repair releases is given.

The customers can also direct their questions to the call-centre via email or fax. The personnel at the call-centre is comprised of three full-time employees with a working time of 35 hours per week and one part-time employees with a working time of 20 hours per week, for each of the three telephone numbers.

The broad product pallet leads to a differentiated customer structure: It is primarily divided into the sectors "retailers" and "end-consumers". This customer structure results in a fluctuation in the frequency of customer inquiries: The end-consumers prefer to take advantage of the call-centre late in the afternoon, whereas specialised trade customers, who represent the largest group, make demands upon the call-centre mainly in the morning or early afternoon. In addition to the daily fluctuations, the call volume exhibits characteristic variations throughout the course of the week or year.

These conditions place various demands on the working time model to be implemented:

- The large proportion of retail customers forces the call-centre to align its business hours with those of the retailers. These are however longer than the individual working times of the call-centre employees.
- In order to provide a high degree of availability at the call-centre the personnel employment must be aligned with the call volume (Figure 1). This ensures, for one, that the customers can be served promptly, even during peak call times, and that they are not lost (e.g. through breaking due to too long waiting times). On the other hand, a high degree of personnel utilisation should also be achieved. Appropriate key figures, which describe the call volume in detail, are a pre-requisite for a customer-oriented alignment of the personnel requirements, e.g. the rate of the incoming calls and the necessary service times.
- Aside from the operational aspects, the individual working time wishes and the stress of the employees resulting from the work tasks must also be accommodated.

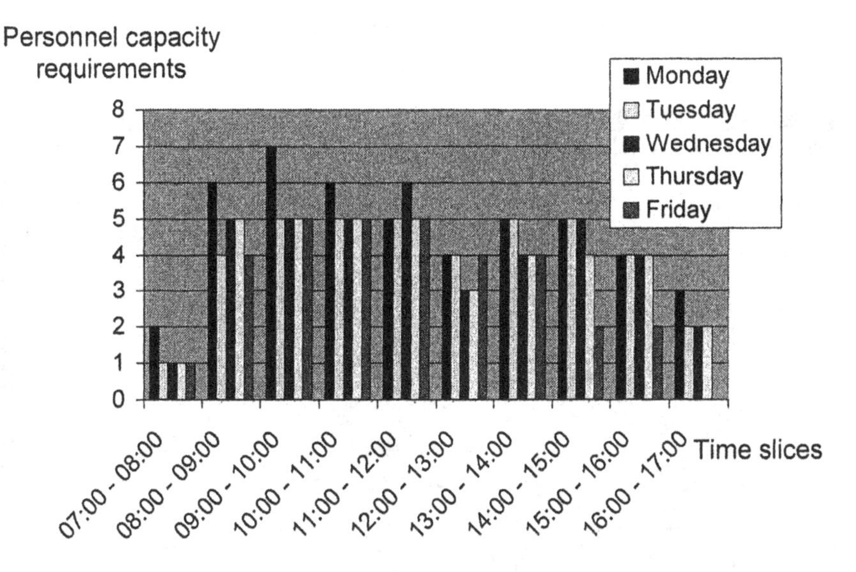

Figure 1. Personnel capacity requirements according to the call volume

3.2 Modelling of the Call-Centre

In the following the initial situation will be described by modelling the customer types, the employees as well as the working time model, in accordance with the modelling concept sketched in section 2. This step is necessary for validating the simulation model by comparing simulation results with real data.

Considering the types of inquiries arising in the call-centre, the following four customer types can be differentiated:

- calls to the order processing call number *AB*,
- calls to the call number EB for spare parts ascertainment,
- calls to the workshop information desk call number *WA* as well as
- written fax inquiries or inquiries over the internet *FI*.

The modelling of these various types of inquiries is carried out using activity networks (JONSSON 2000, pp. 55). An activity network shows, in a type of net representation, the temporal logical dependencies of the sequentially succeeding, parallel processing or alternatively occurring activities of inquiry processing. Figure 2 presents an example activity network for the processing of the written fax and internet inquiries. In this example an activity for the distribution of the written inquiries is modelled. This task can be carried out by all of the call-centre employees. One employee from each of the call numbers is responsible for the incoming faxes. In contrast to telephone customers, the fax enquiries are not associated with a specific waiting time.

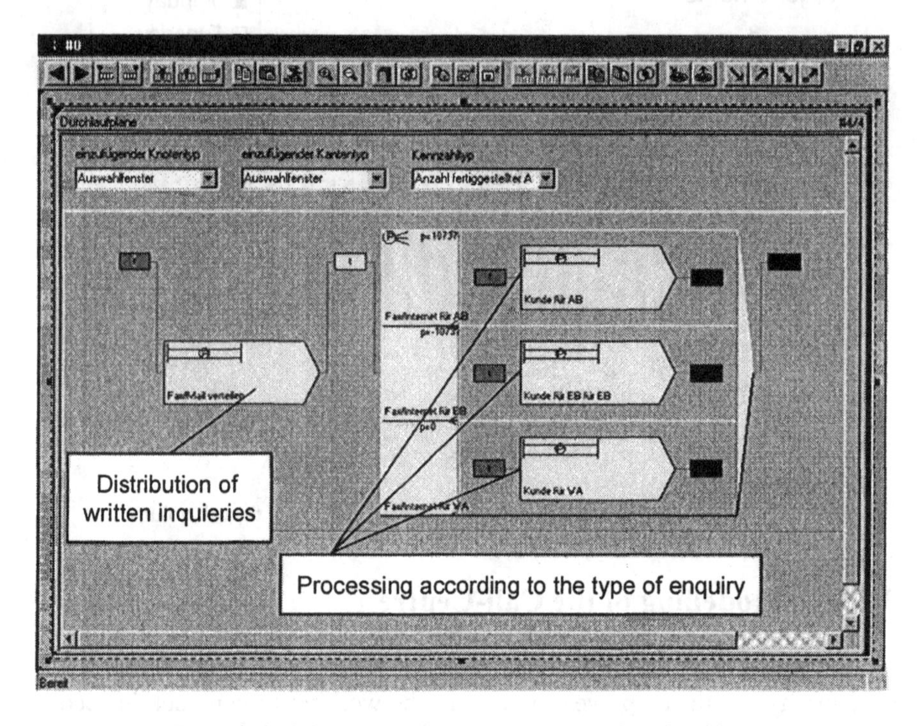

Figure 2. Activity network for customer type "written inquiries"

In the initial situation the call-centre's regularly present personnel is composed of three full-time employees, with a weekly working time of 35 hours, and one part-time employee, with a working time of 20 hours, for each of the groups responsible for customer types *AB*, *EB* and *WA*. This results in a personnel inventory of 12 persons with an available personnel capacity of 75 hours per day. Each employee is modelled along with his respective weekly working time, minimum and maximum daily work duration as well as the fixed and variable employment costs.

Subsequently, the persons and activity networks are linked with one another through the definition of individual qualifications. Figure 3 shows the resource links for the activity network "written inquiries" in a so-called functions-equipment-matrix. The columns and rows of the matrix describe all arising functions and available employees respectively. The qualifications of each employee are modelled through a selection of matrix elements, the so called function elements.

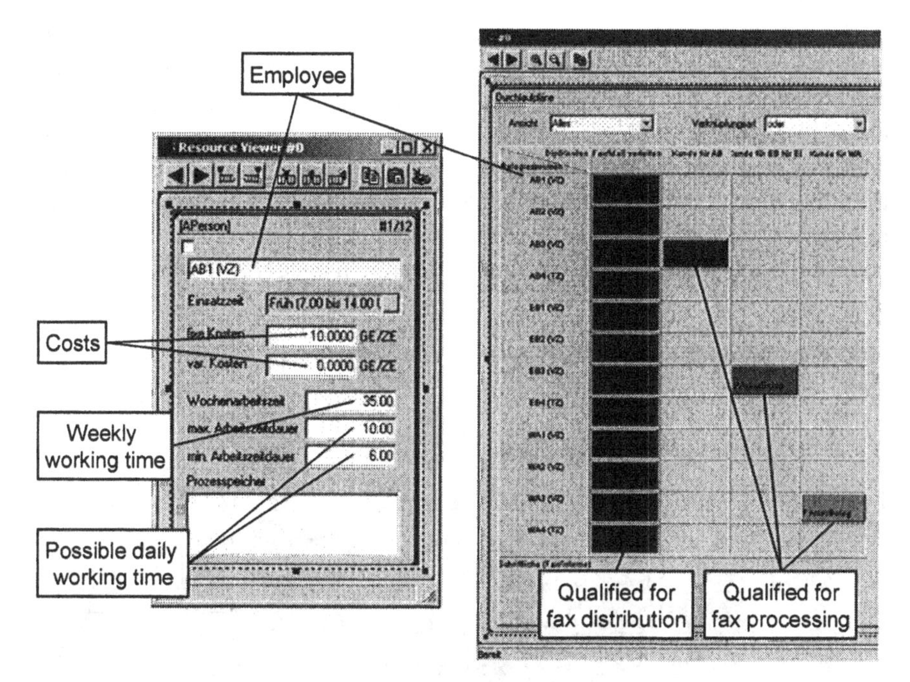

Figure 3. Modelling the employees and their qualifications

In the initial situation various working time models exist, differentiated according to work forms:

- The weekly working time of the full-time personnel comprises 35 hours with a minimum work duration of 6 hours per day and maximum work duration of 10 hours. The full-time personnel work in 4 shifts with a cycle time of 3 weeks.
- The weekly working time of the part-time personnel comprises 20 hours with a minimum work duration of 3 hours per day and a maximum work duration of 10 hours per day. The part-time personnel work form 10 am to 2 pm daily.

Once the working time models of the various work forms have been modelled they are assigned to the individual persons in the employment plan for the entire simulation time period. Figure 4 shows example employment times for the working time model in the initial situation.

The validation of the simulation model results with real data shows that the service degree of the simulation is approximately 10 % lower than the real service degree. This deviance can be traced back to the fact that the existing real data do not possess the needed differentiation, necessary for the simulation, regarding the distribution of incoming inquiries as well as regarding the processing times of the individual functions. This is in turn due to the fact that these were, for enterprise regulatory and data security reasons, not registered. Therefore, the required simulation data first had to be calculated using the generalized key figures of the call-centre. Furthermore, no precise data regarding the length of time the customers were willing to wait was available for the simulation study. The waiting time of 20 s used in the simulation is an optimistic estimate from the call-centre, meaning that, in reality, the customers would probably wait longer. One result that can be deduced is that there are new demands on time management which must be met.

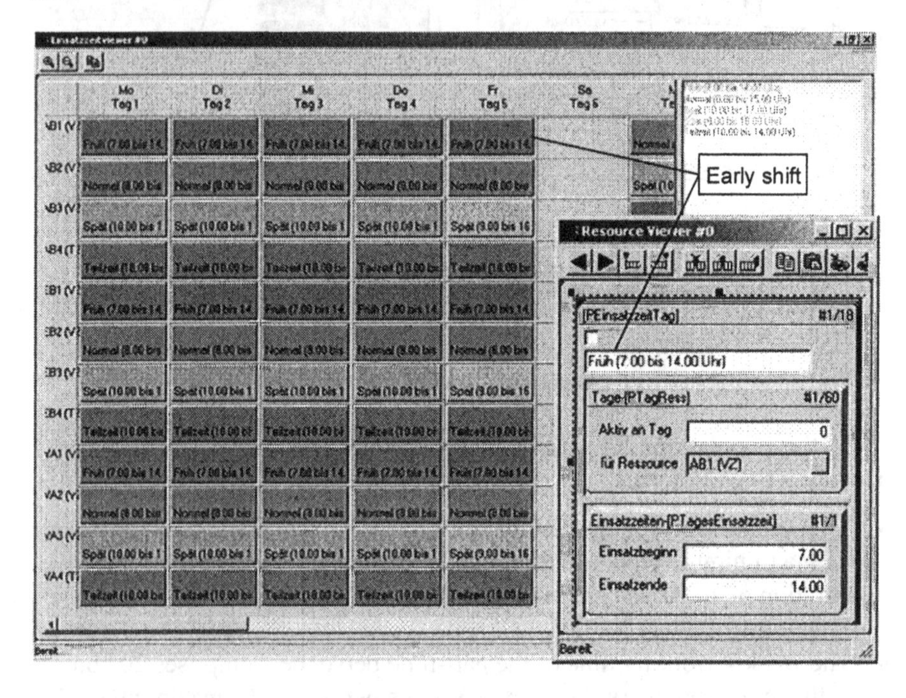

Figure 4. Modelling the working times of the employees

3.3 Development of the Simulation Alternatives

Based on the initial situation, various scenarios were developed for the working time systems as alternative configuration solutions. In the simulation study at hand the following alternatives were examined:

- *Alternative 1:*
  The personnel structure remains constant (meaning the same number of full-time and part-time personnel and in total the same personnel capacity as in the initial situation), and both the full-time and the part-time personnel work in a flexible working time model.

- *Alternative 2:*
  The personnel structure remains constant and the full-time personnel continue to work in a shift model, while merely the part-time personnel work in a flexible working time model.
- *Alternative 3:*
  The existing personnel structure is changed in that, with a constant personnel capacity, exclusively part-time personnel are employed, with flexible working times.

In order to define the concrete employment times the corresponding heuristic implemented in *OSim-GAM* is used for the simulation of these alternatives. In general this can lead to relatively complex employment plans, e.g. the employment plan for alternative 1 contains more than 50 different employment times.

3.4 Results of the Simulation Study

For the execution of a simulation study it is not sufficient to simulate the modelled working time systems one time only. Rather, it is necessary to repeat the simulation several times in order to eliminate random variations of the results (cf. NEUMANN, MORLOCK 1993, p. 702). Therefore, the initial situation as well as each of the three alternatives were simulated and evaluated with twenty different random number seeds. The degree of goal achievement for "workload", "service degree", "lead time", "physical stress", "time stress", "used capacity costs" und "process costs" are used for the evaluation of the alternatives (in particular in comparison with the initial solution).

In Figure 5 the results of the simulation study are presented as a comparison of the working time system alternatives and initial situation. Very differing consequences, dependent upon the simulated working time system, can thereby be seen. In some assessment criteria an alternative led to a marked improvement, but to an extent, also to a deterioration in others. The results are at least significant on the .01 level, with the exception of the degrees of goal achievement for lead time in alternative 1 and 2 as well as for process costs in alternative 2.

With respect to the logistical criteria, alternative 1, with its consistent flexible working time regulation, led to an improvement in the degree of goal achievement for workload and service degree, but to a deterioration of the lead time. The monetary assessment criteria attained the highest value here in comparison with all other alternatives. The higher customer service rate can be noticed in an increased degree of physical stress (seen here as a negative change to the corresponding degree of goal achievement).

In alternative 2 the restriction of the flexible working time to merely the part-time personnel brings with it a great improvement in the degree of physical stress of the workers in general. Since the minor deterioration of the lead time and the customer process costs are barely noticeable this is the alternative for which implementation can be suggested.

In contrast, alternative 3, with its total abandonment of full-time personnel, is not feasible for the service department at hand. Nevertheless, this alternative is worthy to be considered for the reorganisation of a further call-centres. This alternative is marked by the shorted lead times, but also by a slightly lower service degree. This in turn has an impact upon the process costs. Considering the personnel related aspects, the slightly increased workload also led to an increase in the physical stress.

Changes compared to the initial situation

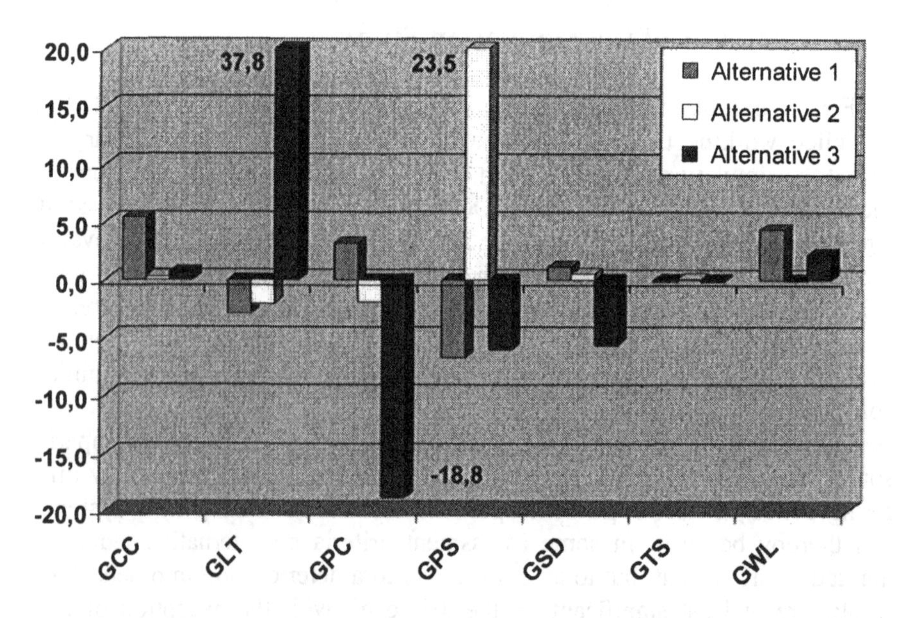

Goal achievement for ...

GCC: Used capacity costs
GLT: Lead Time
GPC: Process costs
GPS: Physical stress
GSD: Service degree
GTS: Time stress
GWL: Work load

Figure 5. Simulation results
(following to ZÜLCH, STEIH, STOCK et al. 2004)

4. CONCLUSIONS

Working time configuration has proven to be a highly complex problem, whose solution is influenced by very differing general conditions. Furthermore, the estimation of the consequences of working time systems is very problematic: Common evaluation methods such as benefit analyses deliver subjective, meaning relatively unreliable, results. Due to the considerable organisational changes, it is generally too laborious to introduce several working time systems into a business in order to prove them. Further difficulties arise from the fact that those responsible for working time configuration possess only limited knowledge regarding existing possibilities.

In particular in service departments, with customer-dependant personnel requirements, business-specific working time models are appropriate in order to provide the customers with the highest level of availability possible. The simulation-based approach has proven to be a suitable means of configuring and assessing working time systems: In the semi-industrial service sector, the expansion of the work content with indirect functions resulted in non-deterministic employment demands, which can only be represented and simulated adequately with probability distributions.

The presented personnel-oriented simulation procedure *OSim-GAM* contains a modelling concept for the generic representation of differing service departments and work systems. Furthermore, capacity-based employment plans in flexible working time models can be developed using an integrated heuristic. In the scenario simulation of the call-centre of a manufacturer of electric devices, various working time systems were simulated and assessed with this approach. The results extend beyond the recommendation of flexible working time regulation for part-time personnel, to the fact that qualification expansion could also lead to an improvement in the degree of goal attainment for various criteria.

ACKNOWLEDGEMENTS

This work was funded within the *FAZEM*-Project "The Effects of Flexible Working Time Models on Personnel Employment and Personnel Strain" by the German Federal Ministry of Education and Research within the framework program "Innovative Arbeitsgestaltung - Zukunft der Arbeit" (Innovative Work Configuration – The Future of Work) (Project No. 01HR9954). A detailed representation of the *FAZEM*-Project can be found in STOCK and BOGUS (2002), BOGUS and STOCK (2002), ZÜLCH, BOGUS, FISCHER (2002) or ZÜLCH, STOCK and BOGUS (2003).

REFERENCES

ACKERMANN, Karl-Friedrich:
Prozeßstandardisierung des Arbeitszeitmanagements. In: Innovatives Arbeits- und Betriebszeitmanagement.
Eds.: ACKERMANN, Karl-Friedrich; HOFMANN, Mathias.
Frankfurt/M., New York: Campus Verlag, 1990, pp. 183-212.

BOGUS, Thomas:
Simulationsbasierte Gestaltung von Arbeitszeitmodellen in Dienstleistungsbetrieben mit kundenfrequenzabhängigem Arbeitszeitbedarf.
Aachen: Shaker Verlag, 2002.
(ifab-Forschungsberichte aus dem Institut für Arbeitswissenschaft und Betriebsorganisation der Universität Karlsruhe, Vol. 31)

BOGUS, Thomas; STOCK, Patricia:
Simulationsbasierte Gestaltung von Arbeitszeitmodellen.
In: Arbeitszeitflexibilisierung im Dienstleistungsbereich.
Eds.: ZÜLCH, Gert; STOCK, Patricia; BOGUS, Thomas.
Aachen: Shaker Verlag, 2002, pp. 141-155.
(ifab-Forschungsberichte aus dem Institut für Arbeitswissenschaft und Betriebsorganisation der Universität Karlsruhe, Vol. 28)

BUNDESMANN-JANSEN, Jörg; GROß, Hermann; MUNZ, Eva:
Arbeitszeit '99. Ergebnisse einer repräsentativen Beschäftigtenbefragung zu traditionellen und neuen Arbeitszeitformen in der Bundesrepublik Deutschland.
Ed.: Ministerium für Arbeit, Soziales und Stadtentwicklung, Kultur und Sport des Landes Nordrhein-Westfalen.
Köln: Institut zur Erforschung sozialer Chancen, 2000.

FERREIRA, Yvonne; LANDAU, Kurt:
Umsetzungen flexibler Arbeitszeitregime am Beispiel eines deutschen Großflughafens.
In: Arbeitsgestaltung, Flexibilisierung, Kompetenzentwicklung.
Ed.: Gesellschaft für Arbeitswissenschaft.
Dortmund: GfA Press, 2001, pp. 245-250.
(Jahresdokumentation 2001)

HORNBERGER, Sonia; KNAUTH, Peter:
Innovative Flexibilisierung der Arbeitszeit.
In: Innovatives Arbeitszeitmanagement.
Eds.: KNAUTH, Peter; ZÜLCH, Gert.
Aachen: Shaker Verlag, 2000, pp. 23-49.
(ifab-Forschungsberichte aus dem Institut für Arbeitswissenschaft und Betriebsorganisation der Universität Karlsruhe, Vol. 22)

JONSSON, Uwe:
Ein integriertes Objektmodell zu durchlaufplanorientierten Simulation von Produktionssystemen.
Aachen: Shaker Verlag, 2000.
(ifab-Forschungsberichte aus dem Institut für Arbeitswissenschaft und Betriebsorganisation der Universität Karlsruhe, Vol. 21)

KNAUTH, Peter:
Was kann das betriebliche Zeitmanagement zur sozialverträglichen Gestaltung von Arbeitszeiten beitragen?
In: Sozialverträgliche Arbeitszeitgestaltung.
Eds.: BÜSSING, André; SEIFERT, Hartmut.
München: Rainer Hampp Verlag, 1995, pp. 209-220.

KRAMER, Ulrich; BURIAN, Klaus; GERBRACHT, Petra et al.:
Wettbewerbsstärke und bessere Vereinbarkeit von Familie und Beruf – kein Widerspruch.
Stuttgart, Berlin, Köln: Verlag W. Kohlhammer, 1998.
(Schriftenreihe des Bundesministeriums für Familie, Senioren, Frauen und Jugend, Band 152)

NEUMANN, Klaus; MORLOCK, Martin:
Operations Research.
München, Wien: Hanser Fachbuchverlag, 1993.

SCHREIER, Jürgen:
Flexible Arbeitszeitregelungen sind kein Privileg von Großunternehmen.
In: Maschinenmarkt,
Würzburg, 101(1995)16, pp. 20-21.

ZÜLCH, Gert; BOGUS, Thomas; FISCHER, Jörg:
Integrated Simulation and Workforce Assignment for the Evaluation of Flexible Working Time Models.
In: System Simulation and Scientific Computing.
Beijing: International Academic Publishers/Beijing World Publishing Corporation, 2002, Vol. I, pp. 353-357.

ZÜLCH, Gert; FISCHER, Jörg; JONSSON, Uwe:
An integrated object model for activity network based simulation.
In: WSC'00 Proceedings of the 2000 Winter Simulation Conference.
Eds.: JOINES, Jeffrey A.; BARTON, Russel R.; KANG, Keebom et al.
Compact disk: WSC'00, 2000, pp. 371-380.

ZÜLCH, Gert; STEIH, Marco; STOCK, Patricia et al.:
Simulationsbasierte Gestaltung von Arbeitszeitmodellen im industrienahen Dienstleistungsbereich. In: Betriebliche Tertiarisierung – Der ganzheitliche Wandel vom Produktionsbetrieb zum "dienstleistenden" Problemlöser.
Ed.: LUCZAK, Holger.
Wiesbaden: Deutschen Universitäts-Verlag, 2004 (in print).
(Schriftenreihe der Hochschulgruppe für Arbeits- und Betriebsorganisation, HAB-Forschungsbericht 15)

ZÜLCH, Gert; STOCK, Patricia; BOGUS, Thomas:
Working time recommendations for the load reduction of employees in retail stores. In: Human Performance and Aging, Proceedings Volume 4. Proceedings of the XVth Triennial Congress of the International Ergonomics Association and the 7th Joint Conference of Ergonomics Society of Korea / Japan Ergonomics Society "Ergonomics in the Digital Age".
Ed.: Ergonomics Society of Korea.
Seoul: Ergonomics Society of Korea, 2003, pp. 227-230.

PART TWO

Human Aspects in the Digital Factory

Impact of the Digital Factory on the Production Planning Process

Eberhard Haller, Emmerich F. Schiller and Ingo Hartel
DaimlerChrysler AG, Mercedes Car Group, D-71059 Sindelfingen, Germany.
Email: emmerich.schiller@daimlerchrysler.com

Abstract: Changing conditions in the automotive industry mean that assembly lines have to be built faster and at an increasingly early stage of the product development process. Since the required efficiency improvements could not be achieved with conventional planning processes, organisational structures and software systems, DaimlerChrysler AG introduced the *Digital Factory* strategic project in mid-2000. At the centre of this project is the complete and integrated digitisation of all areas of production planning. The following article describes the digitisation process – from the selection and introduction of suitable software and IT infrastructure, through the development of methods for efficient planning, to the changing and restructuring of the production planning process and the organisation.

Key words: Digital factory, Production planning process, Digital production planning

1. CHANGING CONDITIONS IN THE AUTOMOTIVE INDUSTRY

In order to maintain and enhance its leading position in an increasingly competitive market, DaimlerChrysler AG launched a wide-ranging model offensive several years ago (SCHILLER et al. 2003). In addition to developing new variants of existing models and vehicle types, a key component of this strategy was the introduction of entirely new vehicle concepts. Alongside this model offensive, DaimlerChrysler initiated a range of measures to significantly reduce product development times. For example, the introduction of CAD systems and the associated shift to computer-aided 3D modelling led to a noticeable shortening in the time needed for the design process. The introduction of a company-wide PDM system also simplified and therefore accelerated the product data acquisition and management processes (in particular those applicable to design data). But the changes went further than simply introducing and implementing new IT systems; the processes and workflows in product development were also restructured - especially those relating to product design.

A few years ago it became apparent that those technical and organisational changes needed to be matched by adjustments within the planning departments that take the designed vehicles and put them into production. Shorter development times, for example, mean that the associated assembly lines have to be planned faster and earlier. Production planning was therefore faced with the challenge of significantly improving the efficiency of its operations in order to keep pace with product development. There are two critical factors to this: Firstly, assembly lines will in future have to be both planned and tooled for an increasing range of vehicle models. Secondly, the introduction of new, innovative vehicles often demands the implementation of new production methods and processes, which must first be made ready for standard production by production planning.

As well as improving the efficiency of existing planning processes, production costs also need to be shaved. In addition to the continuous (in the sense of CIP) drive to reduce these costs, significant potential has to be realised with every type change in a vehicle programme. This is the only way for production at existing sites to remain profitable and competitive in the long term. In order to achieve real cost savings, production needs must be considered early on in the product development process. Much more so than in the past, production planning therefore has be involved in the product in a DFMA (Design for Manufacture and Assembly) sense from the early stages. It is well known that the cost of making changes in the later stages is disproportionately high. Volume production conditions can only be taken into account in the early phases of product design, however, if design engineers and planners work closely together in a *concurrent engineering* approach (HALLER, SCHILLER 2003). The aim of this type of co-operation is to achieve a win-win situation for both design engineering and planning, in other words shorter product development and production planning times combined with improved planning quality and greater maturity of product and production process.

In order to successfully implement *concurrent engineering*, communication and co-operation between the departments concerned are essential. However, this requires both design engineers and production planners to be operating on an equal footing in terms of both technology and processes. As far as the introduction and use of IT systems is concerned, the intensive digitisation process undertaken in design engineering over recent years has put it a good way ahead of production planning. Figure 1 shows how significant progress has been made in establishing digital methods in design engineering, whereas digitisation in conventional production planning is far less advanced. Here two-dimensional, alphanumeric planning supplemented by a number of individual simulation tools, such as material flow simulation, still dominates.

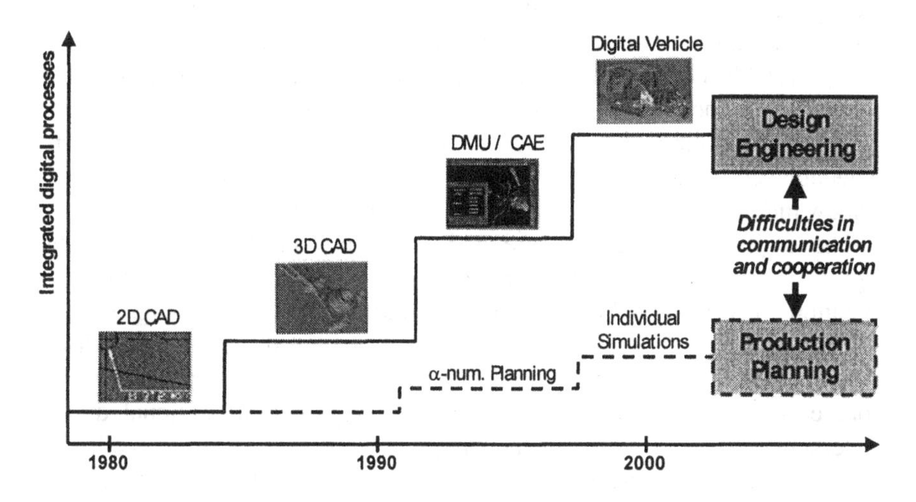

Figure 1. Digitisation in design engineering and production planning

In order to close this technology gap between design engineering and production planning, DaimlerChrysler AG launched the *Digital Factory* (DiFa) strategic project in mid-2003 (SCHILLER, SEUFERT 2003). The aim of this project is to digitise production planning in a way that is comparable to the introduction of CAD in design engineering.

2. DIGITISATION OF PRODUCTION PLANNING

A prerequisite for the digitisation of production planning was the choice of suitable software and the IT infrastructure needed to run it. As well as including the actual *planning functions* such as process modelling, planning validation, analysis/reporting and controlling, the software also had to allow the *integration of data* and applications between design engineering, process planning (e.g. for body-in-white or assembly) and plant planning (e.g. of the building or equipment). In data integration in particular, direct access by the planning departments to product geometry data and parts lists produced by design engineering is critical. The digitisation of production planning at DaimlerChrysler began at a good time since it coincided with a move by software companies that had previously developed CAD systems to start expanding and integrating planning functions into their existing systems.

On the IT infrastructure side, standard technology has been used systematically right from the start. In other words, the digital planning software has to run on conventional PCs with upgraded memory and graphics processing and display capabilities. This is the only way of ensuring that it can subsequently be distributed to several hundred employees. As well as providing production planners with digital planning capability at their workstations, a "power wall" has been set up at Mercedes-Benz' central production planning and design engineering facility in Sindelfingen specifically for the visualisation of major planning issues such as entire lines or plant halls. The 13 m^2 projection screen can be used to display large-scale production areas 1:1 and to show moving 3D simulations of production processes.

The digital planning software and IT infrastructure provided by the *Digital Factory* form the basic building block for integrated, digital co-operation between production planning and design engineering. Giving planners access to digital product data directly within the planning environment enables them to check the data against production requirements far more easily and thoroughly than before. Using three-dimensional digital models, planners can work out their requirements and suggested modifications – particularly regarding production-oriented product design under standard production conditions – at a very much earlier stage and on a much sounder basis than before (SCHILLER, SEUFERT 2004). In addition to the actual software and IT infrastructure, however, this approach also requires the provision of appropriate methods for efficient planning. Figure 2 shows that alongside intensive measures to *standardise* production resources and processes, such methods must include detailed definitions of *workflows* which set out the operational planning steps, including roles and responsibilities.

The following account demonstrates how these prerequisites were established in the *Digital Factory* using as an example the testing of welding guns as part of the planning process for body-in-white production. This test involves checking the geometrical accessibility for welding guns of all weld points on a vehicle. In conventional planning this used to be done using real welding guns on the first prototype vehicles to be assembled, but in the *Digital Factory* the test can be carried out on the digital vehicle. Data integration allows the planner to access not only the geometrical data for the product, but also the relevant weld point information (e.g. position and alignment of the weld point, associated components, weld point diameter). The functions included in the planning software allow the planner to check the weld points after entering predefined parameters. This simulation can be performed either manually or automatically and is based on the manufacturing processes (joining sequences) defined by the planner and displayed in the system. The welding guns used in the digital test come from an approved standard catalogue used by the Mercedes Car Group. The result of the test is a list from which the planner can determine which weld points are accessible with the selected standard guns and which are not, along with the minimum number of guns that are required. To ensure that in future this type of test is no longer restricted to specialist operators, a detailed workflow with clearly defined roles has been developed. In addition, production planners have received training in the steps involved so that simulation experts now rarely need to be called in.

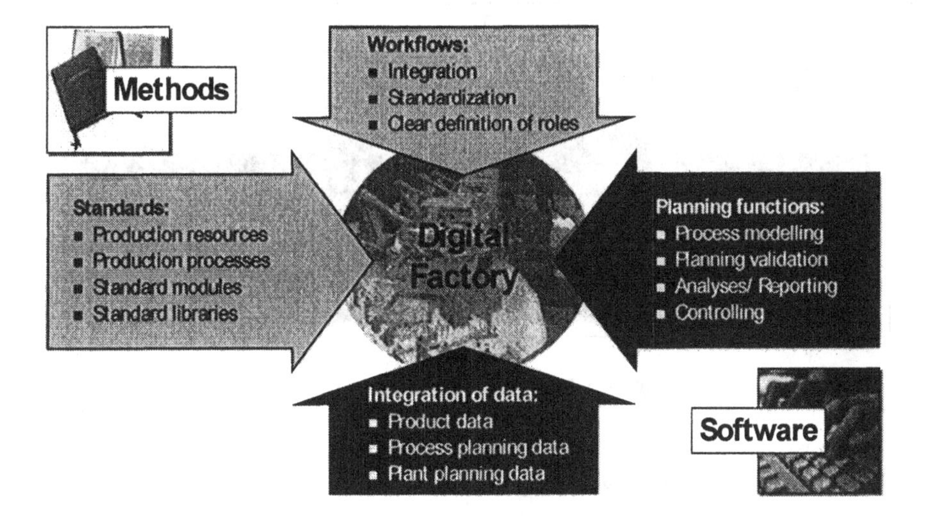

Figure 2. Prerequisites for the Digital Factory

The welding gun test and other methods developed in various vehicle projects that are currently in the development and planning phase have since been successfully used at an operational level. In order to achieve this, not only the software and methodology requirements outlined above had to be fulfilled but also the existing processes and organisational structures that were based on conventional planning had to be adapted for digital production planning. Thereby the new production planning process described below was developed and introduced as part of a comprehensive restructuring initiative

3. NEW PRODUCTION PLANNING PROCESS

The starting point for the design of the new planning process was the need arising from the changed conditions for a stronger product input from planning in the early stages of vehicle development. In line with the development process, which comprises a concept phase and a vehicle development phase, the new planning process is divided into a *standard planning phase* and a *real planning phase* (see Figure 3).

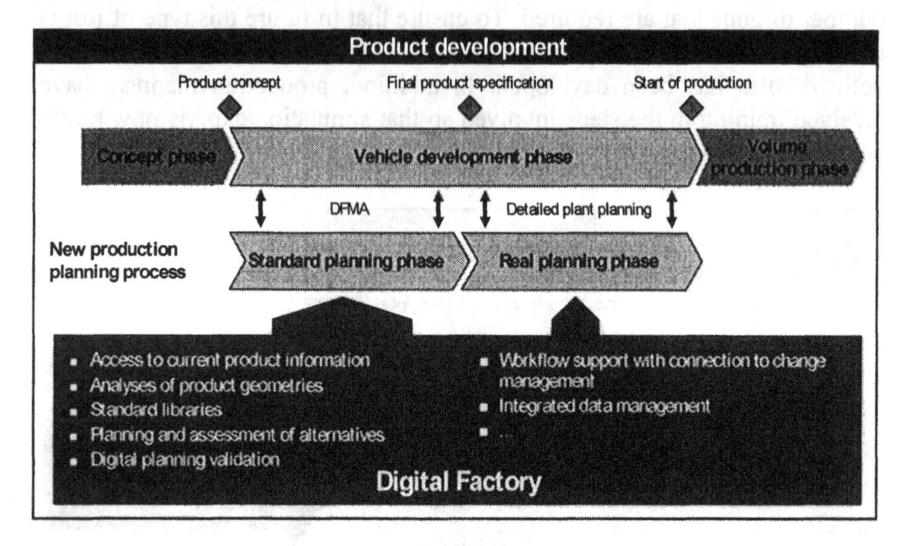

Figure 3. New production planning process

3.1 Standard Planning Phase

The emphasis during the standard planning phase is on working on the product – in other words production-oriented product design. Defined product statuses that exist at specific milestones (quality gates) in vehicle development are tested here under standard production conditions, using the latest product and 3D geometry data for defined vehicle configurations. This data can then be used as a basis for the start of digital planning for a vehicle programme. Suggested improvements and tentative solutions worked out by the planners during this phase can then be discussed with the design engineers on an ongoing basis.

The standard planning phase begins with the specification of the product concept around five years before the start of volume production. The end of the standard planning phase roughly coincides with the creation of the final product specification by design engineering. During this period the production planning activities are broken down into areas such as digital body-in-white, assembly, logistics/equipment and plant planning. Production issues in the various planning areas are taken into consideration using standard modules. By using individual elements from a range of standard libraries, alternative production lines (for body-in-white for example) can be very quickly constructed, analysed and modified if necessary, and also costed out and compared with one another. The results from the digital planning validation in particular are taken into consideration here. In the standard planning phase, the digital planning validation is responsible for the conceptual testing of the vehicle's production feasibility. For body-in-white production, this includes checking the geometric accessibility of the various weld points and the manufacturing sequence and concept or layout. Similarly, in assembly planning the digital planning validation provides results about the safe and ergonomic assembly of components under standard production conditions, and about assembly sequences at stations and on production lines.

3.2 Real Planning Phase

In the real planning phase the valid planning status from the standard planning phase is then transferred to a real, specific plant situation. The starting point for this phase is the "design freeze" which occurs when the final product specification for the entire vehicle is agreed. This gives the product and production concept a much greater stability than it has in the standard planning phase. In the real planning phase the production systems are planned in increasing detail as time goes on, taking account of physical conditions at the production site (e.g. existing buildings), technical conditions (e.g. degree of automation) and specified planning parameters (e.g. number of vehicles to be produced).

3.3 Benefits of the New Production Planning Process

This new planning process, with the *Digital Factory* as an essential element, enables production planners to find optimal solutions at a very early stage using digital data. This has been confirmed in a number of vehicle programme projects since the introduction of the new process. Furthermore, empirical evidence gathered over the course of these projects shows that the new production planning process offers real benefits in the following areas in particular:

- *Increased planning efficiency and planning speed* through the use of standard libraries, data integration and automation. Improvements of over 20 % are achievable here.
- *Improved quality of planning results* through systematic digital testing and plausibility checks. This can be seen in line capacity utilisation, for example, which has been increased by 10 to 15 % with the help of the *Digital Factory*.

- *Improved processes of agreement between engineering, planning and production* through integrated workflows, the ability to test alternatives and the visualisation of planning results. This has allowed optimisation discussions between design engineers and planners to be conducted on a much sounder footing, leading directly to a marked reduction in subsequent design changes.

This combination of benefits has brought about the required increase in the maturity of product and production process. The key factor here is the flow of requirements and findings from planning and production into the product development process. Problems can be identified far earlier than before and resolved jointly with appropriate measures. The digital planning validation is a particularly good example of this. With the new production planning process the joining and assembly sequences are also tested under standard production conditions using DMU. Previously this had to wait until the first real assembly of the vehicle in the zero series at the end of the vehicle development phase. The *Digital Factory* with the digital planning validation mean that any necessary changes can be made roughly 17 months earlier. Analyses comparing conventionally planned vehicle models that are still in production with new vehicle models that were planned and tested with the aid of the *Digital Factory* show that approximately 30 % of the problems that arise at the start of vehicle volume production can be avoided.

4. IMPACT OF THE DIGITAL FACTORY ON ORGANISATION AND CO-OPERATION BETWEEN DESIGN ENGINEERING AND PRODUCTION PLANNING

Early vehicle projects clearly showed the benefits of the *Digital Factory* through specific examples. Now the potential arising from the groundwork done in terms of software and methods and the new production planning process has to be realised across the board for future vehicle models. If the *Digital Factory* is to be established on a sustainable footing, a far-reaching and comprehensive process of change is required. The new production planning process and the potential offered by digital planning software will only bring lasting benefits if the organisation is also adapted accordingly or restructured where necessary. The first organisational changes to have been implemented with the aim of establishing the *Digital Factory* on a broad basis are shown in Figure 4. We will then consider two specific examples:

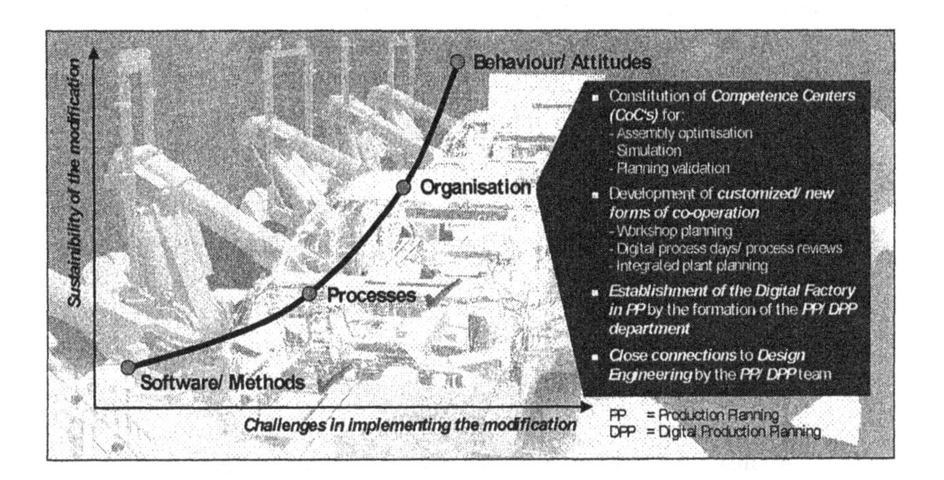

Figure 4. Organisational changes for the purposes of establishing the Digital Factory

- *Competence centre for assembly optimisation:* The *Digital Factory* offers expert tools for optimising assembly processes. To enable these to be used more widely, a special competence centre for assembly optimisation has been established. The centre comes under the production planning umbrella in the Mercedes Car Group. Its principal task is the planning, control and in some cases delivery of interdisciplinary workshops on the optimisation of assembly processes, e.g. balancing lines and stations. This has involved looking not only at vehicle programmes that are currently in the development and planning stage, but also at those that are in production today. Experts from assembly and logistics planning usually take part in the workshops. As they work together to develop and assess alternatives and find optimal overall solutions, staff from the competence centre offer help in the use of the various planning and testing tools from the *Digital Factory*. This procedure has increased capacity utilisation in the assembly process by up to 10 percent.
- *Digital process days and process reviews:* The purpose of digital process days is to visualise production planning statuses digitally, allowing them to be tested at an early stage. Whereas the competence centre provides continuous support for assembly planning, digital process days are one-off events. They are planned according to the progress of vehicle projects. In advance of each digital process day the necessary test parameters and scenarios are decided. The individual tests are then performed with the planning software using the latest product data. On the process day itself, the detailed results of the tests are then presented to representatives from design engineering and production planning with responsibility for

the vehicle projects. The current planning status is discussed within the group and assessed on the basis of checklists. Measures are devised to resolve any problems, in the form of design changes or process adjustments, for example. At the end of a digital process day the measures that have been devised are presented in a special management session to decision-makers from design engineering, production planning and production. In this way individual measures can be agreed immediately, and people nominated to ensure that they are carried out. In between digital process days there is also a mechanism for holding digital process reviews if required, where partial results from the digital planning validation can be presented to a smaller selected group of people, and preliminary measures decided upon together.

The organisational changes described here represent a major first step towards establishing the *Digital Factory* within DaimlerChrysler AG. Further steps are still needed, however. If the *Digital Factory* and Digital Production Planning (DPP) are to be established for the long term, a change in the behaviour and attitude of everyone concerned, at every level of the organisation, will also be required (see Figure 4).

To this end, a number of targeted activities have been undertaken since the start of the project. One of the most important of these involves training staff to use the *Digital Factory*. This begins with process training courses, where staff learn about the new processes, methods and procedures and about the changes they mean to the way they work. Only when planners have learnt about the new processes (workflows) and their consequences do they move on to systems courses on the use of the digital planning software.

Their subsequent hands-on work with the planning software is accompanied by psychology oriented work analyses. Interviews and activity logs are used to measure improvements in planning quality and the impact of the *Digital Factory* on planning contents and their time spread in comparison to conventional planning (SIMON, SCHILLER, HACKER 2004). Design recommendations for further developments to the methods and software in the *Digital Factory* are then derived from these results.

A new co-operation model, in which pilot users from the various departments work with staff from the *Digital Factory* project on current planning issues, has proven to be another useful way of broadening acceptance of the *Digital Factory*. Responsibility for the technical planning result still lies with the department, however. This new way of working has proven to be particularly popular with planners.

5. NEXT STEPS

After more than three years, it is clear that the *Digital Factory* project represents a major step along the road towards true concurrent engineering. With the introduction of digital planning software and the implementation of methods for digital planning and validation production planning have now made comparable progress towards universal digitisation like design engineering. Through the digitisation of production planning, the new planning process and the organisational changes that have been made, design engineers and planners are now able to work together and communicate "eye-to-eye" and with far less friction. This improved co-operation between design engineers and planners means that they can work together to realise their full joint potential. In future, production systems will not be planned, built or run until the product, the production processes and the production facilities have been digitally tested on the computer (SCHILLER, SEUFERT 2003). In contrast to conventional production planning, not only can solutions be reliably evaluated at a much earlier stage but alternative planning scenarios can also be developed quickly and at little cost. In this way, the *Digital Factory* enables the most promising alternative to be selected at a very early stage, thereby reducing the risk of subsequent changes and of wasted investment.

Despite – or possibly even because of – the advanced status of the *Digital Factory* at DaimlerChrysler, there are still things that remain to be done (HALLER, SCHILLER, SEUFERT 2003; HALLER 2003):

- In terms of *software and methods*, issues such as the integration of planning and layout data into existing EDM/PDM systems, integrating and managing large amounts of data, improving software performance, stability and user-friendliness, and extending the existing software planning functions and planning methods are paramount.
- In the area of *processes*, attention will be directed towards complexity management. To this end, roles, responsibilities, information flows and schedules need to be defined more precisely, and also described more clearly and simply.
- The changes to the *organisation* that have been made now need to be extended. More competence centres need to be built for specific areas, for example, and the new co-operation models must increasingly be made to apply across vehicle programmes. For example, as well as design engineers and planners, production staff (supervisors and operators) will increasingly take part in digital process days. In addition, organisational models allowing greater integration of suppliers into digital production planning need to be developed and implemented.

- With regard to changing the *behaviour* and *attitude* of staff to the *Digital Factory*, the goal is to increase motivation through participation, which means that all staff concerned must now be included in training programmes and courses.

REFERENCES

HALLER, E.:
„Planungszeit um bis zu 40 Prozent verkürzen".
In: Automobil Industrie,
Würzburg, (2003)7-8, pp. 46-47.

HALLER, E.; SCHILLER, E. F.:
The Significance of Digital Manufacturing within the New Planning Process.
In: Human Aspects in Production Management.
Eds.: ZÜLCH, Gert; STOWASSER, Sascha; JAGDEV, Harinder S.
Aachen: Shaker Verlag, 2003, pp. 18-23.
(esim – European Series in Industrial Management, Volume 5)

HALLER, E.; SCHILLER, E. F.; SEUFERT, W.-P.:
Herausforderungen beim Hochlauf der Digitalen Fabrik.
In: Innovation in der Automobilproduktion und Produktionsplanung – Ausgewählte Beiträge aus der industriellen Praxis.
Eds.: SCHILLER, E. F.; HALLER, E.
Aachen: Shaker Verlag, 2003, pp. 1-15.

SCHILLER, E.F.; ROLL, K.; WÖHLKE, G.; WIEGAND, K.:
Digital Manufacturing in Press Part Production.
In: Innovation in der Automobilproduktion und Produktionsplanung – Ausgewählte Beiträge aus der industriellen Praxis.
Eds.: SCHILLER, E. F.; HALLER, E.
Aachen: Shaker Verlag, 2003, pp. 63-75.

SCHILLER, E. F.; SEUFERT, W.-P.:
Bis 2005 realisiert – Digitale Fabrik bei DaimlerChrysler.
In: Automobil Produktion,
Landsberg, 16(2002)2, pp. 20-30.

SCHILLER, E.F.; SEUFERT, W.-P.:
Virtual Automotive Production.
In: Business Briefing-Global Automotive Manufacturing and Technology,
London (2003)2, pp. 68-71.

SIMON, A.; SCHILLER, E.F.; HACKER, W.:
(Digitale) Unterstützung komplexer Tätigkeiten in der Produktionsplanung: Anforderungen und erste Ergebnisse.
In: Arbeit + Gesundheit in effizienten Arbeitssystemen.
Ed.: Gesellschaft für Arbeitswissenschaft.
Dortmund: GfA Press, 2004, pp. 291-294.
(Jahresdokumentation 2004)

Integrating Human Aspects into the Digital Factory

New Tools for the Human-oriented Design of Production Systems

Gert Zülch
ifab-Institute of Human and Industrial Engineering, University of Karlsruhe, Kaiserstrasse 12 D-76131 Karlsruhe, Germany.
Email: gert.zuelch@ifab.uni-karlsruhe.de

Abstract: Currently, comprehensive tools are being developed which shall improve the process of factory planning. These tools are referred to as the "Digital Factory" and comprise diverse features, from capacity planning of production resources to visualisation and simulation of a virtual workshop. In a micro-ergonomic view, human-centred functionalities and ergonomic workplace design and assessment can also be included. From a macro-ergonomic point of view, however, the integration of personnel-oriented simulation is still missing. This paper describes the main functionalities of these features and hints at first pilot software for their integration into the Digital Factory.

Key words: Digital factory, Ergonomic workplace design, Personnel-oriented simulation

1. THE DIGITAL FACTORY: A COMPREHENSIVE TOOL FOR FACTORY PLANNING

1.1 Graphic Simulation of Future Production Systems

Currently, computer-aided tools are developed under the name "Digital Factory" and are intended to help improve and accelerate the planning of new production systems. Meant as a set of comprehensive tools for designing, visualising and even running of future production systems in a computer model, the Digital Factory promises to reliably prognosticate dimensions and performance of the future production system long before its realisation. For this purpose, graphic 3D-modelling and visualisation tools are combined with functionalities of event-driven simulation and evaluation tools.

The development of the Digital Factory started in the automotive industry (see SCHILLER, SEUFERT 2002), and further attempts are also being made in the aircraft industry. It can be expected that this new technology will gradually be applied in other industrial sectors as well.

Already existing tools of the Digital Factory are mainly directed towards technical and organisational issues. Their focus lies primarily on procedures for navigating in a virtual building, methods for capacity planning, balancing of production lines and features for animating physical manufacturing processes.

1.2 Micro- and Macro-ergonomic Simulation

Another development approach towards the Digital Factory has its origin in the study of ergonomic, which can be divided into the two different branches of modelling and simulation of humans at work: In a micro-ergonomic view, human postures and body motions are investigated, taking anthropometric and biomechanics basics into account (Figure 1). By using detailed virtual man models, firm insights into the practicability and reasonableness of working tasks shall be gained. Usually, only short-cycled tasks in a spatially limited work system are regarded.

In the macro-ergonomic view, the stress and strain of working humans during longer work cycles are investigated (for the distinction between stress and strain in ergonomics see LANDAU, BRAUCHLER, ROHMERT 1999). Here, specific questions like environmental working conditions during the shift, the personnel structure within complex assembly systems or physiological and psychological stress effects of manufacturing processes are answered. For this application field, personnel-integrated and personnel-oriented simulation tools have also been developed (section 3.1).

Micro- and macro-ergonomic tools should be combined and integrated into the Digital Factory. However, deeper insights show that both kinds of tools are not yet linked to one another. If at all, only micro-ergonomic tools are included in Digital Factory systems. Therefore, a different sets of tools have to be used for macro-ergonomic investigations, causing multiple data input and modelling, each of which have to be done by a different specialist.

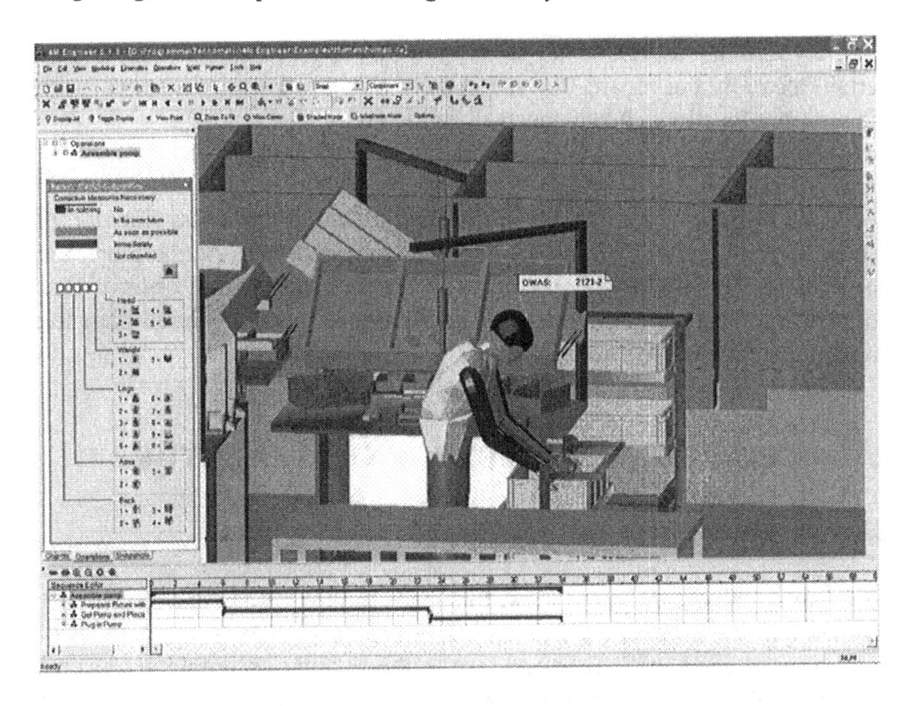

Figure 1. Ergonomic evaluation of a workplace with *eM-Human* by TECNOMATIX

1.3 Vision for Integrating Human Aspects

A prerequisite for overcoming this situation is the development of an adequate data model (see section 4.1). This data model can then serve as a common database for the relevant applications for integrating human aspects into the Digital Factory.

From a micro-ergonomic view, tools for workplace design are needed, such as man models for the verification of human postures and body motions. Such tools are already available (e.g. HALLER 1982 as an early example; *eM-Workplace* from TECNOMATIX 2003). However, these tools are not yet linked to simulation procedures for virtually running the production system in order to evaluate its performance in a production logistics sense (e.g. *Wittness* from LANNER GROUP 2003).

There exists an even more extended vision as to how human aspects should be integrated into the Digital Factory: Beyond well-known anthropometrical and biomechanical aspects of humans, also stress factors, originating either from the work task itself or from the environmental situation, should be prognosticated. Some of these factors, like illumination, can be regarded as constant, while others are influenced by the dynamics of the work process, e.g. the labour energy expenditure. Furthermore, the acoustic noise level as an environmental factor is in first approximation constant, at a certain location, but regarded in the context of a work task of a human with his different locations, it becomes a dynamic factor.

Therefore, from a macro-ergonomic view, the combination of an event-driven simulator for work processes and an evaluator for stress factors is needed in order to prognosticate in the Digital Factory the stress situation of a human during a shift or an even longer working period. Furthermore, the adequate assignment of workers to workplaces and the various forms of their cooperation must be considered. For this purpose, human-centred planning and simulation tools are needed.

2. PLANNING THE PERSONNEL STRUCTURE

2.1 Qualitative and Quantitative Assignment

From a macro-ergonomic view, this leads to the qualitative and quantitative assignment problem. The quantitative problem refers to the number of employees needed in a workshop, while the qualitative problem concerns their possibly divergent qualifications. The starting point for solving this problem is usually given by the pre-defined configuration of the future machinery and equipment of a workshop and a given manufacturing programme which is regarded as representative for a certain planning period. The solution is the personnel structure of the regarded workshop.

With respect to the planning horizon of personnel assignment, different problem areas can be distinguished (Figure 2). The greatest scope of possibilities for the assignment of persons to functions and workplaces in manufacturing systems is found in the case of planning a new factory or in the long-term planning of personnel structures. In this case, the number of employed persons and their abilities can be defined without considering an existing personnel structure. In contrast to the planning of a personnel structure from scratch, the middle-term reorganisation of personnel assignment has to consider the restrictions of an existing personnel structure. In this case, reorganisation means the modification of a personnel structure through training, hiring, and dismissal.

The chosen assignments of workers to functions and workplaces have an important effect on labour costs per manufactured unit and the achievement of logistical and financial objectives. In any case, the aim is to find an optimal transition from the existing personnel structure and to improve the organisational structure by minimising adjustment and personnel costs and maximising logistical goal achievements. New evaluation approaches also include human-oriented criteria (see section 3.2).

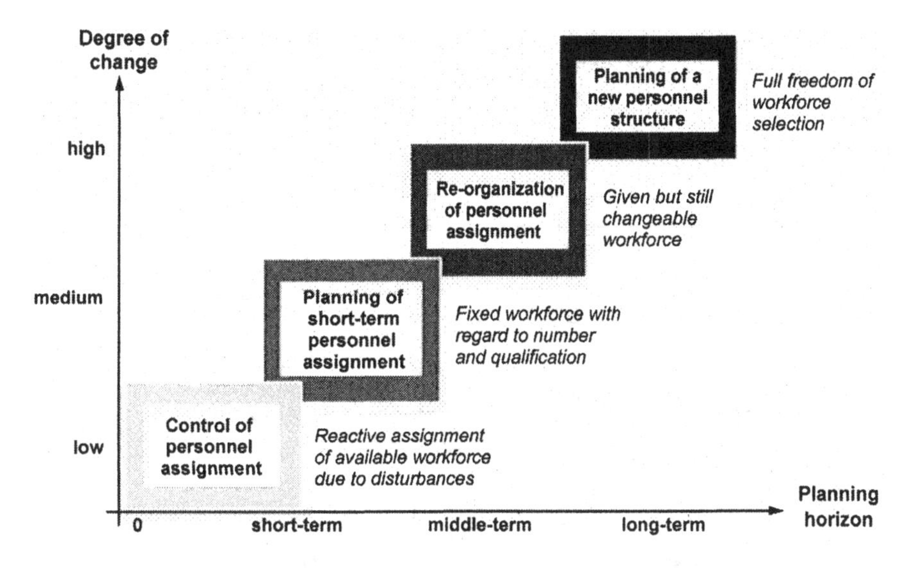

Figure 2. Time horizons for planning and scheduling personnel assignments

Such personnel structure problems can be modelled as a transportation problem, and therefore they ought to be solved by using well-known Operations Research algorithms (Figure 3). However, these algorithms give only a static result, without looking at the dynamics of the production system. The complexity of the problem derives from the numerous events which have to be considered.

There are many other methods for finding an optimal personnel assignment, but none of these incorporate the optimal number-of-workers problem or the optimal staff-skill-composition problem (COCHRAN et al. 1997, pp. 3393). In order to consider the plurality of possibilities for personnel assignments and to exploit the flexibility of human resources, effective planning tools are needed to solve such planning problems. For the time being, the problem can only be solved by investigating various scenarios and evaluating them with an appropriate simulation tool.

2.2 Simulation as an Evaluation Tool for Scenarios

For planning and evaluating potential personnel assignment solutions, the ifab-Institute of Human and Industrial Engineering of the University of Karlsruhe exploits the possibilities of using simulation as a planning tool. Figure 4 shows the general structure of one of them, namely *FEMOS* (German abbreviation for "Manufacturing and Assembly-Simulator"). Dependent upon the type of personnel assignment problem, various other simulation tools have been developed. Some of them are closely connected with planning games for educational purposes (ZÜLCH, BRINKMEIER 1995).

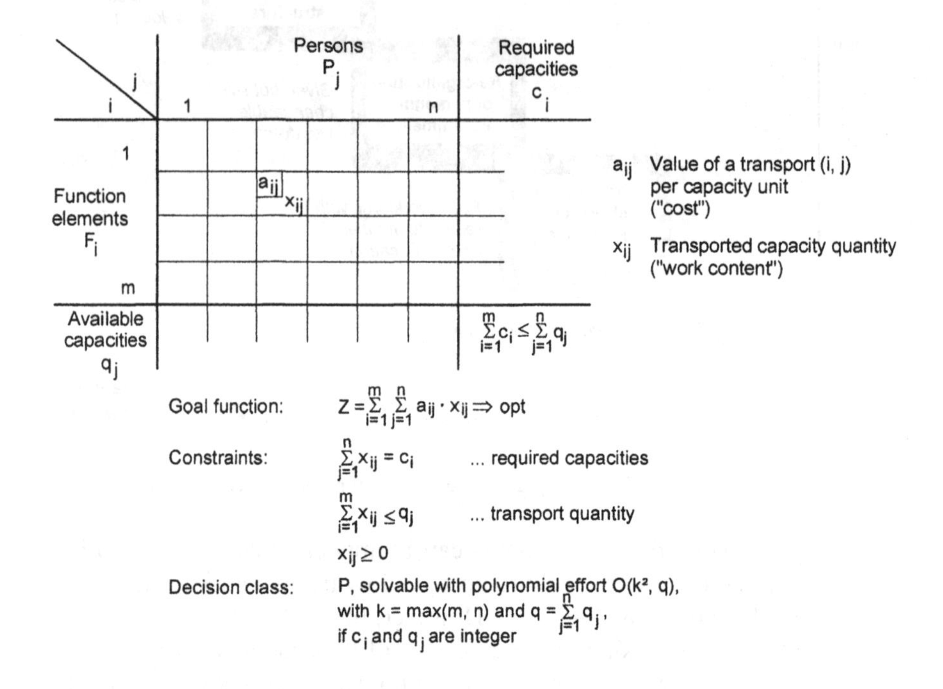

Figure 3. Modelling a personnel structure as a transportation problem (following ZÜLCH 1996)

All these procedures posses the same general approach to model personnel qualifications (ZÜLCH 1995, pp. 114): Different functions - meaning a set of similar activities - which are defined in the manufacturing programme have to be assigned to machines, or more generally to machine or workplace groups or even only to departmental units. Hence, there exists a relation between these functions and machines, which is referred to as feasibility. Every function requires a person or, more generally, a personnel type for its fulfilment, so that at least one person must be able to do this job. This relation is called ability/requirement. As a third relation, this person must be competent on the machine on which the job is performed. Hence, there are three relations connecting a function, a machine and a person which establish one qualification element of this person.

In order to model these relations, the simulation tool *ESPE* (German abbreviation for "Bottleneck-oriented Simulation of Personnel Structures") uses a so-called function-equipment-matrix (Figure 5). This matrix contains all needed functions as rows and the already set configuration of machines as columns. The matrix elements contain all function elements which appear in the production programme and which should be operated at the given machines or manual workplaces. Assigned to a person or personnel type, they are called qualification element (see the grey elements of the matrix). Usually, a set of function elements is assigned to a single worker or type of worker; this is then the qualification of this person or personnel type. As important logistical information, this matrix also contains the needed capacity for machines and personnel in the modelled workshop.

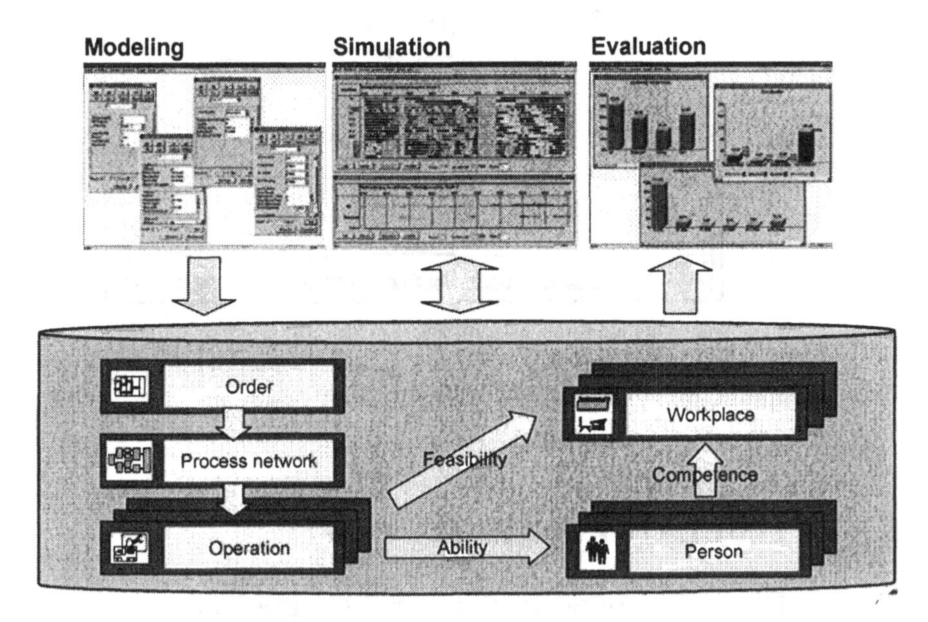

Figure 4. Components of the simulation tool FEMOS (following ZÜLCH, BRINKMEIER 1995, p. 94)

The mentioned simulation tools *FEMOS* and *ESPE* are discrete, event-driven, deterministic simulators with the possibility to model stochastic events like machine disturbances and personnel absenteeism. The given manufacturing programme consists of orders which are modelled as activity networks. Beyond the possibility to model various kinds of personnel assignment, these simulators also allow for changes related to the intended order programme of a workshop as well as for the reconfiguration of the machinery equipment. Thus, various scenarios can be investigated in a simulation study.

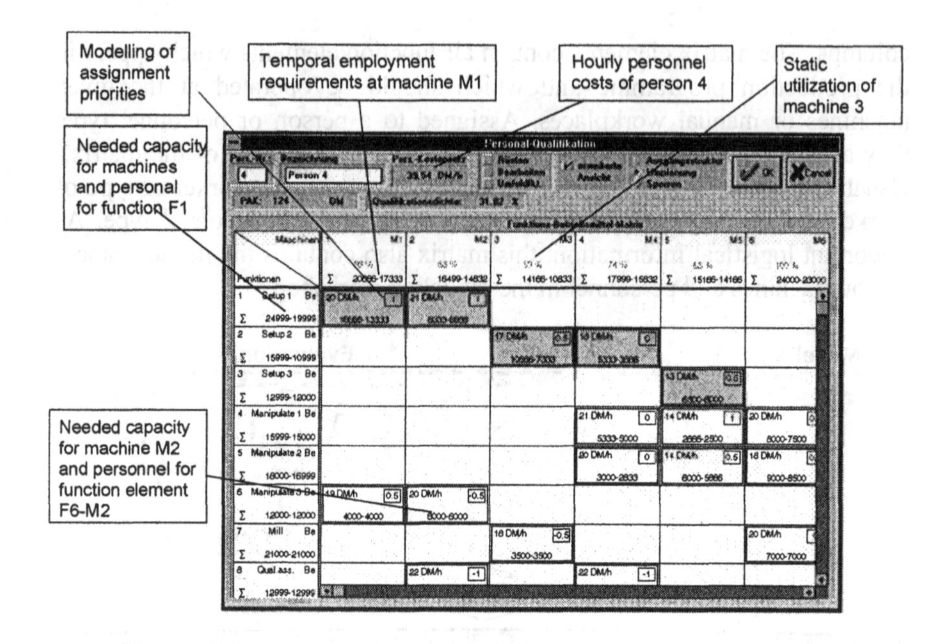

Figure 5: Function-equipment-matrix in the simulation tool *ESPE* (following HEITZ 1994, p. 112)

3. SIMULATION OF HUMAN ASPECTS

3.1 Distinction between Personnel-integrated and Personnel-oriented Simulation

In order to clarify the principles of human-centred simulation, the Association of German Engineers (Verein Deutscher Ingenieure - VDI) published a guideline for the modelling of personnel in simulation models (VDI 3633, part 6, 2001; ZÜLCH, VOLLSTEDT 2000). This guideline concerned with applications in the field of production and logistics only. With respect to the degree of detail in human models, different kinds of simulator are distinguished. The related tools are divided into personnel-integrated and personnel-oriented simulators. Personnel-integrated simulation tools regard the significant properties of the staff employed in production and logistics systems. Their main focuses are applications, in which interactions between workers and machines play an important role.

There are some minimum requirements which personnel-integrated simulation models and tools have to fulfil: First of all, a distinction between the capacities of the personnel and the machines must be possible. Figure 6 shows the animation output of a Gantt-chart which contains these different views of machines and persons.

Furthermore, the determination of the effects of changed work structures, e.g. the personnel capacity requirements for a multi-machine operation or for production cells, must be possible. Therefore, personnel-integrated simulation tools have to offer features for modelling co-operation and group work. Additionally, individual modelling of the working times of the personnel and the operation times of the machines is a main prerequisite for a personnel-oriented simulation tool.

Compared to personnel-integrated simulation tools, personnel-oriented simulators possess a higher degree of detail for answering special human-related questions. Here, the analysis of certain organisational forms and work conditions with respect to the related effects on the production processes and outputs are of particular interest.

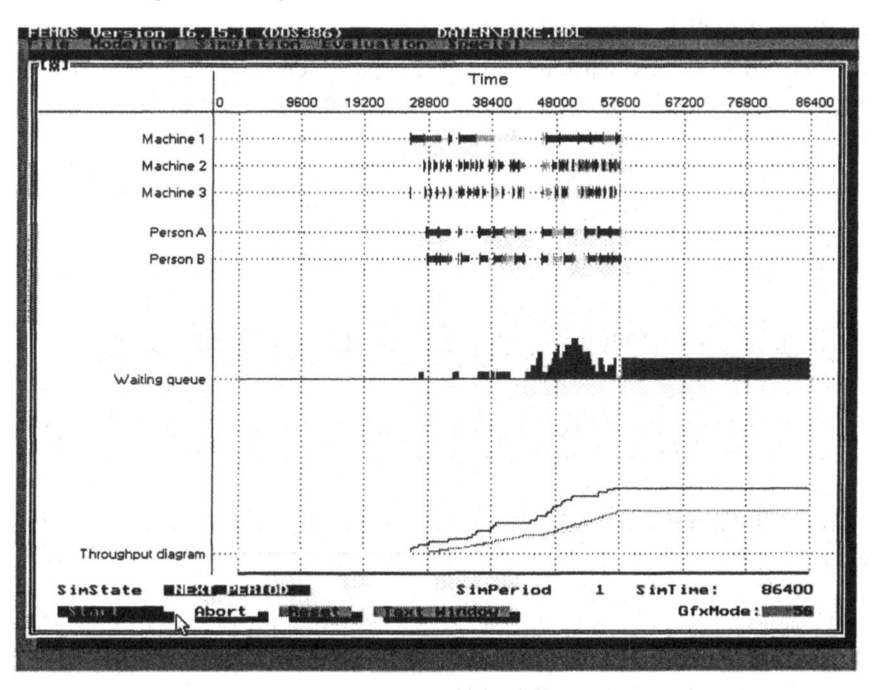

Figure 6. Gantt-chart of order processing, waiting queues and throughput-diagram of a production system in the simulation tool *FEMOS*

For example, emphasis may be put upon on the analysis of the work load and stress for humans, the over- or under-utilisation of workers, learning and unlearning effects, or even the human reliability depending upon the workplace design and work organisation. In comparison to personnel-integrated simulation, these tools require additional information about the work tasks, the environmental conditions and the characteristics of the modelled personnel.

3.2 Human-centred Evaluation

Personal-oriented simulation tools should be used if the effects of human behaviour on the logistical performance of the production system are a central point of interest. This may concern logistical key data like production output, utilisation of resources, lead time degree or tardiness of orders. Furthermore, financial aspects may also be of interest, especially manufacturing costs and process costs.

Beyond this, human-centred evaluation criteria may be used in order to characterise the working conditions for the individual workers or type of worker under investigation. Examples of such studies from an ergonomic point of view are (see VDI 3633, part 6, pp. 7):

- the staff work load with respect to the capacity demand,
- the physiological stress with respect to the working times system,
- the individual energy expenditure, dependent upon the performed activities and the occupied work posts,
- the influence of the qualifications of a worker on his human error probability,
- the degree of overstrain, fatigue, monotony and wear-out dependent upon the performed work tasks, and
- the sequential completeness of performed tasks, i.e. their coverage of planning, operating and controlling aspects.

Even from an industrial sociology point of view, investigations questioning the autonomy, decentralisation and integration of tasks within a departmental unit become possible (Figure 7). Most of these criteria seem to be reasonable, but, from a scientific perspective, a broad field of studies for their validation is opening up.

4. PROGNOSIS OF HUMAN ASPECTS IN THE DIGITAL FACTORY

4.1 Object-oriented Data Model

Some of these modelling and evaluation approaches are ready to be integrated into the Digital Factory. However, the amount and complexity of data and methods demand an adequate system for production data management. The critical question as to which data base architecture is the right one to guarantee consistency, actuality and flexibility for expansion, beyond all the other criteria of data management.

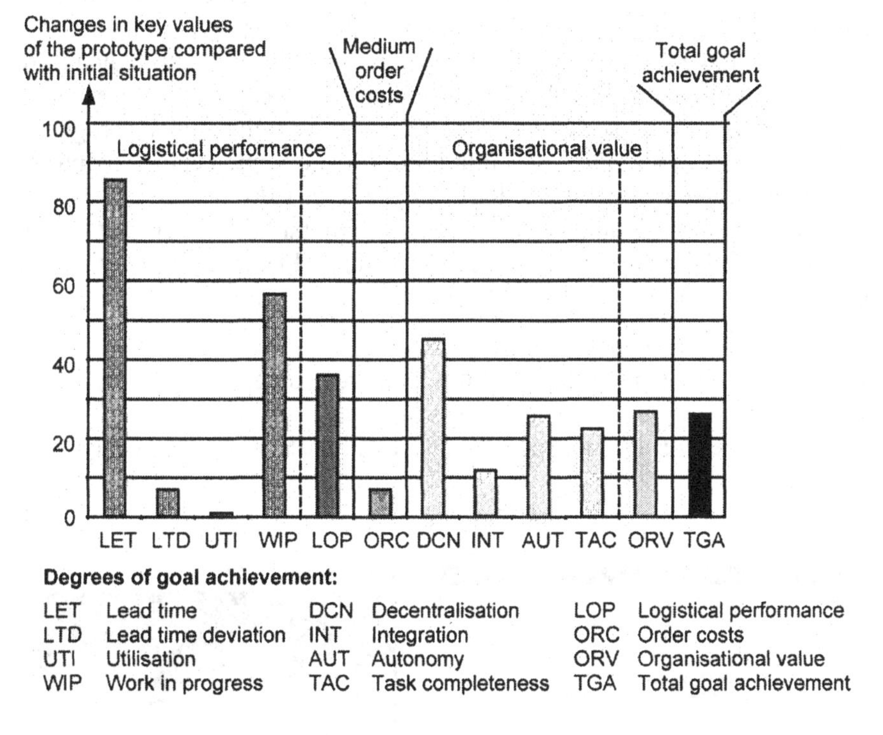

Figure 7. Simulation results of organisational changes in a mechanical component production (following BRINKMEIER 1998, p. 221)

A feasible path can be seen in object-oriented modelling techniques and related tools like *UML* (Unified Modelling Language; FOWLER, SCOTT 1999). The basis of the approach presented here is an object-oriented Product/Production Model. This data model has been developed as basis for the integration of various tools for product design, planning of production systems and operations planning in parts production (ZÜLCH, BRINKMEIER 1998).

4.2 The Information and Planning System *ADAMO*

As a pilot system for the integration of human aspects into the Digital Factory, the ifab-Institute of Human and Industrial Engineering of the University of Karlsruhe is developing a workshop modelling and evaluation tool called *ADAMO* (German abbreviation for "Arbeitsschutzdaten-Modellierer"; KELLER 2001, pp. 142). This information and planning system is designed to incorporate aspects of occupational health and safety (OHS) in the CAD-representation of a workshop.

The pilot system is based on the abovementioned object-oriented database-structure in which the objects of a workshop, together with their OHS-relevant attributes and applicable calculations and visualisation methods, are stored. The integration of calculation methods in this database has been supported by the use of sequence-diagrams. As a first application, prognosis methods for acoustic noise and illumination were implemented (Figure 8) and produced promising results (KELLER 2001, p. 161).

The integration of human aspects into the Digital Factory has just started and is yet producing a number of questions and problems. However, the goal is clear:

- A comprehensive tool for factory planning,
- offering consistent, redundancy-free data processing,
- which delivers human-related prognoses data,
- beyond the already well-known production logistics and cost-related key figures.

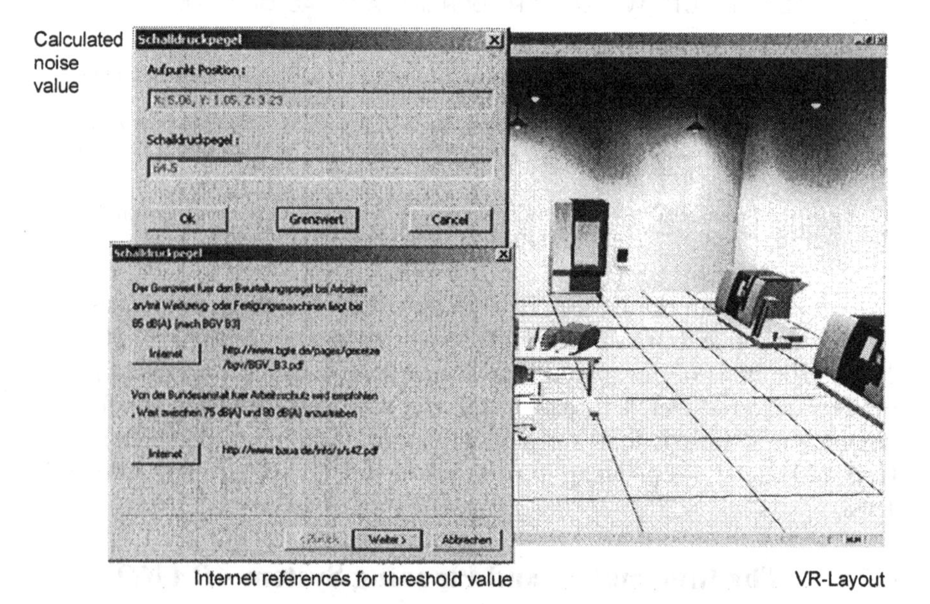

Figure 8. Calculation of the acoustic noise level and the respective threshold value in the OHS-planning tool *ADAMO*

REFERENCES

BRINKMEIER, Bernd:
Prozeßorientiertes Prototyping von Organisationsstrukturen im Produktionsbereich.
Aachen: Shaker Verlag, 1998.
(ifab-Forschungsberichte aus dem Institut für Arbeitswissenschaft und Betriebsorganisation der Universität Karlsruhe, Band 17)

COCHRAN, J. K.; CHU, D. E.; CHU, M. D.:
Optimal staffing for cyclically scheduled processes.
In: International Journal of Production Research,
London, 35(1997)12, pp. 3393-3403.

FOWLER, Martin; SCOTT, Kendall:
UML Distilled.
Boston MA: Addison-Wesley Professional, 2nd ed. 1999.

GROBEL, Thomas: Simulation der Organisation rechnerintegrierter Produktionssysteme.
Karlsruhe Univ.: Institut für Arbeitswissenschaft und Betriebsorganisation, 1992.
(ifab-Forschungsberichte aus dem Institut für Arbeitswissenschaft und Betriebsorganisation der Universität Karlsruhe, Band 3)

HALLER, Eberhard:
Rechnerunterstützte Gestaltung ortsgebundener Montagearbeitsplätze, dargestellt am Beispiel kleinvolumiger Produkte.
Berlin et al.: Springer-Verlag, 1982.
(IPA Forschung und Praxis, Band 65)

HEEL, Jochen:
Reorganisation des Personaleinsatzes mit Hilfe der personalorientierten Simulation.
Aachen: Shaker Verlag, 1999.
(ifab-Forschungsberichte aus dem Institut für Arbeitswissenschaft und Betriebsorganisation der Universität Karlsruhe, Band 18)

HEITZ, Max-Jürgen: Ein engpaßorientierter Ansatz zur simulationsunterstützten Planung von Personalstrukturen. Karlsruhe Univ., Diss. 1994. (ifab-Forschungsberichte aus dem Institut für Arbeitswissenschaft und Betriebsorganisation der Universität Karlsruhe, Band 7 - ISSN 0940-0559)

KELLER, Volker:
Ansatz zur objektorientierten Modellierung betrieblicher Arbeitsschutzdaten. Aachen: Shaker Verlag, 2002.
(ifab-Forschungsberichte aus dem Institut für Arbeitswissenschaft und Betriebsorganisation der Universität Karlsruhe, Band 25)

LANDAU, Kurt; BRAUCHLER, Regina; ROHMERT, Walter:
The AET Method of Job Evaluation.
In: The Occupational Ergonomics Handbook.
Eds.: KARWOWSKI, Waldemar; MARRAS, William S. Boca Raton, FL: CRC Press, 1999, pp. 355-370.

LANNER GROUP (Ed.):
WITNESS Simulation Software.
http://www.lanner.com/corporate/technology/witness.htm, 24.06.2003.

SCHILLER, E.F.; SEUFFERT, W.-P.:
Bis 2005 realisiert - Digitale Fabrik bei DaimlerChrysler.
In: Automobil-Produktion,
Landsberg/Lech, 16(2002)2, pp. 24-32.

TECNOMATIX TECHNOLOGIES (Ed.):
eMPower Products: eM-Workplace NT.
http://www.tecnomatix.com/showpage.asp?page=653, 24.06.2003.
VDI 3633, part 6:
Simulation von Logistik-, Materialfluss- und Produktionssystemen.
Blatt 6: Abbildung des Personals in Simulationsmodellen.
Berlin: Beuth-Verlag, 2001.
ZÜLCH, Gert:
Aspekte weiterführender Forschungsarbeiten zur Simulation der Arbeitsorganisation.
In: Neuorientierung der Arbeitsorganisation.
Ed.: ZÜLCH, Gert.
Karlsruhe Univ.: Institut für Arbeitswissenschaft und Betriebsorganisation, 1995, pp. 109-132.
(ifab-Forschungsberichte aus dem Institut für Arbeitswissenschaft und Betriebsorganisation der Universität Karlsruhe, Band 8)
ZÜLCH, Gert:
A Heuristic Solution to the Personnel Structure Problem.
In: ORBEL10, Proceedings of the Tenth Conference on Quantitative Methods for Decision Making.
Ed.: JANSSENS, Gerrit K.
Brussels: Belgian Operations Research Society, 1996.
ZÜLCH, Gert:
Arbeitsschutz zwischen Umsetzungsdrang und Forschungsbedarf. In: Arbeitsschutz-Managementsysteme.
Eds.: ZÜLCH, Gert; BRINKMEIER, Bernd.
Aachen: Shaker Verlag, 2000, pp. 185-201.
ZÜLCH, Gert; BRINKMEIER, Bernd:
Simulation Aided Planning of Work Structures.
In: Simulation Games and Learning in Production Management.
Ed.: RIIS, Jens O.
London et al.: Chapman & Hall, 1995, pp. 91-104.
ZÜLCH, Gert; BRINKMEIER, Bernd:
Object-oriented Product/Production-Model.
In: Strategic Management of the Manufacturing Value Chain.
Eds.: BITITCI, Umit S.; CARRIE, Allan S.
Boston, Dordrecht, London: Kluwer Academic Publishers, 1998, pp. 577-584.
ZÜLCH, Gert; VOLLSTEDT, Thorsten:
Personnel-integrated and Personnel-orientated Simulation - A New Guideline of the German Association of Engineers.
In: Information and Communication Technology (ICT) in Logistics and Production Management.
Eds.: STRANDHAGEN, Jan Ola; ALFNES, Erlend. Tromsø, 2000, pp. 185-192.
ZÜLCH, Gert; BRINKMEIER, Bernd:
Introduction to the CAESAR Planning Games.
In: Production Management Simulation Games.
Eds.: ZÜLCH, Gert; CANO, Juan Luis; MULLER(-MALEK), Henri.
Aachen: Shaker Verlag, 2001, pp. 3-16.
(esim – European Series in Industrial Management, Volume 4)

ZÜLCH, Gert; BRINKMEIER, Bernd:
Prototyping betrieblicher Organisationsstrukturen.
In: Innovative Organisationsformen.
Ed.: WOJDA, Franz. Stuttgart: Schäffer-Poeschel Verlag, 2000, pp. 435-462.
(HAB-Forschungsberichte der Hochschulgruppe Arbeits- und Betriebsorganisation, Band 12)

Human Aspects in Manufacturing Process Management

Manuel Geyer and Stefan Linner
Tecnomatix GmbH, Richard-Reitzner-Allee 8, D-85540 Haar bei Muenchen, Germany.
Email: manuel.geyer@tecnomatix.com

Abstract: The globalisation of world economy, constantly rising cost pressure in international competition and ever shorter product life cycles create new challenges for production planning. The concept of Manufacturing Process Management (MPM) provides an integrated working environment for production planning and a comprehensive platform to optimise production planning throughout the extended enterprise. The core of the concept is a database hosting a process model that describes manufacturing operations as well as required resources and affected product components. Various planning tools are connected to this database. Ergonomically safe workplaces can be defined using an integrated 3D human model. Ergonomic standards and best practices for manual working operations can be defined and implemented in a very effective way. Hence the planning of production facilities can be done faster with higher quality.

Key words: Production planning, Digital factory, MPM, Ergonomics, Human model, Material flow simulation

1. CHALLENGES FOR PRODUCTION PLANNING

The environment for production planning has become more and more complex. High quality standards and the cost pressure in today's global economy require high quality planning results. At the same time shortening product life cycles, the need to reduce time-to-market and time-to-volume as well as trends like build-to-order put tremendous pressure on the planning.

To respond to such challenges, planning has to start earlier and the planning process itself needs to get more efficient. Starting earlier means, that the product design is still 'unstable' and the planners need to handle many product changes that complicate their planning work. Simultaneous engineering creates the need for improved collaboration and communication both within planning teams as well as with other disciplines within a company or even the supply chain. On top of all that globalisation changes the way how companies operate and creates new challenges like working in geographically distributed teams, transferring proven processes from one plant to another, standardising processes and resources.

2. MANUFACTURING PROCESS MANAGEMENT

Manufacturing Process Management (MPM) of Tecnomatix offers a business strategy for the collaborative development and optimisation of manufacturing processes across the extended enterprise. MPM allows multi-users at multi-sites to collaborate as one single enterprise throughout the entire development of a manufacturing process. MPM leverages specific technologies and methodologies to create a collaborative environment for authoring, simulating and managing manufacturing processes.

MPM addresses the area of manufacturing that has historically been the most neglected area of the industrial process – the actual planning and creation of the manufacturing process. While CAD and PDM address "what" products to manufacture, and MES (Manufacturing Execution Systems) and ERP (Enterprise Resource Planning) address "when" and "where", MPM addresses "how" (Figure 1).

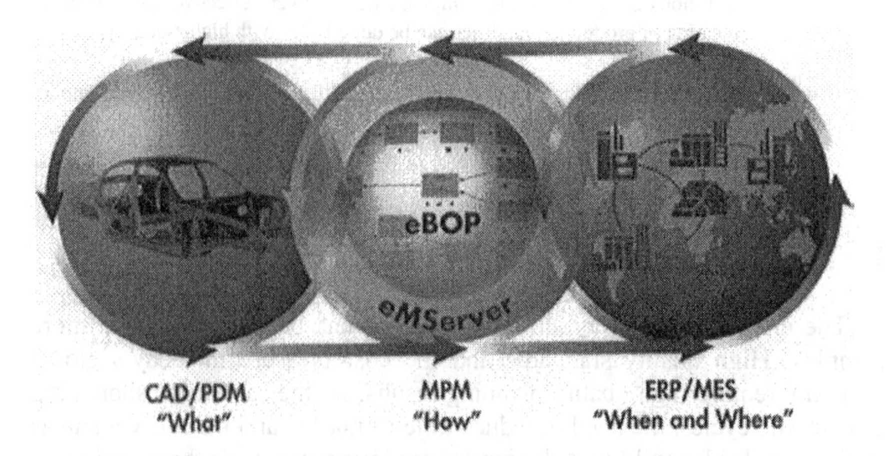

Figure 1. MPM integrates with CAD/PDM and ERP/MES systems (Source: LINNER, GEYER 2003, p. 38)

MPM is a critical component of e-Manufacturing, a broader business strategy that addresses the full industrial process, from the initial concept of a new product until it is delivered to the customer. MPM enables manufacturers to define how the product is going to be manufactured and then to deliver these processes to the shop floor.

While the product design process defines an electronic bill of materials - the "what" - MPM defines an electronic bill of processes (eBOP) - the "how." This information is stored on a server that allows easy access and management of data throughout the enterprise. MPM contributes a common language for the description of manufacturing processes and offers a collaborative environment that facilitates the exchange of information. The eBOP serves both as a clear and defined way of how to describe the manufacturing process and as an information carrier that allows for packaging all information about a process or a manufacturing line in order to send it to someone else.

The concept is realised through a database hosting a process model that describes manufacturing operations as well as required resources and affected product components. Various planning tools are connected to this database - spanning everything from digital mock-up analysis, to defining the production steps, optimising the workplaces with a human model and creating documentation (Figure 2).

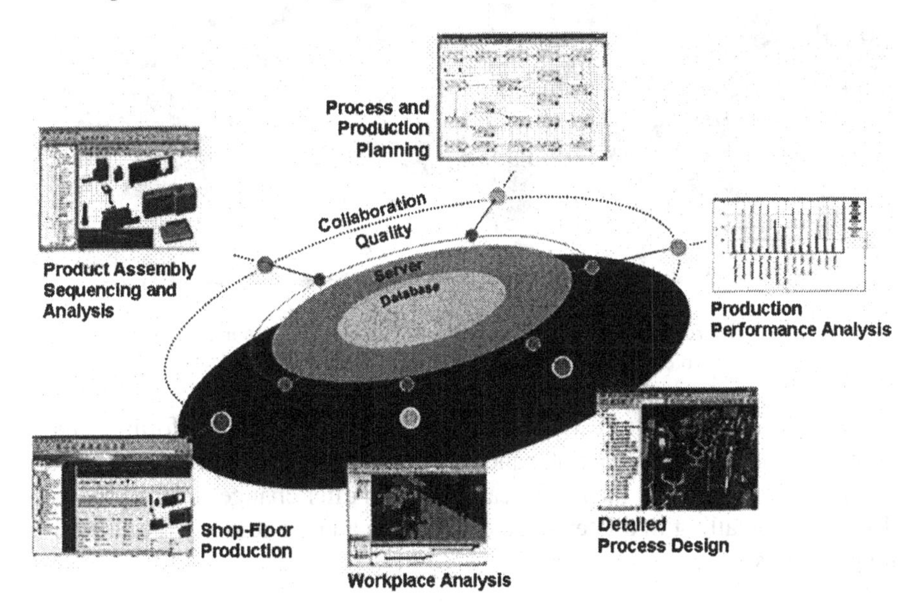

Figure 2. End-to-end Manufacturing Process Management

3. HUMAN ASPECTS IN PRODUCT DESIGN

The design of the product itself obviously has a large impact on the required manufacturing processes. In the early stages of product design, the product is only available as a virtual CAD model, so a virtual environment for simulating the manufacturing aspects of the product design is necessary.

Verifying the manufacturing aspects of the product requires sound domain knowledge and experience and thus must be done by manufacturing planners. The 3D software therefore is specifically tailored to the needs of planners. In this virtual environment the planner will simulate the future assembly sequences and processes to find out problematic areas and to enable with his feedback to the designer an improvement of the product.

Typically used methods are simulation of the assembly paths of the parts, attaching a human hand or the high-end human model Ramsis (Figure 3). The checks in many cases will also include the future production equipment. This simultaneous engineering process enables a cost-effective, ergonomically safe and high-quality manufacturing process.

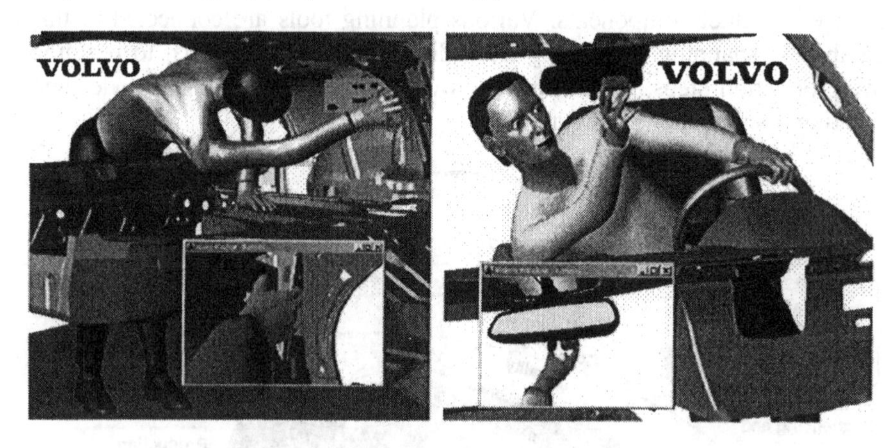

Figure 3. Ergonomic studies of assembly tasks using the 3D human model (pictures provided by Volvo Corporation, Sweden)

In complex products (cars, aircrafts) the planner has to perform a high number of assembly and maintenance studies. The studies can be saved in a structured way in the database for easy retrieval after changes of the product design, for detailing in a later stage of planning and for reuse in future planning projects.

4. HUMAN ASPECTS IN WORKPLACE DESIGN

After rough planning of the layout structure and material flow the details of the workplaces are planned. Optimising manual workplaces and processes is usually driven by the need to minimise production costs. However it is clear, that skilled, healthy and motivated employees are representing a high value for the enterprise. Thus ergonomics and human simulation is a key point in MPM.

As described before, the human model is integral part of the 3D-planning environment (Figure 4). The user loads the human model into his current planning environment to evaluate the workplace configuration according to several ergonomic analysis methods. Most data (e.g. layout configuration, operation plans) required to perform a human simulation is already present on the server. This greatly improves the acceptance of the planners in performing an ergonomic evaluation.

In addition to the standard methods of ergonomic analysis (NIOSH, OWAS, Burandt-Schultetus, Rula) typically various company- and country-specific ergonomic standards are used. A key point for acceptance of such a tool is the ability to customise the relevant methods with low effort to the regulations of the local production site.

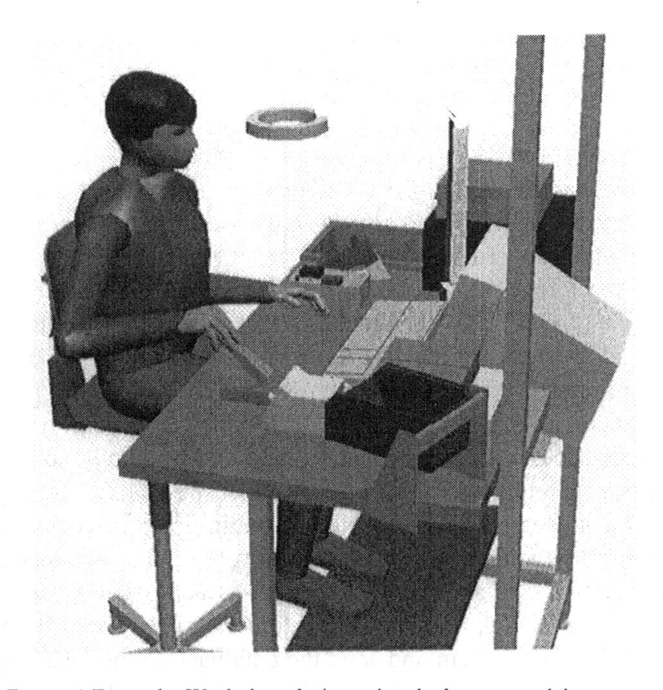

Figure 4. Example: Workplace design using the human model

5. HUMAN ASPECTS AND MATERIAL FLOW SIMULATION

While the 3D simulation with human model mainly applies to analysing and optimising the ergonomic quality and cycle times of individual workplaces, the interaction and performance of the entire production is today analysed and optimised using material flow simulation.

This can also take into consideration aspects of human labour, such as stress, fatigue, learning and unlearning phases, pauses, shift cycles and the qualification of the worker. These aspects can thus be verified early in the planning process.

6. IMPLEMENTATION ON THE SHOPFLOOR

Finally MPM will ensure the implementation of the optimised work practices and workplaces. Various reports can be created that aggregate information out of the process database. From dimensioned 3D workplace drawings to electronic work instructions all necessary data can be generated and communicated.

7. COLLABORATION AND COMMUNICATION

Different levels of tools support different needs of collaboration and range from pure viewing to real collaborative process planning. They range from generating paper-based reports to online-access to the current planning database with standard Web-Browsers.

They may, for example, be used to deliver up-to-date information of the current planning state to management, project managers or other team members. In addition the ability to use a common language for describing production processes enables not only information, but a true collaboration: Typical scenario is a company that locally plans an assembly line for a new product, but does not have specialised knowledge on ergonomics. The local manufacturing engineer will do the layout of the line and conceptual design of equipment. He will now send all relevant information including the bill of materials, parts geometries, work content to be done, workplace configuration, as an eBOP to a corporate ergonomics competence centre in another location with the click of a button. The ergonomic expert will do the optimisation of the work system and send the data back to the local engineer (Figure 5). Of course, the same idea will work for the extended enterprise and external companies.

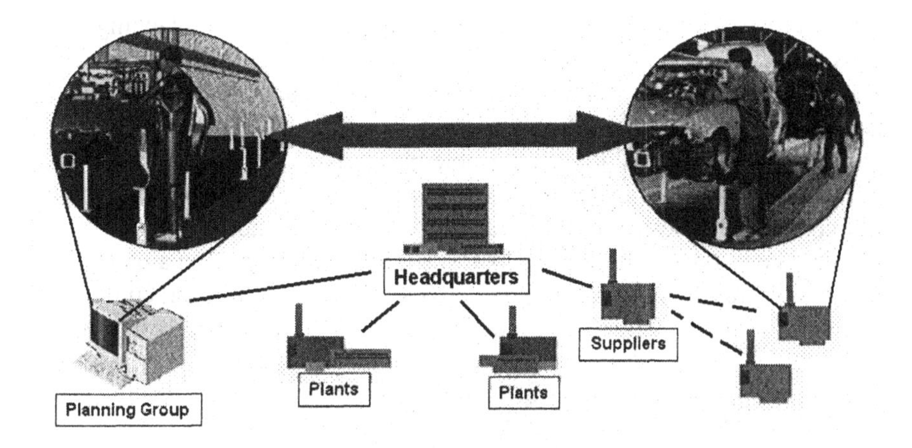

Figure 5. Communication of ergonomically safe procedures and workplaces (Source: LINNER, GEYER 2003, p. 41)

The integrated planning environment and the ability to store best practice processes in a database and reuse them for future projects reduces the planning effort significantly.

Another aspect of MPM is the possibility to define and enforce standards for the enterprise and to easily communicate them. Using the standard eBOP format a company may define libraries of predefined ergonomically safe and proven workplaces and work procedures for certain tasks (Figure 6). Web tools allow for making this information available to all relevant people in the entire organisation.

8. SUMMARY

Whereas both in product design (CAD/PDM) and production control (ERP, MES) fairly developed IT systems with a comparatively high degree of integration can be found, the IT landscape in production planning is primarily characterised by a large number of independent non-interacting software tools. This results in considerable overhead including multiple-entry of the same data, interface problems, etc.

The concept of Manufacturing Process Management (MPM) gives a solution by providing an integrated working environment for production planning and a comprehensive platform to optimise production planning throughout the extended enterprise.

Planning times can be shortened through usage of a common process database and concurrent engineering. The core of the concept is a database hosting a process model that describes manufacturing operations as well as required resources and affected product components. Various planning tools are connected to this database. Ergonomically safe workplaces can be defined using 3D technology using a human model. Additional human aspects will be considered using a material flow simulation.

Web-based collaboration tools enable planners to cooperate between several locations and to transfer proven production processes to other locations. The concept of a best practice database for processes and resources as well as integrated simulation tools help to increase productivity, maintain high quality standards and optimise product cost. Ergonomic best practices may be defined and implemented for the company, as well as for the suppliers.

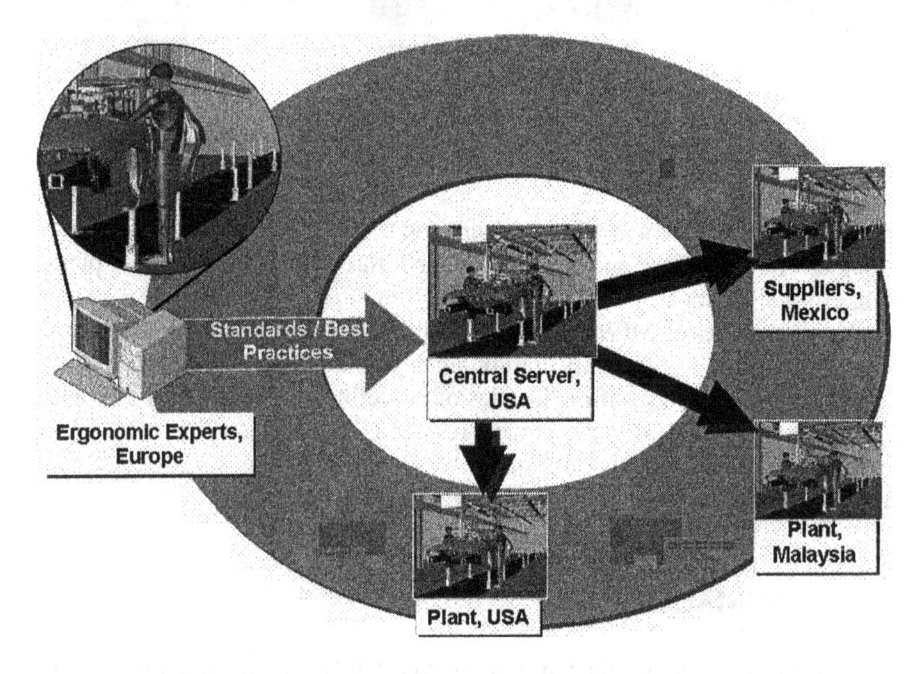

Figure 6. Standardisation and using best practices across the extended enterprise

REFERENCES

BAUR, Cornelius; KAAS, Hans-Werner; KASSNER, Sven et al.: Profitable Wachstumsstrategien in der Automobilzulieferindustrie. Düsseldorf: McKinsey & Company, 2000.

Ernst & Young LLP (eds.):
Profile of Tomorrow's Automotive Supplier.
Atlanta, GA et al.: Ernst & Young LLP, 1998..
http://www.autoindustria.com/encuentro/documentos/
automotive_supplier_capgemini.pdf, 29.07.03.

GEYER, Manuel:
Flexibles Planungssystem zur Berücksichtigung ergonomischer Aspekte bei der Produkt- und Arbeitssystemgestaltung.
Berlin, Heidelberg et. al.: Springer Verlag, 1997.
(iwb, Forschungsberichte Band 112)

JELTSCH, Michael; BAIER, Andreas; WAHRENDORFF, Matthias:
Auto 2010. Eine Expertenbefragung zur Zukunft der Automobilindustrie.
Sulzbach/Taunus: Accenture, 2001.
http:// automobile.karrierefuehrer.de/auto2010.pdf, 29.07.03.

LINNER, Stefan; GEYER, Manuel:
Human Aspects in Manufacturing Process Management.
In: Human Aspects in Production Management.
Eds.: ZÜLCH, Gert; STOWASSER, Sascha; JAGDEV, Harinder S.
Aachen: Shaker Verlag, 2003, pp. 37-43.
(esim – European Series in Industrial Management, Volume 5)

PART THREE

Human Aspects in Production Planning and Control

Human Factors in Production Planning and Control

How to Change Potential Stumbling Blocks into Reliable Actors

Hans-Peter Wiendahl, Gregor von Cieminski, Carsten Begemann and Rouven Nickel
University of Hannover, Institute of Production Systems and Logistics (IFA), Callinstrasse 36, D-30161 Hannover, Germany.
Email: {cieminski, wiendahl}@ifa.uni-hannover.de

Abstract: The influence of human actors on production planning and control (PPC) systems is significant. This paper describes a number of ways, in which human interaction with PPC systems affects the logistic performance of production. It demonstrates how human decisions and behaviour can act as stumbling blocks for PPC. Logistic models and PPC procedures that remove these stumbling blocks are presented. Moreover, the paper proposes the concept of 3-Sigma PPC as a holistic approach to PPC. 3-Sigma PPC recognises the influence of human factors on PPC and incorporates methods that improve human decision-making so that a better logistic performance can be achieved.

Key words: Production planning and control, Disturbances, Human factors, 3-Sigma PPC

1. CLASSICAL STUMBLING BLOCKS OF PPC

So far, progress in the field of production planning and control (PPC) has not been able to change the fact that lead times and inventory levels remain not fully controllable in many manufacturing enterprises and thus hinder compliance with delivery due dates. Relying on the continuing advances in information technology, companies often attempt to solve these problems solely with new PPC software systems. However, the actual causes of the shortcomings of production logistics do not normally reside in inadequate software only. Numerous studies of the Institute of Production Systems and Logistics (IFA) and the Institute of Manufacturing Engineering and Automation (IPA) show that the consequences of a wide range of economic, organisational and behavioural factors are the true stumbling blocks of PPC

(WIENDAHL et al. 2003; WIENDAHL 2003). Figure 1 provides an overview of these factors that originate from within production enterprises as well as from the market environment.

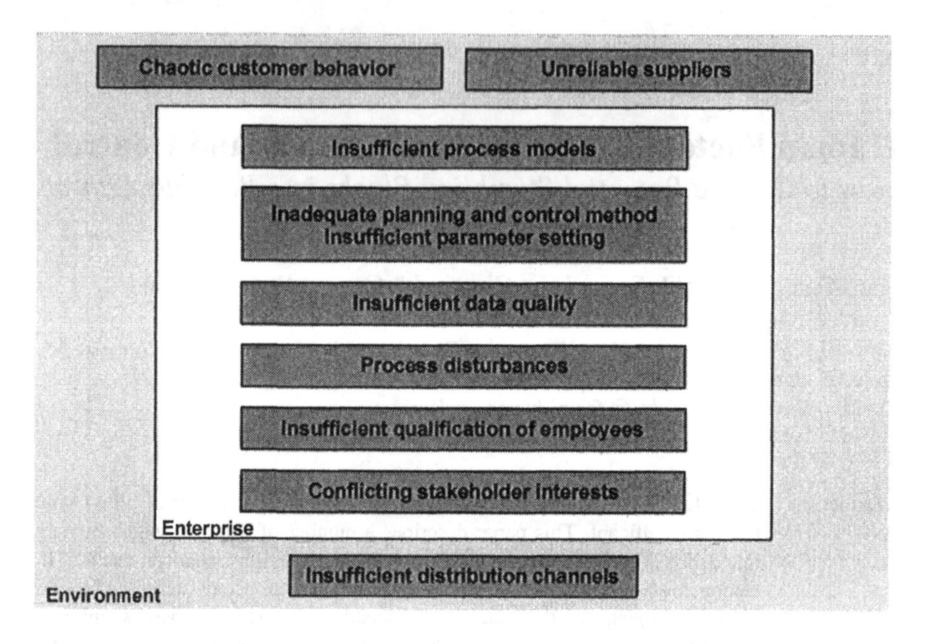

Figure 1. Causes of the classical stumbling blocks of PPC

In the external environment manufacturing enterprises can only control certain factors – e. g. their distribution channels – while other factors remain outside their direct control – e. g. the behaviour of their customers or suppliers. Due to this lack of control, manufacturers are often forced to accept the external factors as stumbling blocks of PPC. Internally however, the enterprises are required to resolve a range of problems that prevent high levels of logistic performance (see Figure 1). These range from insufficient process models and inadequate PPC methods as well as insufficient parameter settings, through insufficient data quality and technical process disturbances, to insufficient qualification of production planners or shop floor operators and conflicting PPC stakeholder interests. The majority of these problems have a strong connection to the activities of human staff working in manufacturing enterprises. Human operators are responsible for carrying out the procedures prescribed by the PPC control methods. They have to ensure high data quality within the PPC system. The impact of their level of qualification or the extent to which they pursue their own interests is self-evident. Hence, human actors have a special significance for the logistic performance of PPC systems, which is the subject of this paper.

The paper uses an overview of the PPC control cycle to generally introduce human influences on PPC systems. It provides detailed descriptions of the specific human stumbling blocks related to PPC systems, the level of qualification of employees and the stakeholder interests of PPC as well as measures to remove them. Furthermore, section six presents the 3-Sigma PPC approach that addresses human aspects of PPC as a key feature for improving the logistic performance of PPC systems.

2. HUMAN INFLUENCES ON PPC SYSTEMS

Employees in production planning departments, operators on the shop floor and all other members of staff who participate in PPC activities are actors of PPC systems. The main responsibility of PPC actors is decision-making. Decisions made by the actors affect the logistic performance of production systems in two ways: If the actors take correct decisions, the PPC system is effectively controlled to achieve the logistic performance objectives even if unplanned disturbances occur. Mistakes or oversights in the decisions made by actors run contrary to the logistic performance objectives. The actors either succeed to guarantee high levels of logistic quality by stabilizing the production process. Conversely, they may reduce logistic performance levels by becoming a cause for disturbances themselves. Figure 2 shows the stages of the PPC control cycle. As the actors are involved in decision-making in every step of this cycle it is useful for highlighting their influence on the logistic performance of PPC.

The first step of the cycle is the setting of strategic logistic objectives and corresponding performance targets. In the next step, planning activities establish the production program and the material and capacity plans. The purchasing, production and distribution functions execute the plan. During fulfilment, production and machine data acquisition systems record data that illustrate the actual throughput of orders through production. Performance measurement systems calculate and analyse the logistic performance of the production system and feed the analysis results back to the planning department. On the basis of the performance data, decisions for the following control cycle are taken.

In addition, Figure 2 indicates that disturbances can lead to deviations of the actual events in production from production plans at several points within the PPC control cycle (WIENDAHL 2003). In these cases, the task of the PPC actors is to either rule out disturbances from the outset by making appropriate control decisions. Alternatively, if disturbances do occur, the actors have to react to them, correct them if possible and take all actions necessary to maintain the fulfilment of the logistic objectives.

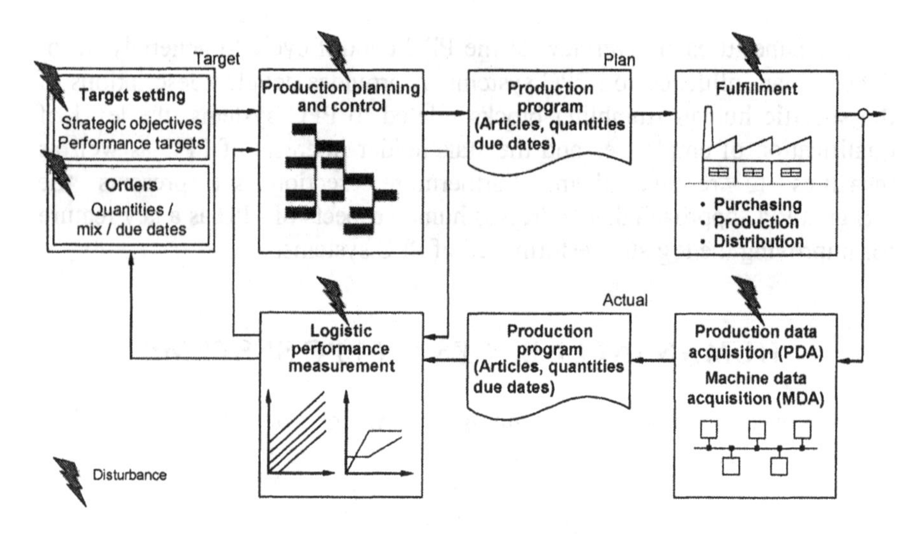

Figure 2. Possible sources of disturbances in the PPC control cycle

Often, target setting is ineffective because production planners try to fulfil conflicting logistic objectives simultaneously. During the execution of production plans, unplanned events like machine breakdowns disturb the original framework of the plans. This requires measures that seek to limit the effects of the disturbances on the logistic performance. A detailed understanding of the logistic effects of certain kinds of disturbances to production processes is required of production planners and operators to rectify the situation. Production and machine data acquisition systems and logistic performance measurement systems depend on accurate data to provide a true view of the actual logistic state of a production system. As these functions are not normally fully automated, the actors who feed data into the systems or who oversee performance analyses have to ensure data accuracy. They have to be aware that incorrect production data distort the actual logistic performance. This can lead to decisions that are based on wrong assumptions about the status of orders.

The three most significant factors that emerge as possible human stumbling blocks of PPC from this discussion are as follows:

1. the PPC systems, which often do not facilitate unambiguous human decision-making in order to achieve specific logistic objectives,
2. the lack of qualification of employees, which prevents actors from understanding the effects of their decisions on PPC systems,
3. the stakeholder interests, which serve individual, localised purposes rather than attempting to achieve the agreed logistic performance targets of an entire production system.

2.1 Stumbling Block: PPC Systems

The planning and control systems applied often represent a stumbling block of PPC. The best-known PPC system is the MRP II (Manufacturing Resource Planning) approach (WIGHT 1984). It incorporates the functions of business planning, production program planning, material requirements planning, scheduling and production control and contains precise procedures for all of these as well as for their connections. MRP II relies on centralised planning procedures that apply the output rate as their control variable and therefore emphasise the logistic objective of achieving high resource utilisation. Although MRP II is still employed in many manufacturing companies and also forms the basis of many PPC software systems (VOLLMANN et al. 1997), the inflexibility of its procedures and the resulting inability to adjust to a dynamically changing environment mean that the system is reaching the limits of its applicability.

The current requirements of manufacturing enterprises render the application of new PPC systems necessary, which are able to adapt to the changing boundary conditions. On the one hand, the systems have to be able to cope with the increasing variety of product variants and have to be applicable to manufacturing systems with highly complex material flows. On the other hand, modern PPC systems must consider the known interrelationships between logistic objectives and control variables (see section 2.2). For this purpose it is desirable that the execution procedures integrate the skills of the machine operators by decentralising as many PPC functions as possible (see Figure 3).

PPC systems like the Decentralised WIP-oriented (DeWIP) control empower the operators of every production work system to rationally pursue its specific logistic performance objectives (LÖDDING 2000). Figure 3 shows the DeWIP control procedure. From a human factors perspective, the most important characteristic of DeWIP are its short control loops. Operators have the unambiguous logistic objective of controlling the work-in-process (WIP) level at their respective work systems. They use direct communication with their colleagues to achieve this target: All operators of the work systems in Figure 3 continuously monitor current and future WIP levels. When an order is released into production from the central production program, work system 1 has to request authorisation to start processing from work system 2. Work system 2 only gives this authorisation if its current WIP level does not exceed capacity. The same logic applies to all other work systems within the production department. The shop floor has the permission to overrule the production program established by the centralised PPC system if the current status of production requires this. Excess WIP levels, which overload manufacturing capacities and increase throughput times, are thus avoided.

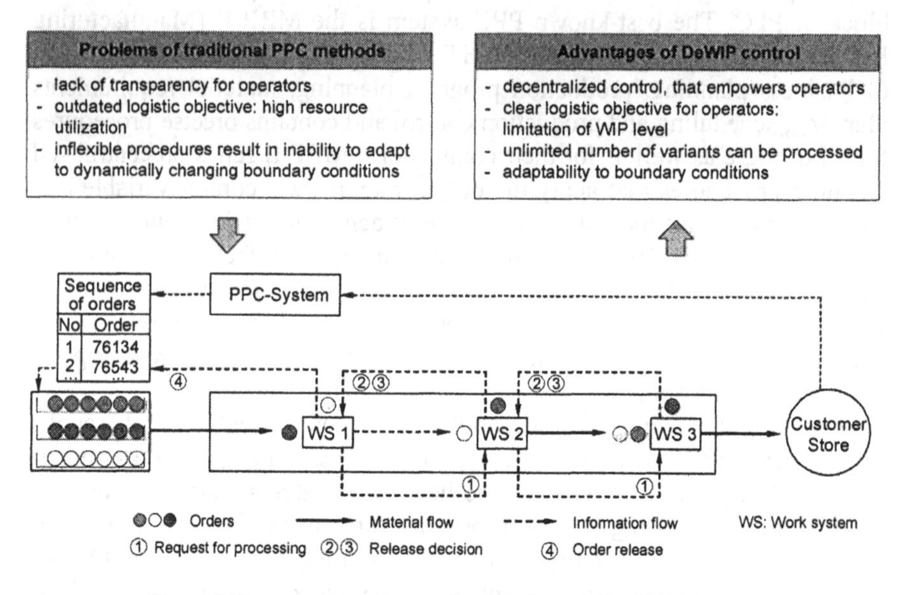

Figure 3. Empowering operators through decentralised WIP control (adapted from LÖDDING 2000)

2.2 Stumbling Block: Qualification of Employees

The level of qualification of the production staff represents a further stumbling block for PPC. In practice, both the employees that are directly involved in the production process as well as the employees that carry out production planning and control tasks may act or react irrationally due to a lack of understanding of dynamic logistic processes. Attempting to achieve a local optimum, they concentrate on the apparently most urgent problem.

A typical behavioural model, which is caused by an insufficient understanding of the planning process, is the vicious circle of errors of PPC (MATHER, PLOSSL 1978) that is shown in Figure 4. Production planners frequently ascribe a poor schedule reliability to short planned throughput times. Consequently, they increase the planned values. Thus, they release production orders earlier and increase the WIP levels in production. Queuing times in production rise and the schedule reliability deteriorates further. Only through the expediting of express orders can the production department manage to deliver orders on time. The employees are unaware that their decisions result in this spiral of events that affect the quality of PPC.

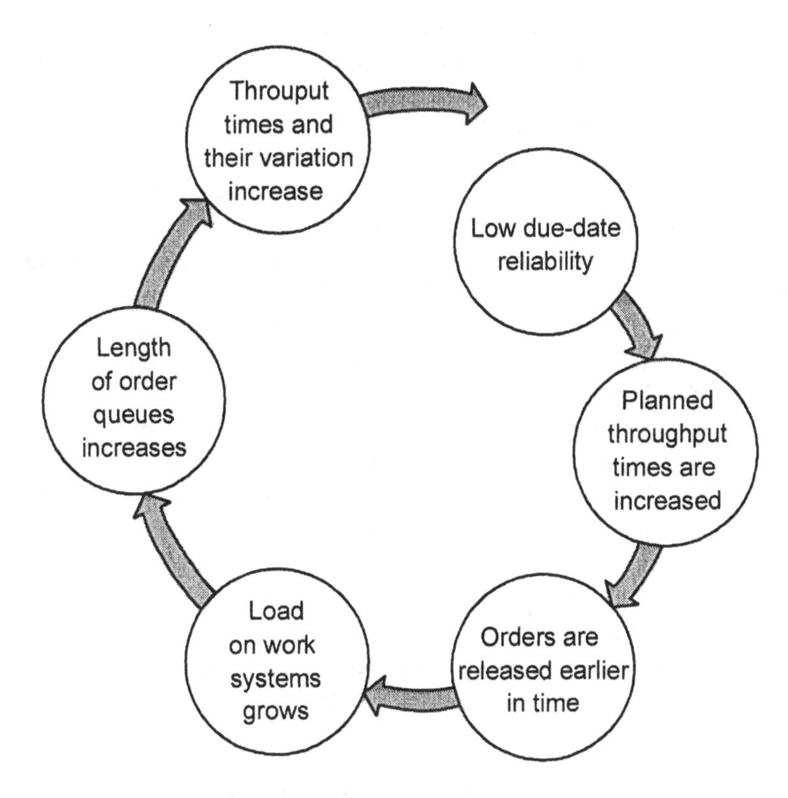

Figure 4. Vicious circle of errors of PPC
(adapted from MATHER, PLOSSL 1978)

Consistent logistic models of production processes can help to break the vicious circle of errors of PPC. The models explain the causal interrelationships between PPC control variables and logistic performance measures. For example, the models can be applied to make human actors aware of the rise in the workload on production resulting from an increase of the planned throughput times. Examples of such logistic models are the funnel model and the throughput diagram (WIENDAHL 1994) as well as the Logistic Operating Curves (NYHUIS, WIENDAHL 2003).

The funnel model is an analogy for the events at a work system during the processing of manufacturing orders. It forms the basis for the throughput diagram that visualises the events and enables the calculation of logistic performance measures from production feedback data. In combination, the two models show the exact consequences of PPC measures on the logistic behaviour of a work system over time.

A phenomenon that can frequently be observed in practice is the arbitrariness with which operators report production feedback data. Industrial studies carried out by IFA show that operators accumulate a WIP level of completed orders at their work systems over the course of a working week. Only at the end of the week or before holidays are these "buffers" dissolved: The operators report the orders as completed and thus release them to the succeeding work system. The WIP buffers serve as an insurance against reductions in output rate caused by possible process disturbances. This behaviour distorts the production feedback data, which can lead to the wrong control decisions being taken. It also increases the work system WIP levels and thus causes increases in the variation of throughput times. The throughput diagram visualises the negative effects of such inaccuracies in the feedback data. The diagram can be used to educate human actors about the consequences of their actions.

On the other hand, the Logistic Operating Curves describe the behaviour of the logistic performance measures output rate, throughput time and delivery reliability depending on variations in the WIP level of production resources. Therefore, the operating curves are a means to identify production planning and control measures that are adequate to achieve specific logistic objectives. They also highlight the conflicts between different objectives. The more human actors are aware of these interrelationships, the better they can contribute to good logistic performance by taking appropriate decisions and setting realistic performance targets. Reiterations of the vicious circle of errors are thus avoided.

2.3 Stumbling Block: Stakeholder Interests

It is essential to bear in mind that natural individual interests of different groups of staff – the PPC stakeholders – coexist besides the formal logistic performance objectives of a PPC system (schedule reliability, throughput time, WIP level, resource utilisation). Commonly, manufacturing companies do not officially acknowledge the existence or nature of these interests. The latter depends on the roles that the stakeholders fulfil in the context of PPC or within the company. Clearly, different groups of staff pursue distinct objectives, which results in a complex structure of individual, possibly conflicting, stakeholder interests and their motivation (MASLOW 1987).

Machine operators constitute a typical group of PPC stakeholders. The main objectives of this group are to protect their workplace and to keep the processing of production orders stable. As has been pointed out before, the operators tend to "hoard" production orders for periods of low resource loading and thus create safety WIP levels. In addition, they arrange the processing sequence of orders in a way that fits their daily and weekly work rhythm. Figure 5 depicts a graphic example of this behaviour that IFA found during a study in a manufacturing enterprise. The company had configured its pull control system for low WIP levels: Six kanban cards were meant to ensure lean production (see Figure 5a). Low production WIP levels caused significant idle times at the work systems in the production department. In order to reverse this situation and to suggest to management that they were busy most of the time, the operators artificially increased the WIP level by adding copies of the original kanban cards to the control cycle (see Figure 5b).

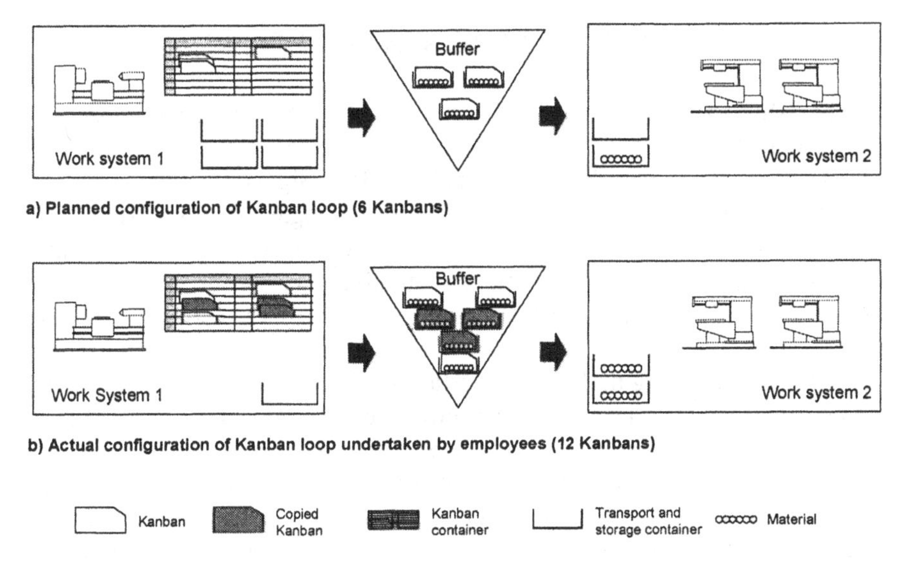

Figure 5. Consequences of the fear of utilisation losses in a Kanban system

A second important group of PPC stakeholders are the expeditors. Their objective is to ensure the on-time fulfilment of priority orders. For this purpose, expeditors "chase" production orders past the production work systems. If successful, they are often regarded as the "heroes of production who make the impossible possible".

Neither the machine operators nor the expeditors have a genuine desire for a PPC system whose procedures "automatically" fulfil the formal logistic objectives in the interest of the company and its customers. The reasons for this are identical in both cases: Both actors fear for their existence. While the fears of the machine operators are mostly unfounded and can frequently be put down to a lack of qualification or an inadequate compensation system, expeditors have a better understanding of the dynamic logistic processes and are aware that more effective PPC systems would render their function superfluous. Stakeholder interests are also caused by a lack of qualification in production logistics. In this context, it is especially important to create an awareness amongst the employees for the consequences of their actions for themselves and their colleagues who are affected. In order to minimise the effects of these personal interests the wage and incentive system has to reward conformance to the overall objectives of the production system. Thus, customer orientation, high logistic performance levels and quality and the conformance of decisions to objectives of the PPC system should all be rewarded.

3. THE 3-SIGMA APPROACH TO PPC

The 3-Sigma PPC approach is based on a holistic view of the impact of the three features human actors, organisation and logistic models on PPC systems (see Figure 6; WIENDAHL et al. 2003). 3-Sigma PPC represents an important extension of traditional PPC methodologies that almost exclusively concentrated on the feature of the logistic models. The concept aims to integrate the solutions to the human stumbling blocks described above with elements from the other features in a single coherent framework for PPC.

The vision of 3-Sigma PPC is to establish PPC systems as quality management systems for the logistic performance of production systems. The concepts of process capability and reliability are transferred to the logistics context and logistic quality becomes measurable in terms of sigma (σ). It stands for the standard deviation, a statistical measure of the distribution of a characteristic around a mean value. For a quality characteristic with a normal distribution and a permissible tolerance equal to 3 Sigma, 99.73 % of all values have to lie within the tolerance range (EVANS, LINDSAY 2001). As its name suggests, 3-Sigma PPC transfers this quality requirement to PPC systems. This means that PPC procedures have to control a production system in such a way that its logistic objectives are met within the specified tolerance in 99.73 % of all cases.

The examples above illustrate the influence of actors within 3-Sigma PPC. For this PPC approach it is essential that all actors have the necessary qualification in production logistics. The actors need to be able to understand the structure of the production-logistic models that the PPC systems use. They have to be aware of the logistic performance objectives and the variables with which they can influence them. Also, they need to know and understand the interrelationships between both, i.e. what consequence the modification of a control variable has on the logistic performance. Besides this qualification, the actors also have to have the competence to apply the PPC system, methods and algorithms, to carry out their task of maintaining the logistic performance levels.

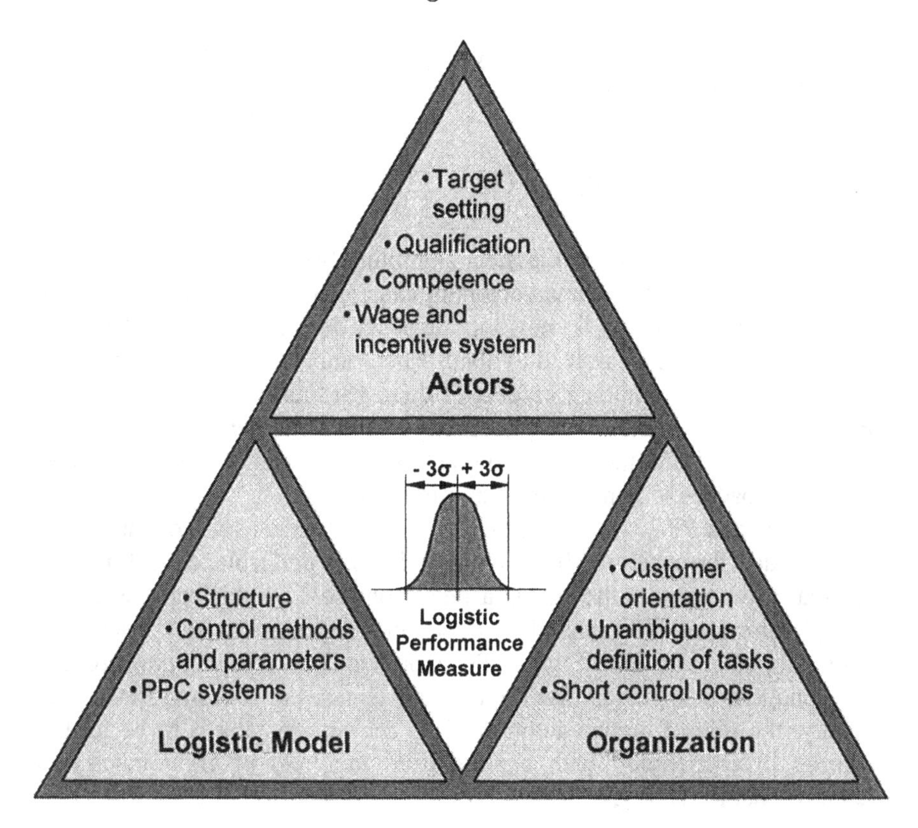

Figure 6. Features of the 3-sigma PPC approach (according to WIENDAHL et al. 2003, p. 91)

Ideally, actors at all levels of the PPC system – production planners and shop floor operators alike – have this qualification and competence. The more the qualification to interpret interdependencies between performance measures and control measures can be decentralised, the more action for performance improvements can be taken rapidly and directly. This applies to all stages in the PPC control cycle in Figure 2. Qualification facilitates the correct interpretation of production-logistic models for appropriate and consistent target-setting. This is followed by the definition of feasible production plans, for which the likelihood of changes occurring is reduced. The group of actors is enabled to rapidly react to disturbances in the purchasing, production or distribution functions so that the effects on logistic performance are minimised. The actors are also aware that for a realistic assessment of the actual logistic situation in a production system, feedback data have to be provided accurately and as early as possible. Moreover, the performance measurements taken from the shop floor should conform to the production-logistic models in the same way as the information that is fed back into the system for control purposes.

4. CONCLUSIONS

The paper introduced the classical stumbling blocks of PPC and emphasised the involvement of human actors in causing and affecting them. Using the PPC control cycle as a framework, the paper demonstrated how decisions taken by human actors can lead to disturbances and adverse effects on logistic performance. The human aspects of three particular stumbling blocks – PPC systems, qualification of employees and stakeholder interests – were detailed further. Here, the paper provided examples from industrial practices as well approaches to remove the stumbling blocks.

The 3-Sigma PPC concept integrates these individual solutions in a general approach for achieving high levels of logistic performance and quality. 3-Sigma PPC combines the human aspects considered in this paper with two further features of PPC, logistic models and organisation. IFA intends to continually elaborate the 3-Sigma PPC approach by developing new production management methods that improve all features of 3-Sigma PPC. Specifically, the role of human actors in PPC decision-making is to be further explored in cooperation with experts from the field of ergonomics and organisational psychology.

REFERENCES

EVANS, James R.; LINDSAY, William M.:
The Management and Control of Quality.
St. Paul, MN: West Publishing, 5th ed., 2001.
LÖDDING, H.:
Decentralized WIP-oriented Manufacturing Control.
In: The manufacturing system in its human context - a tool to extend the global welfare.
Stockholm: CIRP College International pour l'Etude Scientifique des Techniques de Production Mecanique, 2000.
(Proceedings of 33rd CIRP International Seminar on Manufacturing Systems)
MATHER, H.; PLOSSL, G.:
Priority fixation versus throughput planning.
In: Production and Inventory Management,
Washington, DC, 19(1978)3, pp. 82-95.
MASLOW, Abraham H.:
Motivation and Personality.
Boston, MA; Addison-Wesley, 3rd ed., 1987.

NYHUIS, Peter; WIENDAHL, Hans-Peter:
Logistische Kennlinien: Grundlagen, Werkzeuge und Anwendungen..
Berlin et al.: Springer Verlag, 2nd ed., 2003.
VOLLMANN, Thomas E.; BERRY, William L.; WHYBARK, David C.:
Manufacturing Planning and Control Systems.
New York, NY: McGraw-Hill, 4th ed., 1997.
WIENDAHL, Hans-Hermann:
Marktanforderungen verstehen, Stolpersteine erkennen.
In: Proceedings of Management Circle Seminar "Flexible atmende Produktion: Auftragsschwankungen in den Griff bekommen".
Eschborn: Management Circle, 2003.
WIENDAHL, Hans-Peter:
Load-oriented Manufacturing Control.
Berlin et al.: Springer, 1994.
WIENDAHL, Hans-Peter; BEGEMANN, Carsten; NICKEL, Rouven:
Die klassischen Stolpersteine der PPS und der Lösungsansatz 3-Sigma-PPS.
In: Expertensystem Logistik
Eds.: Wiendahl, Hans-Peter; Luczak, Holger.
Berlin et al.: Springer Verlag, 2003.
WIENDAHL, Hans-Peter; BEGEMANN, Carsten; NICKEL, Rouven; CIEMINSKI, Gregor von:
Human Factors as Stumbling Blocks of PPC.
In: Current Trends in Production Management.
Eds.: ZÜLCH, Gert; STOWASSER, Sascha; JAGDEV, Harinder S.
Aachen: Shaker Verlag, 2003, pp. 87-93.
(esim – European Series in Industrial Management, Volume 6)
WIGHT, Oliver W.:
Manufacturing Resource Planning: MRP II.
Indianapolis, IN: John Wiley & Sons, Revised edition, 1984.

Human Factors Aspects in Set-Up Time Reduction

Dirk Van Goubergen[1] and Thurmon E. Lockhart[2]
1) Ghent University, Department of Industrial Management, Zwijnaarde B-9052, Belgium. Email: dirk.vangoubergen@UGent.be
2) Grado Department of Industrial and Systems Engineering, Virginia Polytechnic Institute and State University, Blacksburg 24061, VA, USA. Email: lockhart@vt.edu.

Abstract: The ability to produce products which continue to meet changing customer expectations requires flexibility within a manufacturing plant. Short set-up times are a key enabler for obtaining these flexible production systems. The purpose of this research is to identify human factors issues in set-up reduction and to develop a framework for assessing and improving the changeover from an ergonomics point of view. A Human Factors System Design methodology will be the starting point to identify how the existing human factors body of knowledge can be incorporated in the body of knowledge of design for fast changeover. This publication describes the first steps of this research. We identify the human factors gaps and deficiencies in the current set of design rules for fast changeover.

Key words: Fast changeover, Human factors system design, Design rules

1. INTRODUCTION

1.1 Why Set-up Time Reduction?

Although the need for short set-up times is not new, recently in all types of industry an increased focus on set-up times is perceived. The need for short set-ups is now bigger than ever. Globalization of the market, customization of products and the continuous effort for better efficiency of the existing production equipment are the main driving forces for this phenomenon. (VAN GOUBERGEN 2000, p. 1)

Many companies around the world are implementing lean concepts and customer-pull based production systems. For these systems short set-up times are a sine qua non (WOMACK, JONES 1996, p. 26).

1.2 Existing Set-up Reduction Methodologies

There are several publications and case studies available on how set-up times can be reduced in existing situations (SHINGO 1985; see also VAN GOUBERGEN, VAN LANDEGHEM 2002). Basically all approaches are derived from the SMED method ("Single Minute Exchange of Die") (SHINGO 1985). According to the SMED method all activities related to a set-up can be divided into internal - which are performed while the machine is down and thus the production process is stopped – and external - which take place while the machine is running activities.

Actually, in practice a lot of set-up problems could have been avoided if the machine designer would have paid proper attention to the set-up problem. Several set-up reduction principles from the SMED method, as well as the use of specific industrial engineering techniques, can be applied during the design stage of machines and equipment. However, very limited research has been published on how to incorporate fast changeover capabilities in machine design.

An initial set of design rules is published by (MILEHAM et al. 1999). Most of these rules were technical in nature. This list was substantially extended by (VAN GOUBERGEN, VAN LANDEGHEM 2002) based on practical experience with more than 60 set-up reduction projects on different types of machines in different types of industry. The resulting list of rules include rules dealing with technical as well as organizational or method aspects.

1.3 Goal of this Research – Human Factors Aspects

The principles and rules that are mentioned in the previous section have been proven to be successful in many situations in different types of industry (for examples see VAN GOUBERGEN, VAN LANDEGHEM 2002, p. 213). However, when they were established, the main focus has been on the "time" aspect: how to reduce the machine downtime or the internal/on-line time.

Human factors aspects are only included in an implicit way and are not studied in depth, although it is obvious that humans play a crucial role in set-ups. We strongly believe that the existing body of knowledge on human factors can offer interesting additional insights and ways to improvement of set-ups. An in-depth search of human factors and ergonomics literature has not given one single publication on this subject.

Therefore, the main overall goal of this research is to assess set-ups from a human factors point of view and to propose additional human factors based rules for finding appropriate improvement proposals in order to shorten set-up times. We identified following research steps to reach our research goal:

1. Identify human factors gaps and deficiencies in current set of design rules (i.e. which human factor elements are not covered yet?).
2. Based on the results of the first step, develop additional comprehensive rules that can be used by the practitioner and that complete the existing list of design rules for fast changeover.
3. Refine the current, existing rules using the human factors body of knowledge. We will identify:
   - Which rules need to be extended?
   - Which rules can be more detailed?
4. Validate the new rules with experiments.
5. Develop an assessment framework based on the human factors body of knowledge in order to identify opportunities to reduce set-up times by improving human factors aspects in case of existing set-up methods.

This publication presents the results of the first step of this research: identify human factors related gaps in the existing set-up reduction approaches and design rules by using a universal human factors system model. Based on these results we will also provide some input for the second research step.

2. HUMAN FACTORS SYSTEM MODEL

Current design rules mentioned in section 1.2 were mapped to the appropriate section of an adapted version of the "Human Factors in Systems Design" model (HELANDER 1997; see Figure 1), to see what parts of the human factors design paradigm were missing from current fast-changeover design guidelines. This model was chosen because of its completeness. For all elements of the model an operational definition was documented that is suitable for use in the context of set-up time reduction. Some examples can be found in Table 1.

Next, every existing design rule for fast changeover was mapped under these elements. In total there were 52 design rules divided amongst 9 categories. Table 2 provides an overview of these categories and the number of rules in each category.

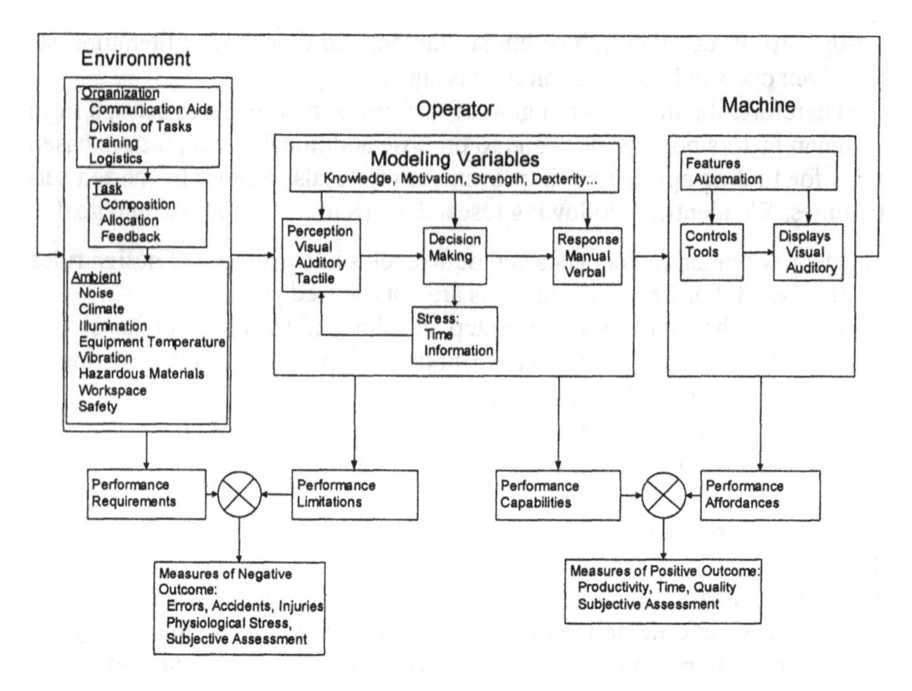

Figure 1. "Human Factors Systems Design" model (HELANDER 1997)

The results of the mapping were visualized using histograms indicating how many rules fall under each element from the "Human Factors System Design" model. Figure 2 shows as an example the histogram portraying the number of rules that are related to the elements related to the operator in the "Human Factors System design" model. Similar diagrams were made for the elements in the categories "Environment" and "Machine".

Table 1. Examples of the operational definition of the elements of the "Human Factors System Design" model adapted for use in the context of set-up reduction

Element	Definition
Division of tasks	Organization of changeover tasks and optimal division among changeover personnel
Task composition	Changing the nature of the task (i.e. changing from a bolt fixture to a clamp fixture)
Equipment temperature	The temperature of parts of the machine being changed, including areas that must be accessed to perform the changeover
Strength	Required muscular strength of the operator needed to perform the changeover
Visual display	Visual information output from the machine

Table 2. Overview of the categories of existing design rules

Category	Number of rules
Less weight	2
Simplification	10
Standardization	3
Securing	5
Location and adjustment	15
Handling	8
Off-line activities	2
Machine lines	1
Method and organization	7

In mapping these rules, the direct human factors impact of the rule was inferred. The human factors literature was consulted to develop general guidelines for incorporating human factors principles in the design.

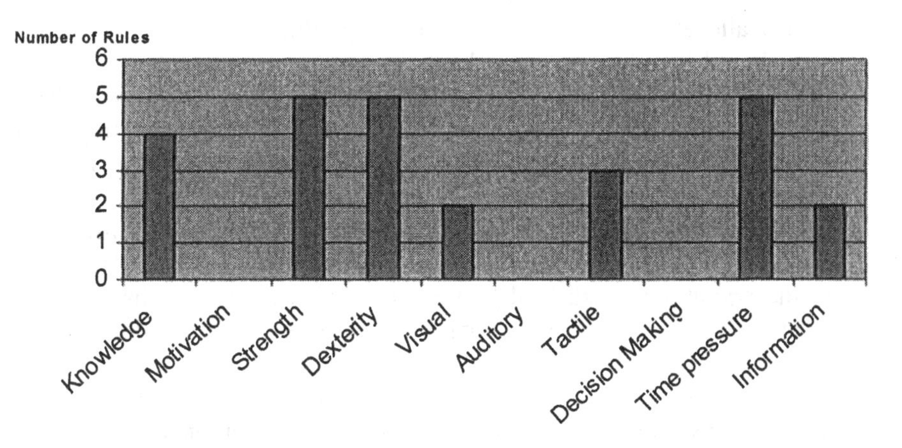

Figure 2. Example of the results of the mapping of design rules: The number of design rules related to the elements of the category "Operator" in the "Human Factors System Design" model

3. RESULTS

Several human factors issues are under represented or not represented at all in the current design rules. Based on evaluation of the current rules, the following human factors issues are not adequately addressed:

1. Communication aids
2. Task allocation
3. Noise
4. Climate
5. Illumination
6. Vibration
7. Hazardous materials
8. Motivation
9. Auditory perception
10. Decision making
11. Automation
12. Visual displays
13. Auditory displays
14. Safety

Short synapses of the results section is presented below:

- *Communication aids:*
  Communication is extremely important for changeovers that require the coordinated effort of a team. In some instances, the use of simple displays, such as andon lights, can aid communication. Additionally, the general rule should be to avoid the requirement for remote support (WHITAKER, FOX, PETERS 1993).
- *Task allocation and automation:*
  Proper allocation of tasks is necessary to produce an optimal system. A rule should be added to remind the designer to consider the options of human and machine operations during machine design. Decision criteria for automation have already been developed and should be employed (OLDER, WATERSON, CLEGG 1997). In practice the decision for automation will be based on a cost/benefit analysis. It is important to not limit the cost analysis to the engineering or purchasing cost. A life cycle cost analysis is more appropriate as in a lot of situations an important part of the operational costs will be the downtime cost due to changeovers. Trade-offs can be made with regard to the degree of automation and the associated design cost versus the reduction of set-up downtime costs. On the other hand, due to the ever more complex manufacturing environment and the flexibility of human operators, complete automation is often infeasible (AYRES, BUTCHER 1993).
- *Noise and auditory perception:*
  As a general rule, noise should be kept to a minimum. Noise in the external environment also contributes to the local noise level, so a more systemic approach is necessary to effectively reduce noise levels.
- *Climate:*
  Some studies have shown that heat decreases mental awareness, increases stress on the human body and can even affect a human operator in psychological ways (NIOSH 1986). VAN GOUBERGEN, VAN LANDEGHEM (2002) propose that motivation is a pillar of producing quick set-ups. Therefore, the psychological affects of working in heat should not be ignored.

- *Illumination:*
  The amount of ambient illumination has several effects on work systems. Visual fatigue can set in if the work area is too bright or too dim. If so, this can reduce the operator's capability to observe accurately and can decrease the desire to perform accurately. This can be detrimental to the set-up process, as can the misreading of visual cues.
- *Vibration:*
  The human body resonates at certain frequencies. It is theorized that physical damage due to vibration occurs when the human body is in a state of resonate amplification. Therefore, in addition to reducing vibration in general, there are certain frequency ranges to avoid (WASSERMAN 1987).
- *Hazardous materials and safety:*
  Current design rules for fast changeover also neglect explicitly to cover many safety issues that face the operator.
- *Motivation:*
  VAN GOUBERGEN and VAN LANDEGHEM (2002) propose that motivation is the foundation of achieving a fast changeover. Although the culture of the organization is crucial here, machine designers can influence motivation by designing the machine to be easily maintained and set-up in a comfortable way.
- *Decision-making aids:*
  On-board diagnostic equipment can greatly reduce the post-set-up adjustment process. Practice shows that even in cases where a standardized set-up method is available and documented in a set-up instruction, insufficient attention is paid to the re-adjustment process. If a trial run shows that there is a non conformance on one (or more) of the product requirements, it can often be seen that each operator tries to solve this problem in a different way. Decision making aids are needed to show the relationship between product requirements and setting parameters. This way, based on how much deviation there is on the product specification, one knows which parameter needs to be re-adjusted and how much.
- *Visual and auditory displays:*
  The design of displays is very important to achieve settings that are right the first time. A lot of machines nowadays are still lacking good adjustment aids such as displays. In that case the setting activity will not be based on measurements, but more on trial and error. Depending on the type of parameter that needs to be set and the required accuracy, an appropriate display needs to be designed. The human factors body of knowledge already contains in-depth information on the design of displays in order to portrayal information. This knowledge can add specific guidelines for this design.

To remedy these gaps in the current set of design rules, new rules need to be added. However, we must keep in mind what the "sphere of control" of the machine designer consists of. He does determine both the concept and the practical realization of the machine. On the other hand, he is not in control of environmental factors and ambient elements of the workplace where the machine is going to be used. So it is of no use to add specific design rules for some of these issues.

Also, some of the missing elements that are discussed in the previous paragraph relate to more general principles and guidelines for machine design and are not directly related to the changeover problem. For example guidelines regarding basic safety issues on machines apply to all machine designs and are usually already available. Therefore they do not need to be added to a specific list of design rules dealing with fast changeover.

4. CONCLUSION

Current design rules for fast changeover neglect the principles of human-centered design. This study has identified the human-related shortcomings of these rules. The next step in this research will be to use the human factors body of knowledge to modify existing rules and add more in-depth rules.

REFERENCES

AYRES, R. U.; BUTCHER, C. D.:
The flexible factory revisited.
In: American Scientist,
Triangle Park, NC, 8(1993)5, pp. 448-459.

HELANDER, M. G.:
The Human Factors Profession.
In: Handbook of Human Factors and Ergonomics.
Ed.: SALVENDY, G.
New York, NY: John Wiley and Sons, 2nd ed., 1997, pp. 3-16.

MILEHAM, A. R.; CULLEY, S. J.; OWEN, G. W.; McINTOSH, R.:
Rapid Changeover – a pre-requisite for responsive manufacture.
In: International Journal of Operations and Production Management,
Bradford, 19(1999)8, pp. 785-796.

NIOSH:
Working in hot environments.
Washington, DC: National Institute for Occupational Safety and Health.
(DHHS (NIOSH) Publication No. 86-112)

OLDER, M. T.; WATERSON, P. E.; CLEGG, C. W.:
Critical assessment of task allocation methods and their applicability.
In: Ergonomics,
London et al., 40(1997)2, pp. 151-171.
SHINGO, S.:
A Revolution in Manufacturing: The SMED System.
Cambridge, MA: Productivity Press, 1985.
VAN GOUBERGEN, Dirk:
Set-up reduction as an organization-wide problem.
In: Proceedings of Solutions 2000 Conference. Norcross, GA:
Institute of Industrial Engineers, 2000.
CD-ROM.
VAN GOUBERGEN, D.; VAN LANDEGHEM, H.:
Rules for integrating fast changeover capabilities into new equipment design.
In: Robotics and Computer Integrated Manufacturing,
Oxford et al., 18(2002)3, pp. 205-214.
WASSERMAN, D.E.:
Motion and vibration.
In: Handbook of Human Factors and Ergonomics.
Ed.: SALVENDY, G.
New York, NY: John Wiley and Sons, 2nd edition, 1997, pp. 650-669.
WOMACK, J. P.; JONES, D. T.:
Lean thinking.
London et al.: Simon and Schuster, 1996.
WHITAKER, L. A.; FOX, S. L.; PETERS, L. J.:
Communication between crews: the effect of speech intelligibility on team performance.
In: Proceedings of the Human Factors and Ergonomics Society. 37th Annual Meeting.
Santa Monica, CA: Human Factors and Ergonomics Society, 1993, pp. 630-634.

Influences of Human Operators on the Logistics of Manufacturing Cells

Gregor von Cieminski and Peter Nyhuis
University of Hannover, Institute of Production Systems and Logistics (IFA), Callinstrasse 36, D-30161 Hannover, Germany.
Email: {cieminski, nyhuis}@ifa.uni-hannover.de

Abstract: Human operators have important influences on the logistic behaviour of manufacturing cells. In many of the possible configurations of a manufacturing cell, operators represent the bottleneck of the cell. Thus they dominate the cell's logistic behaviour. This paper explains how the theory of logistic manufacturing cell operating curves considers human influences on manufacturing cell logistics. The operating curves are a tool to analyse the logistic performance of manufacturing cells. Therefore they represent a means for determining logistically suitable configurations of the human-machine interactions between the human operators and the work systems in manufacturing cells.

Key words: Manufacturing cells, Logistic operating curves, Human aspects

1. HUMAN OPERATORS IN MANUFACTURING CELLS

Human operators fulfil important roles in manufacturing cells. When cellular manufacture first came to widespread use in the 1980s, one of the foci of scientific research and practical applications lay on their human aspects (GREEN, SADOWSKI 1984). A substantial body of work was produced on the ergonomic and organisational aspects involving human operators. Job and workplace design were important themes as well the introduction of team working principles (SCHONBERGER 1990). Manufacturing cells were the first type of manufacturing system that integrated elements of management functions in shop floor operations: Decentralised production planning and control, quality management, maintenance and other management functions all form part of the responsibilities of the operators on the shop floor. Furthermore, numerous studies on manufacturing cells concerned themselves with job enlargement and job enrichment as well as employee empowerment (SLACK, CHAMBERS, JOHNSTON 2000). An analysis of the logistic behaviour of manufacturing cells, however, has to intentionally neglect the practicability of ergonomic and organisational aspects and has to purely concentrate on the effects of human action on the logistic performance of the cells.

This paper introduces the logistic manufacturing cell operating curves as a model of the logistic behaviour of manufacturing cells (WIENDAHL, CIEMINSKI 2003). It describes how the general modelling methodology is capable of considering the particular logistic characteristics of manufacturing cells whose logistic behaviour depends on human operators. General observations on the human influence on manufacturing cell logistics are presented. Also, the paper provides guidelines as to how industrial companies can most effectively configure cells staffed by human operators.

2. LOGISTIC MODELLING OF MANUFACTURING CELLS

Until recently, the lack of models of the logistic behaviour of manufacturing cells meant that the precise human influence on their logistic performance could not be determined (MILTENBURG 2001). Recommendations exist as to how human operators best contribute to a smooth material flow in manufacturing cells. These consider ergonomic issues on a practical level rather than examining measures of logistic performance. Industrial companies therefore experience difficulties in finding logistically adequate set-ups of manufacturing cells when implementing these. The theory of the logistic manufacturing cell operating curves represents a valid model for the logistic behaviour of manufacturing cells (WIENDAHL, CIEMINSKI 2003). The logistic operating curves describe the behaviour of the logistic performance measures output rate and throughput time as depending on the work-in-process (WIP) level of a manufacturing cell. The mathematical model underlying the operating curves incorporates the particular logistic characteristics of manufacturing cells. These serve as boundary conditions for the mathematical model. Figure 1 depicts the schematic view of a manufacturing cell that conforms to the boundary conditions. The most important logistic characteristic of manufacturing cells is that the parts of batch orders arriving at, and departing from, the cell are transported in one-piece flow in between its work systems (see Figure 1).

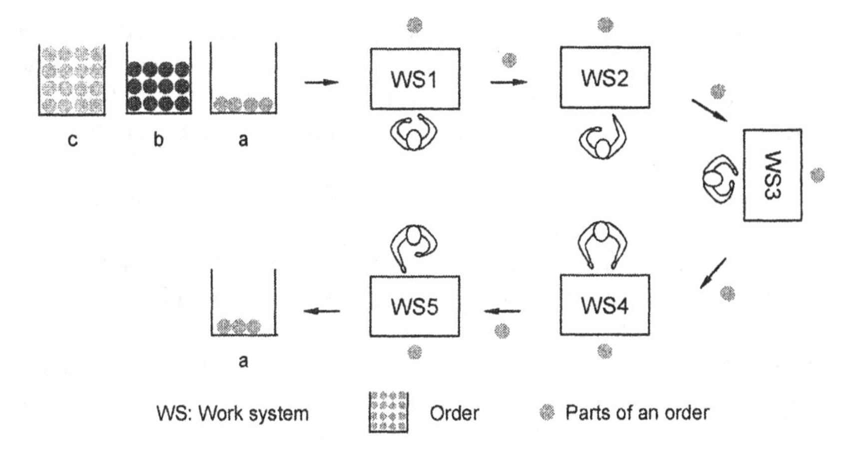

Figure 1. General structure of a manufacturing cell

WIENDAHL and CIEMINSKI (2003) cover the modelling methodology, its logistic boundary conditions and its mathematical algorithm as well as the formulae in detail. They explain how approximation equations for the aforementioned interdependencies between the WIP level and the output rate and the throughput time of a manufacturing cell can be developed. Figure 2 shows the shape of the approximate operating curves.

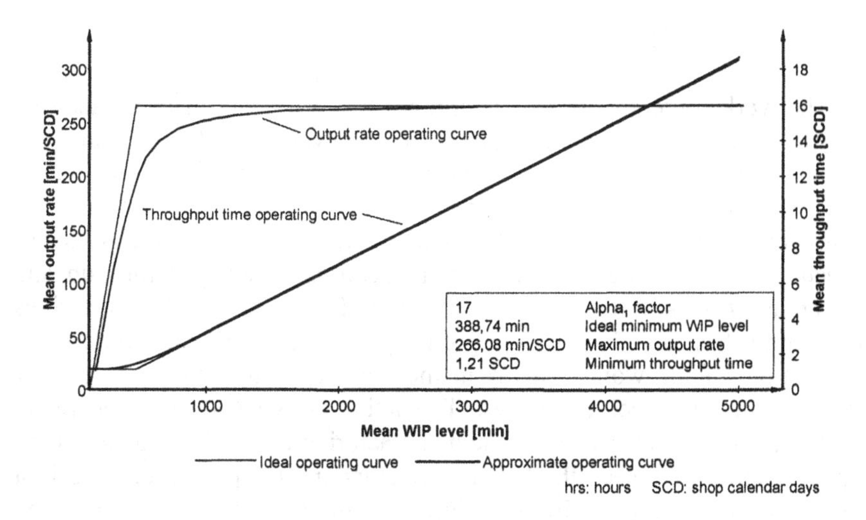

Figure 2. Logistic manufacturing cell operating curves

The mean output rate of a manufacturing cell increases proportionately to rising WIP levels until the capacity limit of the cell is reached. The mean throughput time remains constant up to this particular WIP level. If the WIP level rises further, queues of waiting orders build so that the mean throughput time increases (see Figure 2).

It is important to note that the general model of a manufacturing cell as depicted in Figure 1 assumes that operators service all work systems within the cell. The human resources therefore have no actual influence on the order fulfilment process. The logistic bottleneck of the cell is determined purely by the work system with the maximum processing time. Manufacturing cells implemented in industry normally differ from this simplified structure. Here, the number of active operators in a manufacturing cell is usually reduced so that the influences of the human operators on the logistic processes become evident. Equation 1 defines the work content of manufacturing cells and can be used to explain the human influence:

$$TO_{MC} = \frac{T1_n + (LS - 1) \cdot TCYCLE}{60} \quad (1)$$

TO_{MC}	order work content of manufacturing cell [hrs]
$T1_n$	initial output time at nth work system in mftg. cell [min]
n	number of work systems in manufacturing cell [units]
LS	lot size [units]
$TCYCLE$	cycle time [min]

The work content of orders determines the WIP level of manufacturing cells as well as their output rate and load. The cycle time *TCYCLE* in equation (1), which is equal to the processing time of the bottleneck of the manufacturing cell, has a dominating effect on the work content of orders. Configurations of manufacturing cells exist, in which the cycle time is determined by the activities of the human operators (loading, setting-up and unloading of work systems). This is the case if the duration of the activities an operator has to carry out during one cycle exceeds the maximum processing time of the work systems in the manufacturing cell. The operator becomes the bottleneck of the cell. The activities assigned to the operators depend on the configurations of the manufacturing cells. Figure 3 shows the typical relations between the human operators and the parts processed in a manufacturing cell (TAKEDA 1996). According to the "caravan" logic operators carry a single part through the entire manufacturing cell (see Figure 3a). Following the "relay" logic, parts are passed between operators as in an athletics "relay" (see Figure 3b). In case of the "pool" logic, operators are responsible for operations at certain work systems but are free to move whichever work system that requires attention (see Figure 3c).

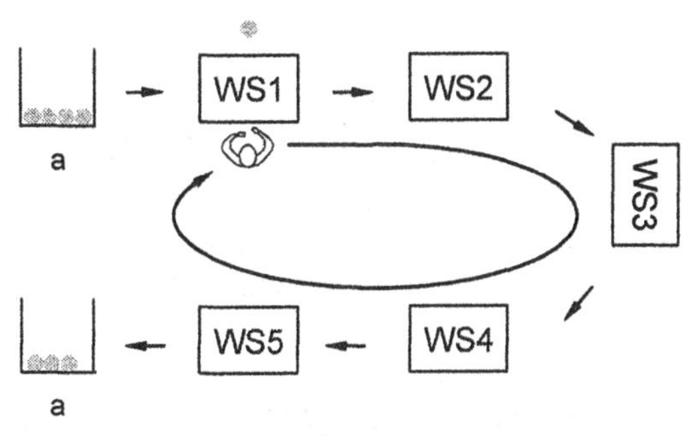

a) Operator movements according to "caravan" logic

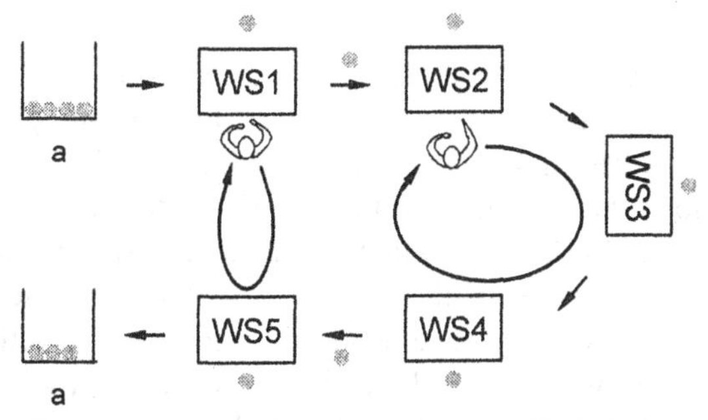

b) Operator movements according to "relay" logic

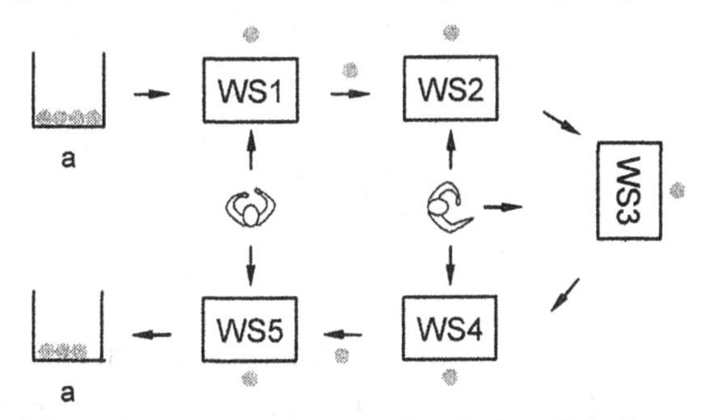

c) Operator movements according to "pool" logic

Figure 3. Operator movements in manufacturing cells
(Source: CIEMINSKI, NYHUIS 2003, p. 143; supplemented)

The number of operators working in a manufacturing cell can vary for all three movement logics. Also, a number of different assignments of operators to work systems and of different specifications of the sequence of activities are feasible. The cycle time determined by the human operators is longest for the caravan logic. In this case, the cycle time is equal to the duration of time it takes an operator to pass all work systems of the cell. The only real difference between the relay logic and the pool logic is that the sequence of activities of the operators is fixed in the relay set-up. Pool configurations make full use of the flexibility of the operators. Therefore, their cycle times are normally lower than for relay configurations. The prerequisite for installing both kinds of configurations, however, is that the operators are qualified to use several, if not all, work systems within the cell.

3. DETERMINING THE LOGISTIC EFFECTS OF HUMAN ACTIVITIES

In order to determine the logistic performance of manufacturing cells with different specifications of the activities of the human operators, simulation experiments were carried out. These examined the behaviour of a simulation model modelling a real manufacturing cell designed for an industrial company. The model had been validated in earlier experiments that did not consider the human aspects of the logistic behaviour. In a series of experiments the logic of operator movements, the sequence of activities and the number of operators working in the cell were varied. All models processed the same random sequence of orders so that the logistic performance measures recorded in the experiments were directly comparable.

The first finding of the simulation experiments was that it is not possible to establish a mathematical equation for the manufacturing cell cycle time if this depends on the human operators working in the cell. Diverse dynamic logistic interactions between operators and work systems occur for different cell configurations. These cannot be expressed mathematically by a single approximation equation. Furthermore, the dynamic processes lead to operator waiting times that form a part of the cycle time. The waiting times can only be predetermined analytically by means of Gantt charts or other project planning methods. Figure 4 shows an example of such a Gantt chart. One operator is responsible for the loading and unloading of all work systems in the manufacturing cell. The manufacturing cell has reached a stable logistic state and the sequence of activities shown repeats itself. It is immediately obvious is that the low number of operators causes a high degree of idle times for the work systems. The machines have to wait until the operator has finished activities at other work systems. Even though the work systems are not fully utilised, there is a period of an operator waiting time. During this work systems 1, 2, 4 and 5 are busy and there are no parts available that could be loaded onto work system 3 (see Figure 4).

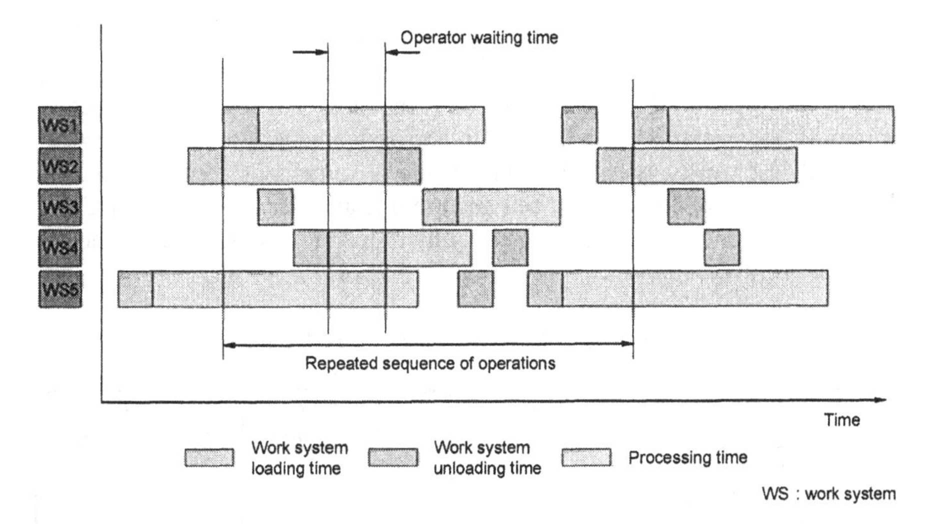

Figure 4. Determination of operator cycle time using Gantt charts

In the simulation experiments, the cycle times were therefore measured rather than predetermined. It was found that using the measured cycle time values, the logistic operating curves could still be applied. It follows, that companies that are setting up new manufacturing cells or reconfiguring existing ones should measure the operator cycle times of the cells in order to be able to determine their logistic performance.

The simulation experiments were secondly used to determine the effects the number of operators had on the logistic performance. These effects can be expressed in terms of the cycle time and the output rate of the manufacturing cells. Table 1 shows the variation of the measured cycle time for different configurations of the manufacturing cell modelled in the simulation experiments.

On the one hand, the experiments proved that manufacturing cell configurations using the pool logic lead to the shorter operator cycle times than configurations of the caravan or relay logic. The utilisation of operators for pool configurations was higher which resulted in reduced operator cycle times. The figures in Table 1 relate to configurations using the pool logic.

On the other hand, variations of the assignments of operators to work systems showed that the operators represent the manufacturing cell bottleneck in the minority of all possible configurations. The likelihood of operators becoming bottlenecks increases if the number of operators working in a manufacturing cell is low. For cost reasons, companies prefer these configurations despite the logistic disadvantages that arise. Even in such circumstances possibilities exist to reduce the extent of these disadvantages. A comparison of the cycle times between manufacturing cells with balanced operator workloads and imbalanced ones shows that it is logistically preferable to assign operators to work systems in such a way that the total processing times per operator are balanced (see Table 1). Such configurations minimise the operator cycle time. Also, the assignment of an operator to the first and last work system of a cell, which is sometimes recommended in literature as a link between the cell and the overall material flow (TAKEDA 1996), was found to lead to particularly high operator cycle times. It should therefore be avoided.

Table 1. Cycle times of different manufacturing cell configurations

Allocation of operators	Cycle time product 1[min]	Cycle time product 2 [min]	Cycle time product 3 [min]
1 operator at each work system	4.8	4.8	4.1
More than 2 operators	4.8	4.8	4.1
2 operators with balanced workload	4.8	4.8	4.1
2 operators with imbalanced workload	5.3	5.3	4.6
1 operator in manufacturing cell	6.1	6.3	5.1

Furthermore, the simulation experiments showed that the output rate of the manufacturing cell depends on the number of operators working in the cell. For reasons of cost minimisation, it is common industrial practice to minimise the number of operators working in the cell. The manufacturing cell analysed consisted of five work systems, the minimum number of operators was one. The manufacturing cell operating curves shown in Figure 5 demonstrate the effect that reductions of number of operators have on the output rate. The curves in the diagram were calculated for a manufacturing cell configuration in which the work systems rather than the operators represent the bottlenecks of the cell. The separate points in the diagram show the results recorded during the simulation experiments. One can see that reducing the number of operators in the cell to one leads to a loss in the mean output rate of c. 20 %. Correspondingly, the mean throughput time of orders exceeds the values that are possible if enough operators are working in the cell. In comparison to configurations with more staff, using a single operator still causes a reduction of the mean output rate of around 8 %. Companies therefore have to reach a suitable trade-off the between the costs of production and the output that their manufacturing cells can produce.

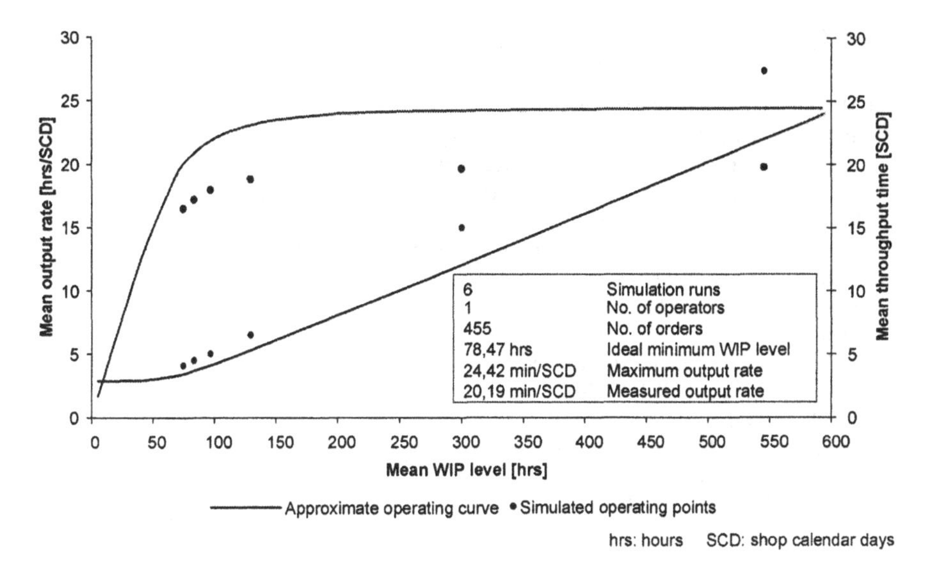

Figure 5. Impact of number of operators on logistic performance

4. CONCLUSIONS

The simulation experiments on a model of a manufacturing cell prove that different configurations of the interactions between human operators and work systems in manufacturing cells have different logistic properties. For some configurations, operators represent the logistic bottlenecks that determine the cycle time of the manufacturing cell and, as a consequence, its logistic performance. If it is possible to determine the operator cycle time, the theory of the logistic manufacturing operating curves serves as a modelling methodology for manufacturing cell logistics, with which the logistic adequacy of operator-work system assignments can be evaluated. The operating curves can be utilised to explain the effects of certain logistic configurations and also to predict the logistic performance of cells that are still at an implementation stage. In this way, industrial practitioners are assisted in finding those set-ups of manufacturing cells, including human operators that allow companies to achieve the logistic performance levels required.

REFERENCES

CIEMINSKI, Gregor von; NYHUIS, Peter:
Human Aspects of the Logistics of Manufacturing Cells.
In: Human Aspects in Production Management.
Eds.: ZÜLCH, Gert; STOWASSER, Sascha; JAGDEV, Harinder S.
Aachen: Shaker Verlag, 2003, pp. 209-215.
(esim – European Series in Industrial Management, Volume 5)

GREEN, T. J.; SADOWSKI, R. P.:
A Review of Cellular Manufacturing Assumptions and Advantages and Design Techniques.
In: Journal of Operations Management,
Amsterdam et al, 4(1984)2, 85-97.

MILTENBURG, J.:
One-piece Flow Manufacturing in U-shaped Cells: a Tutorial.
In: IIE Transactions,
London, 33(2001)4, pp. 303-321.

SCHONBERGER, R.:
Building a Chain of Customers.
London: Hutchinson Business Books, 2nd ed., 1990.

SLACK, N.; CHAMBERS, L.; JOHNSTON, R.:
Operations Management.
London: FT Prentice Hall, 3rd ed., 2000.

TAKEDA, H.:
Das System der Mixed Production.
Landsberg: Verlag Moderne Industrie, 1996.

WIENDAHL, H.-P.; CIEMINSKI, G. von:
Logistic Manufacturing Cell Operating Curves – Modelling the Logistic Behavior of Cellular Manufacture.
In: Production Engineering,
Berlin, 9(2003)2, 277-284.

Simulation of Disassembly and Re-assembly Processes with Beta-distributed operation Times

Jörg Fischer, Patricia Stock and Gert Zülch
University of Karlsruhe, ifab-Institute of Human and Industrial Engineering, Kaiserstrasse 12, D-76131 Karlsruhe, Germany.
Email: {joerg.fischer, patricia.stock, gert.zuelch}@ifab.uni-karlsruhe.de

Abstract: In the case of designing or improving industrial disassembly and repair shops, a special characteristic to be considered is the stochastic nature of disassembly, diagnosis and re-assembly times. Because of this special characteristic, the re-organisation of such work systems shall be supported by using a simulation method which considers distributed operation times. A question to be answered is the type of distribution which should be used. The problem is that during the data collection phase of a simulation project, the parameters of the often suggested Normal- or Gamma-distribution are normally not available. In contrary the parameters of the Beta-distribution are easy to get, because it could be characterised by the optimistic, the pessimistic and the most common time value. The following investigates in which way the different distributions effect the quantitative evaluation of a modelled repair shop compared with fixed operation times. Finally disadvantages and advantages of the use of different distributions in a simulation study will be discussed.

Key words: Simulation, Stochastic operation times, Beta-Distribution, Disassembly

1. INTRODUCTION TO THE PROBLEM

For the design and control of production systems simulation has proven to be a powerful tool. When designing or improving industrial disassembly and repair shops, a special problem to be considered is the feasibility of these operations and the stochastic nature of disassembly, diagnosis and re-assembly times.

In order to take the latter into account, the re-organisation of these kinds of work systems shall be supported by using a simulation approach which can realise stochastic execution times (with respect to feasibility of disassembly operations refer to SCHILLER (1998, pp. 25). Furthermore, it is necessary to consider which type of distribution is particularly suitable for the simulation of execution times.

2. OBJECT-ORIENTED SIMULATION

The simulation tool *OSim* which is used here, was developed at the ifab-Institute of Human and Industrial Engineering of the University of Karlsruhe (JONSSON 2000, p. 181; ZÜLCH et al. 2000; ZÜLCH, FISCHER 2002). The modelling of enterprise processes based on activity networks were studied by Johnson (JONSSON 2000). Activity networks are graphs with a logical sequence of activities, which can be used in the description of production and service processes (GROBEL 1992, p. 26). The production or service system to be modelled can thus be considered as a collection of activity networks, with whose help all processes occurring in the system can be modelled. An exemplary activity network is shown in Figure 1.

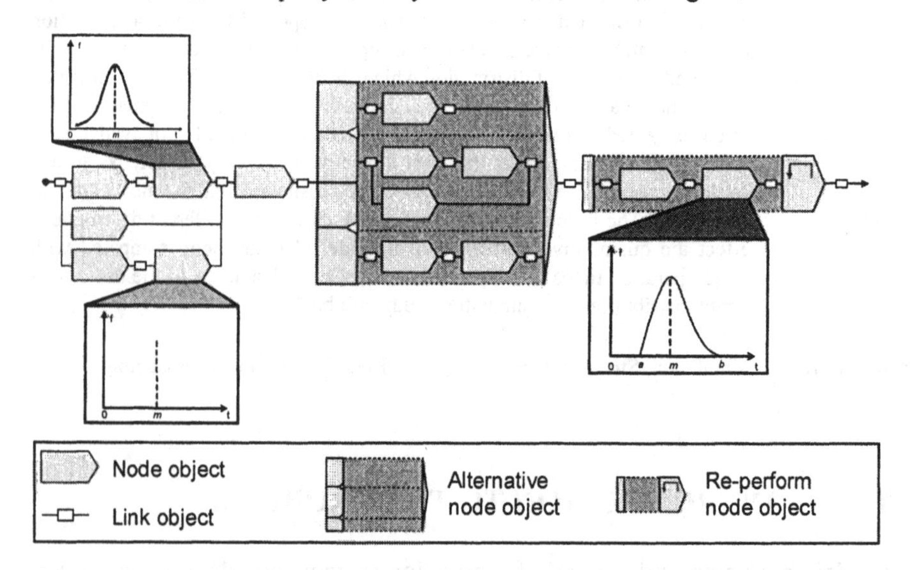

Figure 1. Example of an activity network with stochastic activity duration

Each node represents an activity executed within the process. Usually the nodes are connected with a conjunction, i.e. every activity will be executed exactly one time (for every triggering). This way of modelling the enterprise processes is quite common for a production process, however another logical linkage of the activities might be necessary (JONSSON 2000, p. 66). For instance there may be several alternatives for the disassembly of an electric device depending of the condition of the device (e.g. rusted screws). *OSim* offers the possibility for modelling such a disjunction of the activities (see JONSSON 2000, pp. 83). Additionally, each node takes up a certain length of time. The generic approach, which forms the basis of the simulation tool *OSim*, allows each node of an activity network to be assigned to an arbitrary distribution. This creates the basics for the representation of processes with stochastic activity execution times.

3. DISTRIBUTED EXECUTION TIMES

The question then arises as to which type of distribution is particularly appropriate for the simulation of such activities with stochastic execution times. NEUMANN (1975, p. 213) recommends, based on practical project experiences, to choose a distribution for execution times within an activity network which fulfils three conditions: First of all, the distribution should be steady; second, the resultant activity execution times should be bounded above and below, and third, the realisation of the activity execution times should be concentrated around a certain value.

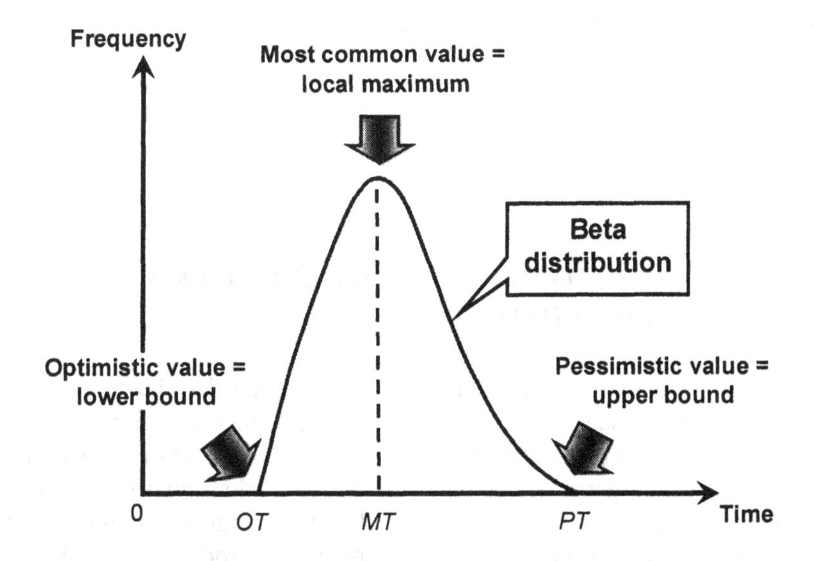

Figure 2. Modelling beta-distribution following the PERT methodology

One distribution which complies with these conditions is the beta-distribution. It possesses an additional advantage: In accordance with the PERT methodology, stochastic activity times can be modelled using an optimistic value (OT), a pessimistic value (PT) and a most common value (See Figure 2; NEUMANN 1987, pp. 165).

These types of values can be determined relatively easily by interrogation or check lists during the data collection and modelling phase of the simulation model. Furthermore, with this kind of Beta distribution modelling, it is possible to get the execution times non-axially symmetric. Furthermore, KRAJEWSKI and RITZMAN (1999, p. 814) suggest that this approach is also suitable for the modelling of operation durations, which are subject to strong scattering.

The probability density function (pdf) of the Beta distribution is (NEUMANN 1975, pp. 213):

$$f(t) := \begin{cases} \dfrac{(t-a)^{\alpha}(b-t)^{\beta}}{(b-a)^{\alpha+\beta+1}B(\alpha+1,\beta+1)} & \ldots \text{for } a \le t \le b \\ 0, & \ldots \text{else} \end{cases} \tag{1}$$

$$a := OT,\ \ b := PT,\ \ m := MT$$

The expected value EV can be estimated by:

$$EV = \frac{OT + 4 \cdot MT + PT}{6} \tag{2}$$

4. GENERATION OF BETA DISTRIBUTED RANDOM NUMBERS

During a simulation run it is necessary to provide distributed random numbers for the simulated activities. The concept applied in *OSim* is based on the random number generator suggested by KLAUKE und PAWLICEK (1981, p. 43), which generates equally distributed random numbers according to the congruence method. In this sort of generator random numbers are created from their direct predecessor. This creates a reproducible list of random numbers for the simulation run, in which the same start number will always lead to the same series of random numbers.

These equally distributed random numbers can be transformed into any statistical distribution. This is usually done through the inversion of probability transformation method, in which the inverse function of the cumulative distribution function (cdf) is used (NEUMANN 1977, pp. 313). For some distributions, however, among others the Beta distribution, it is very difficult to determine an elementary inverse function (LIEBL 1995, pp. 36).

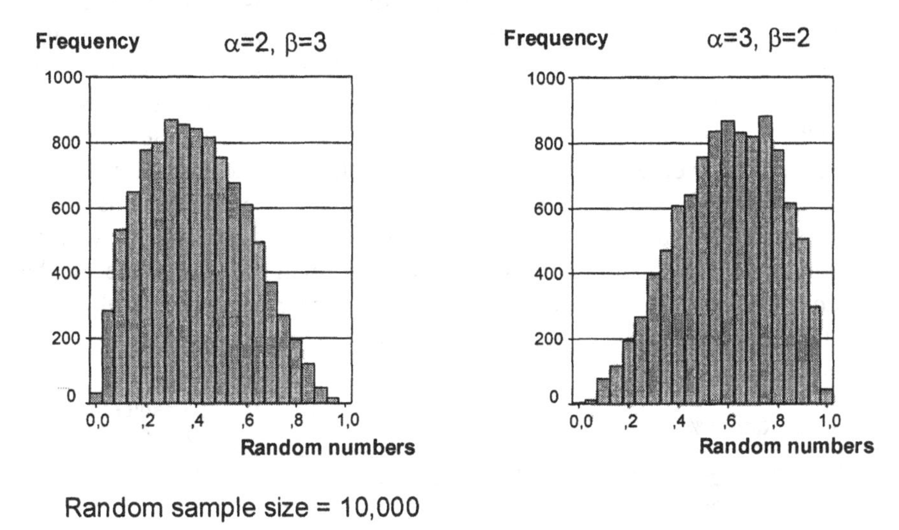

Figure 3. Generated beta-distributed random numbers in the interval [0,1]

To nonetheless be able easily to generate beta-distributed random numbers, *OSim* uses the rejection method for the transformation of equally distributed to beta-distributed random numbers (SCHMEISER 1977, p. 51). This method sorts out those random numbers which are not adequate for the corresponding distribution. In Figure 3 one can see different examples of beta-distributed random numbers in the interval between 0 and 1, generated with the used beta-distributed random number algorithm suggested by DEVROYE (1986, K. 9.3-6).

5. APPLICATION MODEL

In the following a model, a real system will be used to investigate which effects the use of various distributions has upon the simulation results, and to what degree the beta-distribution is suitable for the modelling of stochastic execution times. The implemented model represents an industrial repair line of an industrial repair shop for electrical devices.

The most important objective for the repairs in this application case is the achievement of a high service rate. This means that all repairs processed on a given day should be shipped back to the customer on the same day. If this is the case the service rate attains the optimum value of 100 %. Taking the type of the electrical device into consideration, a total of approx. 400 activity networks with variously distributed activity times were found. In the model, 6000 orders, the order programme for an entire month, were considered.

Optimistic, pessimistic and most common values were already available for each of the activity times to be modelled. Various distributions, which were later analysed in several simulation runs with various starting values from the random number generator, were modelled based on these values.

Assuming fixed execution times for the activities, the same key figures were achieved in every simulation run if the order programme is fixed. However, if one deposits distributions various key figure values will be determined for each simulation run, depending upon the start value for the random number generator. Based on these differing results, a result range can be derived for each key figure (Figure 4).

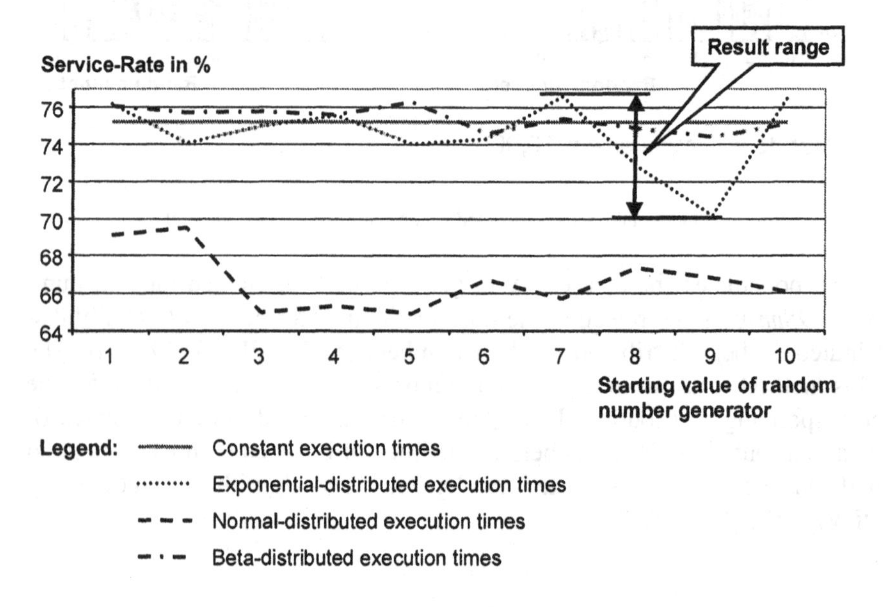

Figure 4. Creation of result ranges

Figure 5 represents the result range for the key figure "service rate". The result area for the beta-distribution is the smallest, implying that these results have the lowest degree of dispersion (standard deviation: for beta-distribution: 0.6; for exponential distribution: 1.6; for normal distribution: 1.9). The reason for this can be found, in the fact that the beta-distribution is bound. Values above or beneath the optimistic or pessimistic value are not present, in contrast to exponential and normal distributions. Incidentally, this bounding of the values corresponds with the experience that was made in the existing industrial repair line for electrical devices.

Furthermore, one can see that the beta-distribution value range encloses the simulated fixed values, whereas e.g. the exponential distribution tends to produce "poorer" results. This can be traced back to the fact that the exponential distribution is not bound above, and thus execution times which are "too long" must also be considered. Within the use of the normal distribution one can see that, generally, "poorer" results are attained for the service rate. Subsequently, it can be assumed that the use of this distribution would create execution times which tend to be too long. Further on, a normal distribution, which is the result of an addition of two independent normal distributions, has a greater statistical spread, i.e. the normal distribution is "broader" (see JACOD, PROTTER 2000, p. 116).

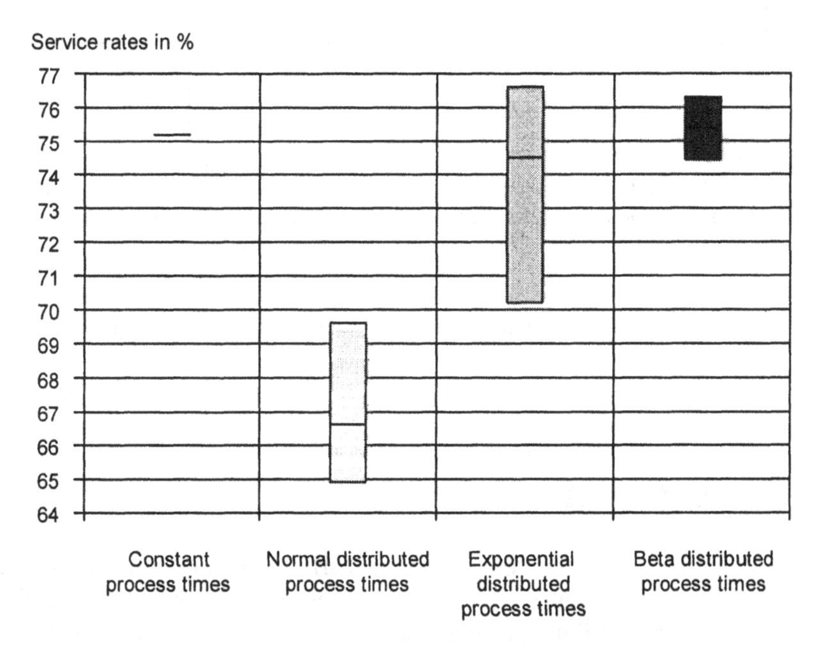

Figure 5. Ranges of service rates when simulating various distributions of execution times

Comparable correlations can also be seen with the key figure "utilisation" (see Figure 6), whereas the result area for the beta-distribution is the again smallest (standard deviation: for beta-distribution: 0.3; for exponential distribution: 0.7; for normal distribution: 0.4). In this case, in particular the normal distribution results area lies considerably higher compared to values determined with constant times, which in turn indicates that a stronger equalisation of the operating times exists.

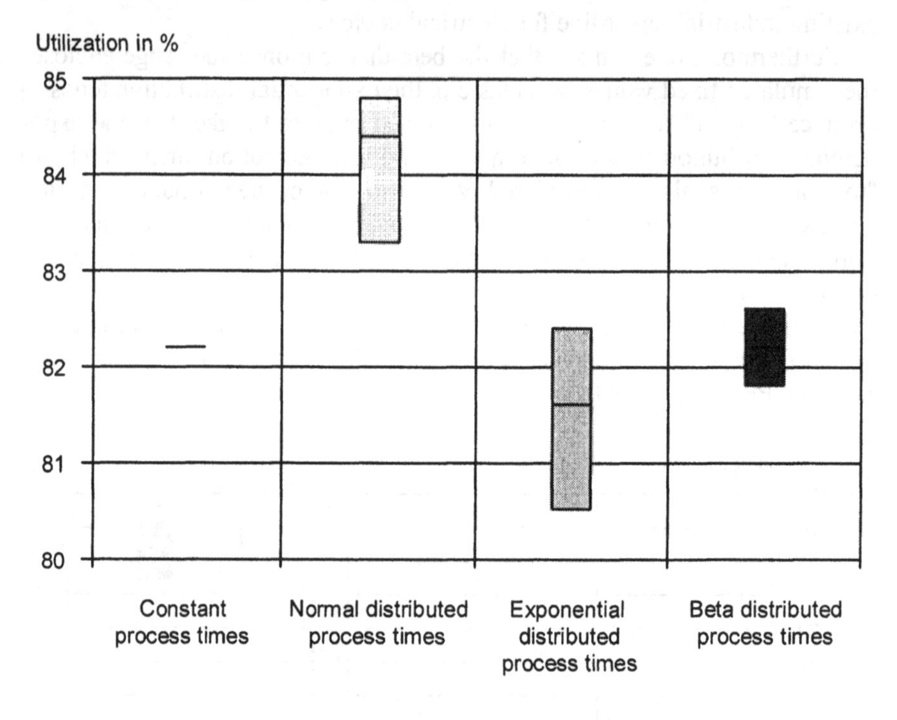

Figure 6. Ranges of utilisation when simulating various distributions of execution times

6. SUMMARY

The examination of the model at hand clearly showed that the beta-distribution is particularly well suited for the modelling of distributed execution times. This can be attributed to the ease of modelling and the resulting simplicity of the data collection. Furthermore, the results are closer to reality since the dispersion is smaller. If it can be guaranteed in a practical case that the optimistic or pessimistic values are neither exceeded nor under-run, it can then be safely expected that the results of the real system will also be in the result range the simulation model has predicted.

ACKNOWLEDGEMENTS

This work was funded within the *SAIDER*-Project "Simulation aided improvement of disassembly and re-assembly processes" in cooperation with the SIM-SERV consortium (see also SAIDER 2004 for further information). SIM-SERV (2004) is financed by the EU under Framework Programme V "Competitive and Sustainable Growth".

REFERENCES

DEVROYE, Luc:
Non-Uniform Random Variate Generation.
Berlin, New York, Heidelberg u.a.: Springer Verlag, 1986.

GROBEL, Thomas:
Analyse der Einflüsse auf die Aufbauorganisation von Produktionssystemen. In: Innovatives Arbeits- und Betriebszeitmanagement.
Eds.: ACKERMANN, Karl-Friedrich; HOFMANN, Mathias.
Frankfurt/M., New York: Campus Verlag, 1990, pp. 183-212.

JACOD, Jean; PROTTER, Philip:
Probability essentials.
Berlin et al.: Springer, 2000.

JONSSON, Uwe:
Ein integriertes Objektmodell zu durchlaufplanorientierten Simulation von Produktionssystemen.
Aachen: Shaker Verlag, 2000.
(ifab-Forschungsberichte aus dem Institut für Arbeitswissenschaft und Betriebsorganisation der Universität Karlsruhe, Vol. 21)

KLAUKE, Adolf; PAWLICEK, Jiri:
Die Erzeugung gleichverteilter Zufallszahlen für die Anwendung in stochastischen Simulationsmodellen.
In: Zeitschrift für Arbeitswissenschaft,
Köln, 35(7NF)(1981)1, pp. 35-44.

KRAJEWSKI, Lee J.; RITZMAN, Larry P.:
Operations Management.
Reading, MA u.a.: Addison Wesley, 1999.

LIEBL, Franz:
Simulation.
München: R. Oldenbourg Verlag, 1995.

NEUMANN, Klaus:
Operations Research Verfahren. Volume 3: Graphentheorie, Netzplantechnik.
München, Wien: Carl Hanser Verlag, 1975.

NEUMANN, Klaus:
Operations Research Verfahren. Volume 2: Dynamische Optimierung, Lagerhaltung, Simulation, Wartschlangen.
München, Wien: Carl Hanser Verlag, 1977.

NEUMANN, Klaus:
Netzplantechnik.
In: Grundlagen des Operations Research, Teil 2.
Hrsg.: GAL, Tomas.
Berlin et al.: Springer, 1987, S. 165-260.
SAIDER:
Sim-Serv - Sucess Stories: Simulation-aided improvement of disassembly and re-assembly processes.
http://www.sim-serv.com/pdf/success_stories/member_22_story_75.pdf, 23.02.2004
SCHILLER, Emmerich F.:
Ein Beitrag zur adaptiv-dynamischen Arbeitsplanung in der Demontage.
Aachen: Shaker Verlag, 1998.
(ifab-Forschungsberichte aus dem Institut für Arbeitswissenschaft und Betriebsorganisation der Universität Karlsruhe, Vol.14)
SCHMEISER, Bruce:
Methods for modelling and generating probabilistic components in digital computer simulation when the standard distributions are not adequate: A survey.
In: Proceedings of the 1977 Winter Simulation Conference.
Eds.: HIGHLAND, H. J.; SARGENT, R. G.; SCHMIDT, J. W.
Piscataway, NJ: The Institute of Electrical and Electronics Engineers, Volume 1, 1977, pp. 50-57.
ZÜLCH, Gert; FISCHER, Jörg:
An integrated object model as a world of model components for an activity network based simulation approach.
In: Simulation in Industry.
Eds.: BREITENECKER, Felix; HORTON, Graham; KAMPE, Gerald et. al.
Delft, Erlangen, San Diego: SCS-Europe, 2002, pp. 74-79.
ZÜLCH, Gert; FISCHER, Jörg; JONSSON, Uwe:
An integrated object model for activity network based simulation.
In: WSC'00 Proceedings of the 2000 Winter Simulation Conference.
Eds.: JOINES, Jeffrey A.; BARTON, Russel R.; KANG, Keebom et al.
Compact disk: WSC'00, 2000, pp. 371-380.

Strategic Analysis of Products Related to the Integration of Human Judgement into Demand Forecasting

Séverine Meunier Martins, Naoufel Cheikhrouhou and Rémy Glardon
Swiss Federal Institute of Technology at Lausanne, Laboratory for Production Management and Processes, Ecublens, CH-1015 Lausanne, Switzerland.
Email: severine.meunier@epfl.ch

Abstract: For companies proposing a large offer of products and services, it is important to identify the products for which the forecast quality is critical. The objective of this paper is to propose a strategic analysis of the products in order to determine priorities in the demand forecasting process: (1) a *high priority* class including the products for which forecast accuracy has a strong impact on the logistic performance of the company and where judgement should be integrated in the forecasting process; (2) a *low priority* class with the remaining products for which forecast accuracy is less important and where forecasts could be established in an automated way. The classification process takes into account different criteria chosen on the basis of cost efficiency reasoning.

Key words: Judgement in forecasting, Strategic analysis, Fuzzy classification

1. INTRODUCTION

Most of the mathematical forecasting models consist of more or less complex extrapolation techniques and are based on the hypothesis that established patterns in the previous data will not change during the forecasting phase and are expected to continue in the future. The context in which companies evolve nowadays is characterised by a high uncertainty level and quick changes in the environment for which this hypothesis is generally insufficiently verified. Under these conditions, integrating judgmental forecasts established by a forecaster on the basis of his/her interpretation of contextual information, his/her knowledge of the market and his/her experience, with mathematical forecasts appears promising; it would in particular allow the anticipation of changes in the demand characteristics and thus to increase the reliability and accuracy of the resulting forecasts (MEUNIER MARTINS 2003).

The integration of human judgement into mathematical forecasts can be very costly if a forecaster has to review every product. Furthermore, this systematic process could be inefficient as the forecaster would become bored with this mass of work. It is therefore crucial to identify the most important products for which the forecast quality is critical and where judgement should be integrated. This paper proposes a strategic analysis of the products in order to define priorities in the forecasting process. Two classes of finished products are determined:

- A *high priority* class including the products for which forecast accuracy has a strong impact on the logistic performance of the company and where judgement should be integrated into the forecasting process;
- A *low priority* class with the remaining products for which forecast accuracy is less important and where forecasts could be established in an automated way, without any human supervision.

This paper is organised in five sections. In section 2, the justification and motivation for a strategic analysis of the products related to the integration of human judgement into demand forecasting are detailed. The classification criteria are also presented in this section with the reason of their choice. Section 3 introduces the classification problem with regards to some known approaches. Then the new classification method based on fuzzy logic and proposed in this paper is presented and justified. The presentation of a case study with data supplied by two companies is given and discussed in section 4. Finally, conclusions are presented in section 5.

2. STRATEGIC ANALYSIS OF PRODUCTS

The classification criteria are chosen on the basis of cost efficiency reasoning that could be illustrated by the question: *how far do the resources invested in judgmental forecasting contribute to improved logistic performances and better service level*? Three types of criteria have therefore been taken into account in the classification process. They are related to the:

- Strategic importance of the products for the company;
- Difficulty to establish forecasts (judgmental or mathematical);
- Influence of the forecasts accuracy on the logistic performance of the company.

The criteria related to the strategic importance of the products for the company are the contribution of a product to the total turnover and its position in the life cycle. The contribution of the products to the total turnover identifies the items making the greatest part of the financial volume. The position of the product in the life cycle is a subjective criterion complementary to the previous one. It makes it possible to identify the items with a strong potential or dying, whereas this information does not appear clearly when evaluating their contribution to the total turnover.

The criteria relating to the difficulty in establishing reliable judgmental or mathematical forecasts are respectively the availability of useful contextual information and the randomness of the time series pattern related to the historical demand for a product. Many studies show that the availability of contextual information is crucial for the forecaster to establish reliable forecasts (SANDERS, RITZMAN 1992; EDMUNDSON et al. 1988). Although the randomness of the time series pattern does not completely evaluate the difficulty to establish accurate mathematical forecasts, this information presents the advantage of simplicity and indicates with a reasonable probability if the hypothesis for the use of extrapolation techniques for forecasting is met. The criterion "availability of contextual information" is subjective whereas the "randomness of the time series pattern" can be evaluated with statistical tests.

The criteria related to the influence of the forecast accuracy on the logistic performance of the company are the lead time of the constituting components of the products and their commonality, i.e. their level of standardisation. Many other criteria could be chosen in considering the role of forecast accuracy on the logistic performances. The choice of the former criteria is based on the fact that the upstream part of the supply chain (supply) depends most strongly on forecasts and requires longer term forecasts for long delivery lead time components. Furthermore, the commonality of the constituting components is important, as products made of specific components require more accurate forecasts than the others (ZIPPER, MEUNIER MARTINS 2002).

3. CLASSIFICATION METHOD

3.1 State of the Art

Single criterion ABC classification based on the financial or material flow is the most popular and used technique for the classification of items in stocks. The great popularity of this procedure is explained by its simplicity;

nevertheless it suffers a major drawback: only one criterion is taken into account.

FLORES and WHYBARK (1986) propose a derived approach from the ABC classification. They use several classification criteria as the delivery lead time of the components, their possibility to be substituted, etc. FLORES et al. (1992) propose the use of a matrix in order to integrate two criteria in the classification process, for example the annual turnover and the lead time. Nine classes of items are determined, requiring nine different policies. This procedure becomes difficult to manage for more than two criteria.

Analytical Hierarchy Process (AHP) developed by SAATY (1980) has been successfully used by FLORES et al. (1992) and GAJPAL et al. (1994) for multi-criteria classification. They use AHP to reduce the different criteria in a unique and quantifiable measure, allowing them to classify the items according to the structure of a simple ABC classification. AHP requires a lot of subjectivity in the pair wise comparisons of the different criteria. It may be difficult for the expert to judgementally compare two criteria and assign them an objective weight according to their relative importance.

GUVENIR and EREL (1998) propose the use of genetic algorithms for the determination of the weights assigned to the AHP classification criteria. They are determined according to the frontier points between the A-B and B-C classes of a pre-classified set of items. The main inconvenient of this method is that it requires a pre-classification of a sample set of items.

3.2 Proposed Classification Method

In real cases of classification, there is often no absolute demarcation between the different classes that generally overlap. Fuzzy classification is appropriate for such situations where elements characterised by continuous data can not be clearly assigned to discrete classes and makes it possible to use linguistic criteria vaguely defined (ZADEH 1965). It consists in evaluating the compatibility of the different products to the two classes on a $[0;1]$ interval.

This operation is done in two steps. Firstly, an elementary compatibility measure is processed for each product according to each classification criterion. The elementary compatibility measures are calculated according to membership functions on a $[0;1]$ interval. The membership functions $\mu_{LP}^{c}(x_c)$ and $\mu_{HP}^{c}(x_c)$ give respectively the elementary compatibility measures of the item x o the sub-classes of low and high priority according to the criterion c and with a value x_c for this criterion.

Secondly, the global membership levels to the two classes (high and low priority) are calculated for each product as a linear combination of their respective elementary compatibility measures. Generally, an average of the compatibility measures is used. This approach is sufficient if the classification criteria are independent and of equal importance.

Once the membership levels have been evaluated, a threshold defining the limit between the items of the low and high priority classes must be defined. This threshold corresponds to the value of the membership level to the high priority class above which the items will be considered as high priority ones. It is defined in order to minimise the number of products in the high priority class and maximise the number of components concerned with the forecasts established for these products.

3.3 Definition of the Membership Functions

3.3.1 Criterion 1: Contribution of the Product to the Total Turnover

This criterion has been chosen so that greater attention in the forecasting process goes to the products generating the more important part of the turnover. Generally, 20 % of the products make 80 % of the turnover (Pareto's law). The membership functions corresponding to the criterion "contribution to the total turnover" are defined according to the values defining the transition interval between the critical and non-critical items. After a classical ABC analysis, the lower and upper values defining this interval, respectively $S_1 - \Delta_1^{\text{inf}}$ and $S_1 + \Delta_1^{\text{sup}}$ are fixed. The corresponding membership functions to the low and high priority sub-classes, respectively $\mu_{LP}^1(x_1)$ and $\mu_{HP}^1(x_1)$, are illustrated in Figure 1 and defined in detail below.

$$\mu_{LP}^1(x_1) = \begin{cases} 1 & x_1 \le S_1 - \Delta_1^{\text{inf}} \\ 1 + \left(S_1 - \Delta_1^{\text{inf}} - x_1\right) / \left(\Delta_1^{\text{inf}} + \Delta_1^{\text{sup}}\right) & S_1 - \Delta_1^{\text{inf}} < x_1 < S_1 + \Delta_1^{\text{sup}} \\ 0 & S_1 + \Delta_1^{\text{sup}} \le x_1 \end{cases}$$

$$\mu_{HP}^1(x_1) = \begin{cases} 0 & x_1 \le S_1 - \Delta_1^{\text{inf}} \\ 1 + \left(x_1 - S_1 - \Delta_1^{\text{sup}}\right) / \left(\Delta_1^{\text{inf}} + \Delta_1^{\text{sup}}\right) & S_1 - \Delta_1^{\text{inf}} < x_1 < S_1 + \Delta_1^{\text{sup}} \\ 1 & S_1 + \Delta_1^{\text{sup}} \le x_1 \end{cases}$$

S_1 — Contribution of a product corresponding to the maximum difference between the slopes of two successive points in the cumulative sum of the individual contribution to the turnover

$S_1 - \Delta_1^{\text{inf}}$ — Contribution of a product corresponding to the inferior limit of the transition zone between the products making the smallest and the greatest part of the turnover

$S_1 + \Delta_1^{sup}$ Contribution of a product corresponding to the superior limit of the transition zone between the products making the smallest and the greatest part of the turnover

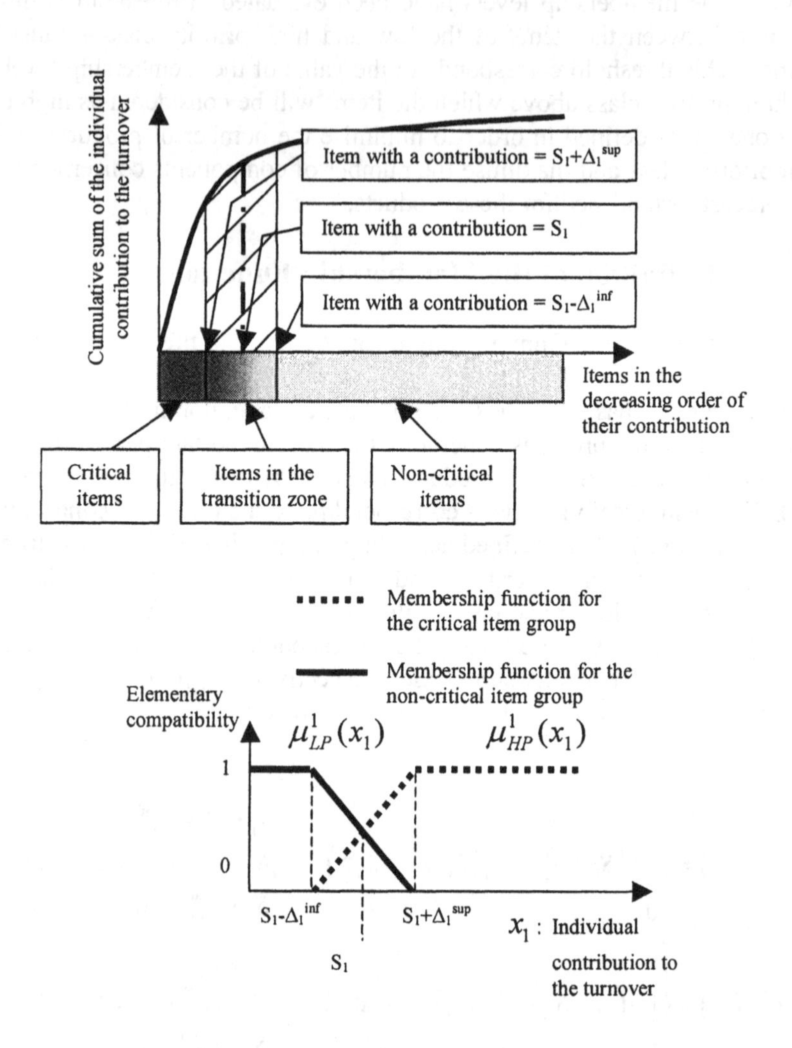

Figure 1. Membership functions for the criterion "contribution to the turnover" (Source: MEUNIER MARTINS, CHEIKHROUHOU, GLARDON 2003, p. 199)

3.3.2 Criterion 2: Position in the Life Cycle

The products in a growing or decreasing phase should receive special attention in the forecasting process in order to limit forecast errors generating high costs (stock outs, over inventories difficult to use/sell, etc.)

The value of this criterion is judgmentally evaluated on a 1 to 5 scale. The value 5 means that the considered product is in a strongly growing phase, whereas the value 1 means that it is dying. The value 3 indicates a stable position of the product. The membership functions corresponding to this criterion are depicted in Figure 2.

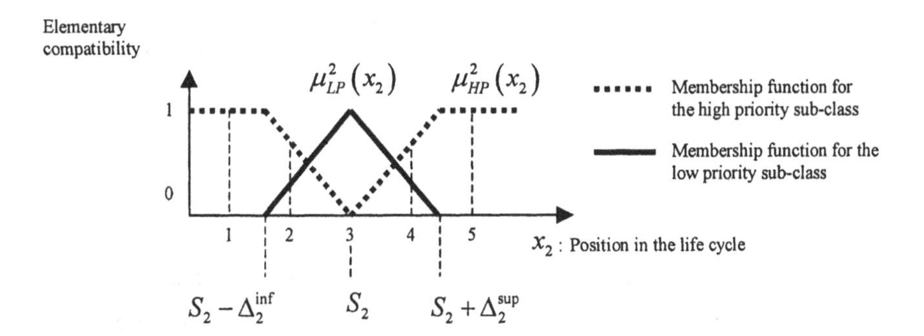

Figure 2. Membership functions for the criterion "position in the life cycle"

S_2	Position in the life cycle corresponding to a stable phase
$S_2 - \Delta_2^{inf}$	Inferior limit of the transition zone between the growing and the stable phase
$S_2 + \Delta_2^{sup}$	Superior limit of the transition zone between the stable and the decreasing phase

3.3.3 Criterion 3: Availability of Useful Contextual Information

Contextual information is very difficult to integrate into the forecasts unless interpreted by an expert and judgmentally combined with the forecasts. The value of the criterion "availability of contextual information" is judgmentally evaluated on a 1 to 5 scale. The value 5 means that a lot of significant contextual information is available for demand forecasting; on the opposite the value 1 means that no useful contextual information is available. Therefore, products for which this information exists should be supervised by an expert in the forecasting process. The membership functions corresponding to this criterion are depicted in Figure 3.

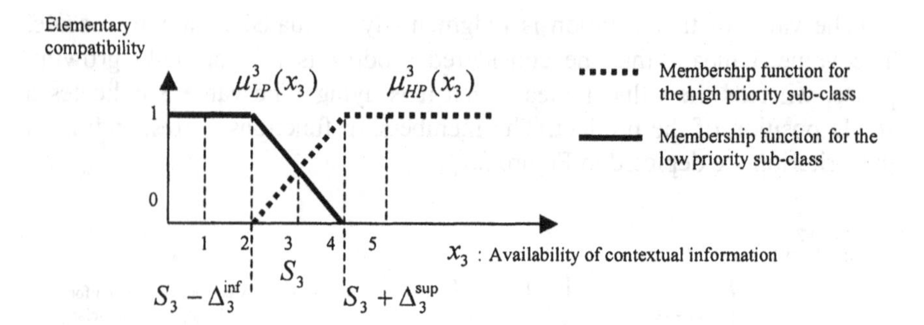

Figure 3. Membership functions for the criterion "availability of contextual information"

S_3	Medium value concerning the availability of contextual information
$S_3 - \Delta_3^{inf}$	Inferior limit of the transition zone between the non-availability and availability of information
$S_3 + \Delta_3^{sup}$	Superior limit of the transition zone between the non-availability and availability of information

3.3.4 Criterion 4: Randomness of the Time Series Pattern Related to the Historical Demand

Extrapolation forecasting models should be applied only to time series having a non random pattern. The turning point test (KENDALL 1990) is used to check whether the time series pattern is random or not. A turning point in a time series is a point where the series changes direction, each turning point represents either a local peak or a local trough. This method is based on the premises that a trended or positively auto-correlated series should have fewer turning points than a random one and a negatively auto-correlated series should have more. If the series is actually a random series, the sampling distribution of the number of turning points U is approximately normal. We define:

$$Z = \left|\frac{U - \mu_U}{\sigma_U}\right|, \mu_U = \frac{2(n-2)}{3}, \sigma_U = \sqrt{\frac{16n - 29}{90}}$$

where μ_U and σ_U are respectively the expected value and variance of U and n is the number of observations of the time series.

The probability that the time series pattern is random is higher than 90 % if $Z < Z_{\alpha/2}$, $Z_{\alpha/2} = 1.282$. The membership functions corresponding to the criterion "randomness of the time series" are depicted in Figure 4.

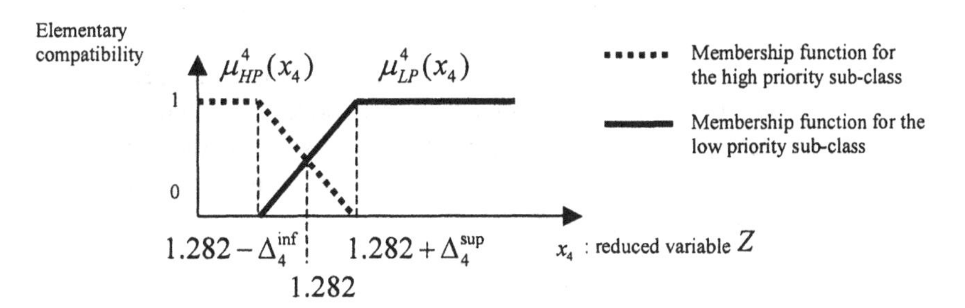

Figure 4. Membership functions for the criterion "randomness of the time series pattern"

S_4	Threshold value above which the probability that the time series pattern is random is higher than 90 %
$S_4 - \Delta_4^{\inf}$	Inferior limit below which the time series pattern is considered as random
$S_4 + \Delta_4^{\sup}$	Superior limit above which the time series pattern is considered as non-random

3.3.5 Criterion 5: Delivery Lead Time of the Constituting Components

The membership functions corresponding to the criterion "delivery lead time of the constituting components" are respectively $\mu_{HP}^{5}(x_5)$ and $\mu_{LP}^{5}(x_5)$ for the high and low priority sub-classes and are shown in Figure 5.

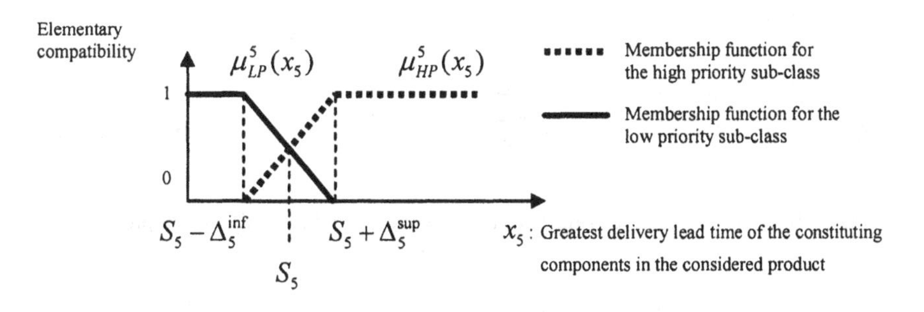

Figure 5. Membership functions for the criterion "delivery lead time of the constituting components"

S_5	Threshold corresponding to the component delivery lead time considered as critical by the company (in regard to the product delivery time for example)

$S_5 - \Delta_5^{inf}$ Inferior limit of the critical delivery lead time

$S_5 + \Delta_5^{sup}$ Superior limit of the critical delivery lead time

3.3.6 Criterion 6: Commonality of the Components

The commonality of a component is defined as the ratio between the number of products including the considered component to the total number of products; the lower the commonality, the more specific the component. The criticality of a product made up of m constituting components with commonality c_i $(i = 1...m)$ is defined as

$$C = 1/m\sum_{i=1}^{m}(1/c_i).$$

The results of a simulation study (ZIPPER, MEUNIER MARTINS 2002) show that the products requiring a particular attention in the forecasting process are the ones scoring a high criticality. In the classification process the number of products belonging to the high priority sub-class related to the criterion "commonality of the components" should be minimal and the population of related components should be maximal. The membership functions corresponding to the criterion "commonality of the components" are respectively $\mu_{HP}^6(x_6)$ and $\mu_{LP}^6(x_6)$ for the critical and non-critical item groups and are shown in Figure 6.

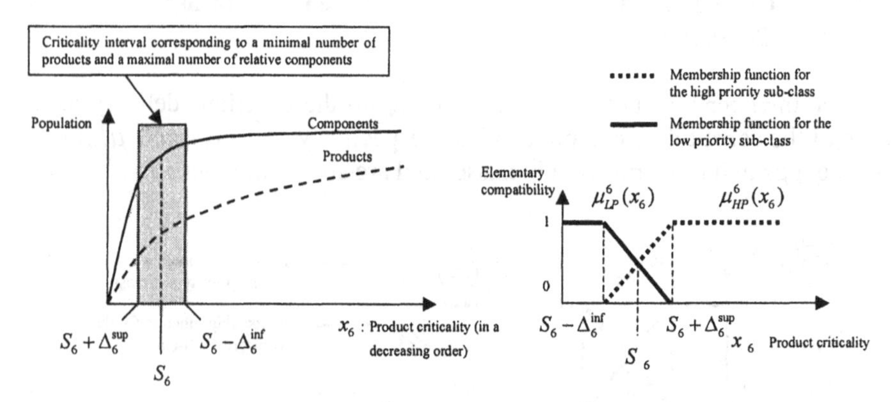

Figure 6. Relation between the criticality of the products and the populations of products and components

S_6 Product criticality threshold

$S_6 - \Delta_6^{inf}$ Inferior limit of the product criticality threshold

$S_6 + \Delta_6^{sup}$ Superior limit of the product criticality threshold

3.3.7 Final Classification

As suggested in section 3.2., a simple average of the elementary compatibility measures is used to process the final membership levels of a product x to the low and high priority classes, respectively $\mu_{LP}(x)$ and $\mu_{HP}(x)$:

$$\mu_{LP}(x) = \frac{1}{6}\left(\mu_{LP}^{1}(x_1) + \mu_{LP}^{2}(x_2) + \mu_{LP}^{3}(x_3) + \mu_{LP}^{4}(x_4) + \mu_{LP}^{5}(x_5) + \mu_{LP}^{6}(x_6)\right)$$

$$\mu_{HP}(x) = \frac{1}{6}\left(\mu_{HP}^{1}(x_1) + \mu_{HP}^{2}(x_2) + \mu_{HP}^{3}(x_3) + \mu_{HP}^{4}(x_4) + \mu_{HP}^{5}(x_5) + \mu_{HP}^{6}(x_6)\right)$$

4. CASE STUDIES

Real data from two Swiss manufacturing companies are used to provide real numerical examples of product classification. Company A designs and produces electric micro-motors. Company B designs and produces electronic sub-assemblies, sensors, etc. For company A, 1465 products are considered made up of more than 2600 components; 21 products made up with 1443 components are used for company B.

The membership levels to the low and high priority classes obtained for some of the products of the company B are presented in Table 1.

Table 1. Example of the results for the membership levels to the low and high priority classes obtained for some of the products of company B

Product	Membership level to the low priority class $\mu_{LP}(x)$	Membership level to the high priority class $\mu_{HP}(x)$
463000	0.053	0.947
713317	0.224	0.776
663288	0.410	0.589
667173	0.734	0.266

Table 2 shows the populations of products and related components with a membership level to the high priority class higher than a specific value. This table is very useful to define the threshold fixing a limit value in the high priority class above which the products will be considered as high priority in the forecasting process. This threshold should be chosen in order to minimise the number of high priority products and maximise the number of components related to these critical products.

Table 2. Relation between the membership level to the high priority class (HP) and the corresponding populations of products and components

	Company A		Company B	
Membership level HP	Population of products	Population of components	Population of products	Population of components
0.5	20.4 %	71.5 %	76.2 %	83.5 %
0.55	18.9 %	70.5 %	71.4 %	83.2 %
0.6	17.4 %	68.9 %	33.3 %	79.8 %
0.65	16.7 %	66.9 %	33.3 %	79.8 %
0.7	16.1 %	66.5 %	28.6 %	27.1 %
0.75	2.9 %	26.9 %	4.8 %	5.2 %

From these results, a good compromise for company A and company B seems to be obtained with a threshold value for the membership level to the high priority class equal to 0.6 (in grey in Table 2). Products with a membership level to the high priority class higher than 0.6 are considered as critical in the forecasting process. In this case the distribution of the population of products for company A and company B is shown in Figure 7.

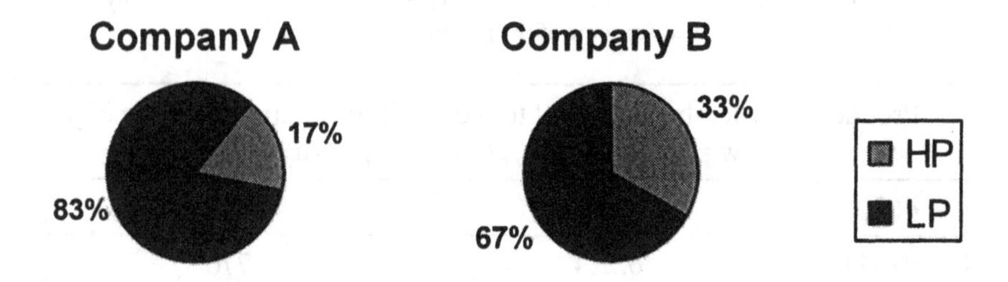

Figure 7. Distribution of the products according to the high and low priority classes

The populations of components related to the products belonging to the low and high priority classes may be compared with the following financial criteria:

- Sum of the unit price of the components used to produce one unit of each product (FV1);
- Sum of the unit price of the components used to produce the products within one-year sales history (FV2).

The results of this comparison are presented in Table 3.

Table 3. Financial volumes of the components related to the low and high priority product classes (LP and HP)

	Company A		Company B	
	LP	HP	LP	HP
FV1: Total component cost related to the production of one unit of each product	25.2 %	74.8 %	85.9 %	14.1 %
FV2: Total component cost related to one-year products sales	14.2 %	85.8 %	32.7 %	67.3 %

The results obtained for the company A confirms that the high priority products hit the components making the greatest part of the total financial volumes described above. For the company B, the most expensive components are essentially included in the low priority products. It explains the poor result concerning the total component cost related to the production of one unit of each product. Nevertheless, when one-year sales history is analysed it appears that the high priority products hit the components making 67.3 % of the supply turnover.

5. CONCLUSIONS

In this paper, a strategic analysis of the products related to the integration of human judgement into demand forecasting has been presented. This strategic analysis is based on a new classification approach based on fuzzy logic. This approach makes it possible to take into account the vagueness of some classification criteria as well as linguistic criteria. Various classification criteria can be taken into account in this process. We chose them in order to emphasise the critical elements for which the quality of the forecasts has a deep impact on the logistic performances of the company. Nevertheless, if the objective of the company is to improve the demand forecast accuracy in order to increase the reliability of supply management at the lowest costs, more specific criteria may be defined. For example the optimal supply management policy for each component may be another criterion for the selection of the high priority products. Once the criteria are chosen, elementary compatibility measures are calculated for each criterion and each product according to the high and low priority classes. Finally the membership levels to the high and low priority classes are determined for each product on the basis of a linear combination of its elementary compatibility measures. The results obtained with the data from two companies show firstly that this method can be applied successfully and secondly that the number of high priority products is limited in a drastic way, without penalising too much the related components. Further research would be needed to find out the influence of the choice of the various membership functions as well as to strengthen the validation of the proposed approach through other real case applications.

ACKNOWLEDGEMENTS

The authors thank the Swiss Agency of Promotion and Innovation for the financial support of the project "Socio-technical system improvement for the reliability of supply forecasts".

REFERENCES

EDMUNDSON, Robert; LAWRENCE, Michael; O'CONNOR, Marcus:
The use of non-time series information in sales forecasting: a case study.
In: Journal of Forecasting,
Chichester, 7(1988)2, pp. 201-211.
FLORES, Benito; OLSON, David; DORAI, V. K.:
Management of multicriteria inventory classification.
In: Mathematical and Computer Modeling,
London, 16(1992)12, pp. 71-72.
FLORES, Benito; WHYBARK, D. Clay:
Multiple criteria ABC analysis.
In: International Journal of Operations and Production Management,
Bradford, 6(1986)3, pp. 38-46.
GAJPAL, Prem; GANESH, L. S.; RAJENDRAN, Chandrasekharan:
Criticality analysis of spare parts using the analytical hierarchy process.
In: International Journal of Production Economics,
Amsterdam et al., 35(1984)1-3, pp. 293-297.
GUVENIR, H. Altay; EREL, Erdal:
Multicriteria inventory classification using a genetic algorithm.
In: European Journal of Operational Research,
Amsterdam, 105(1998)1, pp. 29-37.
KENDALL, Maurice; ORTH, J. Keith:
Time series, 3rd edition.
London: Arnold, 3rd ed., 1990.
MEUNIER MARTINS, Séverine:
Approche stratégique pour la prevision de données de gestion: méthode ciblée integrant des elements subjectifs et mathématiques.
Lausanne: EPFL, PhD diss., 2003.

MEUNIER MARTINS, Séverine; CHEIKHROUHOU, Naoufel; GLARDON, Rémy:
Strategic Analysis of Products Related to the Integration of Human Judgement into Demand Forecasting.
In: Current Trends in Production Management.
Eds.: ZÜLCH, Gert; STOWASSER, Sascha; JAGDEV, Harinder S.
Aachen: Shaker Verlag, 2003, pp. 196-203.
(esim – European Series in Industrial Management, Volume 6)

SAATY, Thomas :
The analytical hierarchy process.
New York, NY: McGraw-Hill, 1980.

SANDERS, Nada; RITZMAN, Larry :
The need for contextual and technical knowledge in judgemental forecasting.
In: Journal of Behavioural Decision Making,
Chichester, 5(1992), pp. 39-52.

ZADEH, Lotfi A.:
Fuzzy Sets.
In: Information and Control,
New York, NY, 8(1965)3, pp. 338-353

ZIPPER, Hélène; MEUNIER MARTINS, Séverine :
Analyse de la criticité des produits.
Internal report.
Lausanne: EPFL-LGPP, 2002.

Workplace Injury Risk Prediction and Risk Reduction Tools for Electronics Assembly Work

A Software Tool for the Non-Ergonomist

Leonard O'Sullivan and Timothy Gallwey
Manufacturing and Operations Engineering Department, University of Limerick, Plassey Technological Park, Limerick, Ireland.
Email: {leonard.osullivan, timothy.gallwey}@ul.ie

Abstract: Ergonomics approaches to risk of musculo skeletal disorders are often reactive rather than proactive, an approach supported by a lack of ergonomics expertise, especially in small to medium size companies. However, the EU framework directive on health and safety at work (89/391/EEC) demands a comprehensive risk assessment by the employer. Further, the machinery directive (89/392/EEC), demands a comprehensive risk assessment at an early design stage. Presented are the results of field studies that have identified suitable risk evaluation methods for electronics assembly and rework tasks. To assist non-expert ergonomists, a risk reduction module is also presented. Combined, the risk evaluation and risk reduction approachs presented here provide a framework for reducing musculsoskeletal disorders in line with EU safety directives.

Key words: Musculoskeletal disorders, Interventions, Risk, Assessment.

1. INTRODUCTION

Work-related Musculo-Skeletal Disorders (WMSDs) are common in the majority of industrial settings. Clinical and epidemiological studies have identified four main factors that help identify high risk tasks, i.e. inadequate postures, repetitiveness, high levels of force, and lack of rest. A lot of industries, for example computer and other high tech industries, involve considerable amounts of assembly work, which by nature is very difficult to automate. For assembly type tasks, a lot of concern focuses on Repetitive Strain

Injuries (RSIs) of the wrist. Although these tasks may be hand intensive, injuries are not isolated to the hand and wrist as arm movement requires continuous activation of the shoulder girdle and the glenohumeral joint (WINKEL, WESTGAARD 1992). GRIECO (1998) in addition to wrist injuries described incidences of shoulder tendinitis, lateral epicondylitis, and tension neck syndrome for repetitive tasks. In a study of an aircraft engine plant, DIMBERG (1987) found that 7.4 % of workers had lateral epicondylitis.

However, many *s*mall and *m*edium size *e*nterprises (SMEs) often cannot afford the services of ergonomics specialists to evaluate workplace risks and make design improvements. This is supported by a review of the practical implementation of EU safety directives (Trade Union Technical Bureau, 2004) that found costs to be an impediment for SMEs. But, new EU directives on workplace ergonomics necessitate the need for safe workplace design such that the risk of injury to employees are minimised. However, the benefits of successful ergonomics interventions are not limited to reduced injury claims, as better workplace layout also improves productivity and product manufacture quality. Therefore the benefits of good ergonomics layout can be of immense benefit to SMEs.

Various workplace evaluation techniques are available that range in application complexity and risk sensitivity. The majority of these are performed by observation and the result is an overall risk rating for a task, often in the form of action levels. The risk ratings are based on task data relating to limb joint angles, repetition, force of exertion and recovery time. Evaluation techniques provide risk ratings for tasks but the results do not indicate the risk rating for each body part. Furthermore, the techniques do not provide the user with suggestions on task redesign for the reduction of risk ratings.

As part of the development of a computerised workplace evaluation suite of tools for an EU funded project, work was undertaken to enhance existing ergonomics evaluation techniques such that non-ergonomists could perform simple risk assessments and follow guidelines on task redesign to reduce the risk of injury associated with tasks. In this respect it is important that guidelines are provided to these users to also help them reduce high-risk levels through administrative and engineering interventions. In the following a risk reduction approach is shown which incorporates a suite of evaluation tools. The current draft is a suggested framework for reducing risk levels in the redesign of work and is a compilation of approaches from various sources within the ergonomics literature.

2. RISK ASSESSMENT

2.1 Need to Include Risk Reduction

In many cases, job analysis can be accomplished by observation and discussing with employees the tasks they are performing. An adequate analysis should identify all risk factors present in each task studied. Figure 1 presents a definition of risk analysis and risk assessment according to EN 1050. However, many SMEs that do contract electronic assembly and rework contracts often do not have access to expert ergonomics judgment. Simple ergonomic evaluation methods that are easy to learn can highlight problematic tasks, but difficulty arises in the risk reduction phase during interventions as the skills required to redesign tasks safety are by its nature a form of tacit knowledge.

The approach in this work was to (a) provide risk analysis tools for electronics industry, and (b) provide a risk reduction approach that is integrated with the risk evaluation stage of a complete assessment.

2.2 Selection of Risk Evaluation Methods

Field studies were performed using various risk assessment methods in electronics industries. Based on the results, four methods were chosen for use by either expert ergonomists or non-experts. Production managers and health and safety personnel were regarded as potential non-expert users in SME electronics companies. These methods were programmed using Visual Basic software for ease of use. The application provides for quick evaluation of workplaces from digital photographs or video recordings.

The final tool set includes *RULA* (MCATAMNEY, CORLETT 1993), *REBA* (HIGNETT, McATAMENY 2000), the *Strain Index* (MOORE, GARG 1995) and *OCRA* (OCCHIPINTI 1998). *RULA* and *REBA* are similar since *REBA* is just an extension of the RULA approach to the rest of the body, instead of being confined to the upper limbs. However, because it was developed later and has achieved less publicity, *REBA* is less well known and so is likely to be used only when newcomers have become accustomed to *RULA*. Also, it was felt that those who were familiar with *RULA* (in many cases the only tool they know) would be upset if it was missing and would therefore be discouraged from using the final version. On this basis both were incorporated. The *Strain Index* is probably even less widely known but it has some attractive features in that it takes into account some measures not included in other tools. The fact that it results in an index number is also an attractive feature although the base of case material used to set the bands of acceptability etc. was rather small for some.

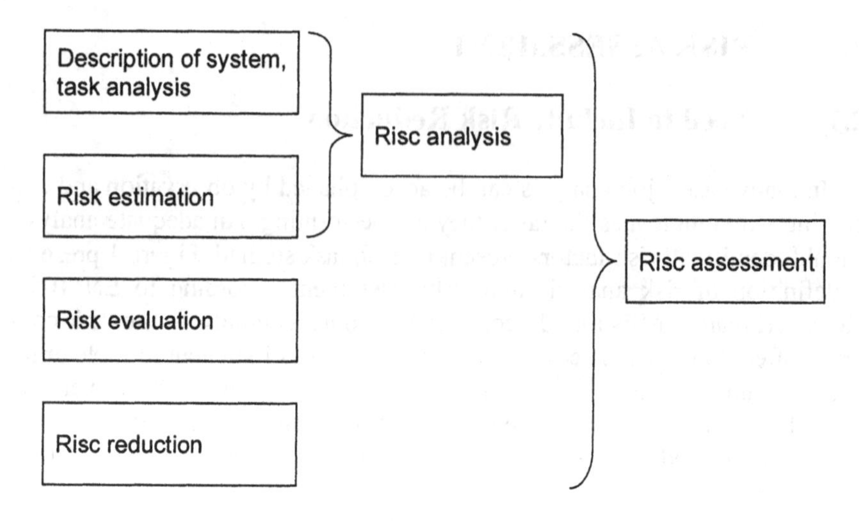

Figure 1. Definition of risk analysis and risk assessment according to EN 1050

Table 1 shows the decision matrix for selection of methods. It is based on level of assessment detail required and the level of ergonomics expertise of the user.

Table 1. Risk assessment methods

Assessment detail ↓ / Ergonomics expertise →	Non-expert	Expert
Lower	*RULA*	*Strain Index*
Higher	*REBA*	*OCRA*

2.3 Field Studies

An expert ergonomist evaluated thirty electronics assembly and reworking tasks using each of the evaluation methods. Sample tasks are show in Figure 2 and Figure 3. A correlation analysis (Spearman's rho coefficient) was performed on the results. The data (*Table 2*) show a pattern of two groups of significant correlations ($p<0.05$).

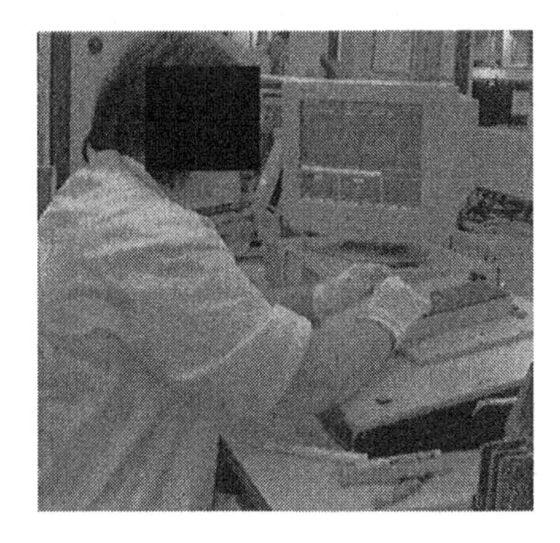

Figure 2. Assembly task

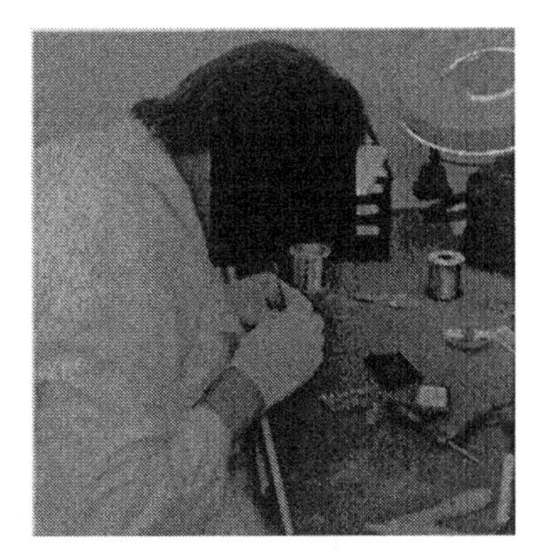

Figure 3. Circuit reworking

The first set is between the *RULA* and *REBA* scores and the second between *Strain Index* and *OCRA* score. This indicates that there is agreement between *RULA* and *REBA* in the estimation of risk for the tasks. But, the grouping of *Strain Index* and *OCRA* estimates together suggest that they are (a) both rating the same tasks as high and low risk, and (b) they are both assessing risk factors more thoroughly and estimating different risk estimates than the broad assessments of *RULA* and *REBA*. Hence, these data support the selection of the tools for the target users in this type of industry.

3. RISK REDUCTION

The risk reduction guidelines were structured in four parts in the order that they are to be used. The approach firstly provides guidance on the initial rectification of the workplace including a framework for making an intervention. General ergonomics guidance from published guidelines are given followed by a more systems approach review of the design of the work. The final stage of the risk reduction approach includes specific points on hardware aids for the worker and issues with static loads.

The technical content of the risk reduction module is also provided as a help file in the Visual Basic software of the risk evaluation tools. Combined, these provide a complete approach of risk analysis, risk evaluation and risk reduction that is convenient to use. These are detailed in the following sections.

Table 2. Correlation of *RULA, REBA, Strain Index* and *OCRA* scores for 30 electronics tasks

	RULA Left hand	*RULA* Right hand	*REBA* Left	*REBA* Right	*Strain Index*	*OCRA* Left hand	*OCRA* Right Hand
RULA Left	1.0	0.586(**)	0.581(**)	-0.016	0.194	0.351	0.205
RULA Right		1.0	0.520(**)	0.566(**)	0.175	-0.027	0.458(*)
REBA Left			1.0	0.569(**)	0.293	0.129	0.071
REBA Right				1.0	0.218	-0.185	0.221
Strain Index					1.0	0.395(*)	0.441(*)
OCRA Left						1.0	0.389(*)
OCRA Right							1.0

* Correlation is significant at $p<0.05$ level

** Correlation is significant at $p<0.01$ level

3.1 Initial Rectification of the Workplace

In many cases it becomes apparent very quickly that there are a number of fairly simple and obvious changes that need to be made. Before carrying out a detailed analysis and improvement programme as described below, it is suggested to rectify these obvious deficiencies first. Doing so may render detailed study unnecessary. Alternatively, the problem(s) may be eliminated or reduced substantially by changes in the organisation of the work. A variety of suggestions are set out below.

3.1.1 Consult the Workers

It is the workers who experience the discomfort and inconvenience involved in doing the job, and these are difficult if not impossible for an observer to detect. The worker can tell the investigator what these are, where the pain/discomfort is felt, and what difficulties are experienced in doing the job. From continued experience of a limited range of activities the worker can often also make useful suggestions for improvement of the workplace design.

3.1.2 Repeat the Observation/Improvement Cycle

Often, once the first set of improvements has been made, it is found that still more changes are needed. Similarly, one set of initial changes may themselves cause some other, new deficiencies. In either case further investigation is needed, and several cycles of this process might result.

3.1.3 Consider Job Rotation or Job Sharing

In some cases it may be very difficult or even impossible to rectify completely some problem aspects. In such cases rotating the job between two or more workers can relieve exposure of the individual. Such a reduction in exposure also reduces the time required for recovery from the pain or discomfort and so productivity will be increased.

3.1.4 Re-allocate Some of the Tasks

Rather than have one worker perform several adverse tasks it may be possible to spread these among several workers instead, i.e. spread the pain around. In this way recovery from one or two adverse tasks may take place while performing several other "easy" tasks. KONZ and JOHNSON (2000) have described this as "working rest" which makes the point rather clearly. Another version of this strategy is to allocate one or more of the adverse tasks to a machine or other piece of equipment. Provided the expense is manageable this can be a very effective measure.

3.1.5 Types of Control

It is common to classify the controls of risk on two levels, namely technical and administrative controls:

1. *Technical controls* (engineering controls) will be used to reduce or eliminate the hazards by technical manoeuvres like design of the workplace, design of the tools and working methods, and design of the product.
2. *Administrative controls* are manoeuvres that effect the organisation of the work or on the ways the work is done. These include, allocating workers to different tasks, working schedules, rotation of workers between different tasks, rules for performing the work, and training of working techniques.

3.2 General Ergonomics Measures

The basic rule is that the job activities, workplace design, and equipment design must be arranged to suit the person doing the job. This means that the worker should perform the tasks that make up the job in ways that are as close as possible to being "natural". All other parts of the work must be changed to fit human capabilities. As a general guide to fulfil these aims CORLETT (1978) enunciated a set of principles.

3.2.1 Corlett's Principles

As the list of Corlett's principles is worked through from top to bottom each level incorporates more of the items in the level above, thus becoming more complex and making it increasingly likely that an adequate design decision will be a compromise. At the same time, principles attached to any items are subject to the overriding effect of principles attached to items at a higher level. By this arrangement each decision in designing a work situation should take account, to some extent, of the complexity of people and avoid the sub-optimisation which sometimes arises.

3.2.2 Typical Disorders and Things to Avoid

The following is adapted from KROEMER (1989). He notes that most of the typical disorders are easily observable: rapid and frequently repeated actions; exertion of finger or arm forces; contorted body joints; blurred outlines of the body owing to vibration; the feeling of cold and the hissing sound of fast flowing air. Training of operators in physiologically correct activities, and the provision of alternating work (to allow "breaks" in otherwise repetitive or maintained activities) are also essential.

3.3 Examine the Design of the Work

Often there are features of the work design that can be changed to reduce the ergonomics deficiencies and thereby reduce or eliminate the postural problems at source. The process requires a highly detailed examination of all elements of the tasks that make up the job. It should be broken down into the set of tasks involved, a task being defined as "the smallest part of the work that can be allocated to a separate operator". This is task analysis.

Task analysis also raises questions as to whether some tasks presently performed by a person should instead be allocated to a machine (including a computer) or the reverse. Changing these allocations offers further opportunities to reduce or eliminate ergonomics problems and/or improve productivity. A convenient way to examine these issues is set out by means of a critical questioning matrix as recommended by KONZ and JOHNSON (2000).

3.4 Specific Workplace Design Recommendations

3.4.1 Use Hardware to Assist the Work

In many manufacturing jobs the designer provides jigs and fixtures to obtain accuracy in machining or assembly. Often, opening and closing them or positioning the work piece, calls for more movements on the part of the operator than are strictly necessary. For example, a tool may have to be used to tighten a nut when a wing nut would be more suitable; or the top of the jig may have to be lifted off to insert a part when the part might be slid into it instead.

3.4.2 Avoid Static Load

This maintains a static contraction of a muscle often from a hold or from maintaining a particular posture. The effect is to impede the flow of blood to and from the muscle in question thus starving it of nutrients and creating a build–up of waste products. But when work is dynamic and rhythmic the alternating contracting and relaxing of the muscle assists both processes.

4. DISCUSSION

The risk evaluation and reduction approach provides details for a suitable framework for both expert ergonomists and non-experts. This allows for screening of tasks within the company with reduction approaches for initial rectification. If desired, the enterprise can follow up on high-risk/problematic tasks that are difficult to improve by buying in ergonomics expertise. This is a lot more cost effective, especially for SMEs that may not be in a position to buy in expertise for a complete evaluation.

Finalisation of the evaluation methods and risk reduction approach was made in conjunction with testing in industrial reference groups. These will also be used for software testing in conjunction with case study compilation as part of a *c*omputer *b*ased *t*raining (CBT) tool development. The final result of the project will be a software tool for use by a wide cross-section of people employed in the design of products and workplaces. It will be offered on the internet with a CBT program and a CD-ROM with a suite of case study applications to assist users. Due to their ease of use and speed, these tools will enable more situations to be analysed, in more detail, more thoroughly, and in less time.

ACKNOWLEDGEMENTS

The research in this paper was funded by the European Union "Growth" Programme (Musculoskeletal Injuries Reduction Tool for Health and Safety, GIRTH-CT-2001-00574).

REFERENCES

89/391/EEG:
European Council Directive 89/391 on the introduction of measures to encourage improvements in the safety and health of workers at work.
12. June 1989.

89/392/EEG:
European Council Directive 89/392 on the approximation of the laws of the Member States relating to machinery.
29. June 1989.

EN 1050:
Safety of machinery. Principles for risk assessment.
1996.

CORLETT, E. N.:
The human body at work: New principles for designing workspaces and methods.
In: Management Services,
Lichfield, 22(1978), pp. 52-53.

HIGNETT S.; MCATAMNEY L.:
Rapid Entire Body Assessment (REBA).
In: Applied Ergonomics,
Amsterdam, 31(2000), pp. 201-205.

KONZ, S.; JOHNSON, S. L.:
Work Design: Industrial Ergonomics.
Scottsdale, AZ: Holcomb Hathaway Publishers, 5th ed. 2000.

KROEMER, K. H. E.:
Cumulative trauma disorders: Their recognition and ergonomics measures to avoid them.
In: Applied Ergonomics,
Amsterdam, 20(1989), pp. 274-280.

KROEMER, K. H. E.; GRANDJEAN, E.:
Fitting the Task to the Human: A Textbook of Occupational Ergonomics.
London: Taylor and Francis, 5th ed. 1997.

MATAMNEY, L; CORLETT, E. N.:
RULA: A survey method for the investigation of work-related upper limb disorders.
In: Applied Ergonomics, Amsterdam, 24(1993)2, pp 91-99.

MOORE, J. S., GARG, A.:
The strain index: A proposed method to analyse jobs for risk of distal upper extremity disorders.
In: American Industrial Hygiene Association Journal,
Fairfax, 56(1995), pp. 443-458.

OCCHIPINTI, E.:
OCRA: A concise index for the assessment of exposure to repetitive movements of the upper limbs.
In: Ergonomics,
London, 41(1998), pp. 1290-1311.

WINKEL, J.; WESTGAARD, R.:
Occupational and individual risk factors for shoulder-neck complaints: Part I – Guidelines for the practitioner.
In: International Journal of Industrial Ergonomics,
Amsterdam, 10(1992), pp. 79-83.

GRIECO, A.:
Application of the concise exposure index (OCRA) to tasks involving repetitive movements of the upper limbs in a variety of manufacturing industries: preliminary validation.
In: Ergonomics,
London, 41(1998) pp. 1347-1356.

DIMBERG, L.:
The prevalence and causation of tennis elbow in a population of workers in an engineering industry.
In: Ergonomics,
London, 30(1987) pp. 573-80.

Trade Union Technical Bureau:
Communication from the Commission to the European Parliament, the Council, the European Economic and Social Committee and the Committee of Regions on the Practical implementation of the Healthy and Safety at Work directives.
Brussels, COM(2004), 62 Final.

PART FOUR

Knowledge Management

Effectiveness of Knowledge Management

A process based survey method

Jürgen Fleischer and Andreas Stepping

University of Karlsruhe, Institut of Production Science, Kaiserstrasse 12, D-76131 Karlsruhe, Germany.

Email: juergen.fleischer@wbk.uka.de

Abstract: Every company uses knowledge management tools – but only some of them are able to determine their effectiveness. This approach based on a survey method was developed in order to fill this information gap. By means of a three way survey, managers and non-managerial personnel were interviewed about the actual employment of knowledge management methods and tools. This information was assessed and in a further step aggregated into one radar diagram per company. Carried out in several companies, it allows the detection of best-practice knowledge management tools among them.

Key words: Knowledge management, Benchmarking, survey

1. INTRODUCTION

In recent years, companies have increasingly realised the importance of their employees' knowledge and its impact on the competitiveness. Especially in industrialised countries, knowledge increasingly defines a company's potential to act and react sufficiently flexible.

But the question is how companies measure whether their way to deal with knowledge is an effective one? Only few companies are able to determine the effectiveness of employed knowledge management (KM) methods – even less in small and medium sized enterprises (SME).

In a research project funded by Stiftung Industrieforschung, Germany, the Institute of Production Science (wbk), Karlsruhe, Germany, developed a tool to benchmark a company's knowledge management along the core process chain.

2. AIM

The aim was to build up a tool to analyse a company's knowledge management with an effort of only one single day. In order to realise surveys in such a short time, an appropriate structure for both the questions and the answers had to be designed. Therefore, a system of questionnaires linked to a database was developed. The relational database stores the questions as well as the answers and the assessments for each interviewed person. It makes a semi-automated evaluation of the survey possible.

The questionnaires were developed to be applied in small and medium sized enterprises (SMEs) so that they take into account the SME-specific conditions like the exceptional overview of employees, short ways and direct communication between the personnel.

3. SURVEY

In order to achieve comparable results in all different companies, a survey by interviews was chosen. This type of survey offers the possibility of detailed explanation and short discussion to minimise issues of comprehensibility by the interviewed person. The survey uses questionnaires on two different hierarchical levels: on the one hand department managers and on the other hand non-managerial personnel.

3.1 Department Managers' Questionnaire

This hierarchical level is used to get a quick and wide overview of the company's conditions and the instruments and methods of knowledge management it uses. The manager's questionnaire is subdivided into 4 chapters (Figure 1).

After collecting data about the interviewed person in chapter one and about the company's properties in chapter 2, the third chapter's questions try to assess the current employment of knowledge management instruments. Therefore, almost all questions of this chapter are linked to a predefined assessment system in the database, with the effect that answers can be evaluated once the data is entered in the system.

The last chapter opens the possibility to collect, compare and aggregate the SMEs' ideas of how knowledge management should work and which scale it should have in SMEs.

Chapter	Content	Pages
Chapter 1	Information about interviewed person like age, number of years in company in order to analyse how much experience the interviewed person has in general and in this company.	2
Chapter 2	Brief overview of company's turnover, organisation and product type in order to classify the company	4
Chapter 3	Overview of the current situation in the company. This is the only assessed chapter in this questionnaire.	13
Chapter 4	How should an optimal knowledge management look like?	6

Figure 1. Chapters of manager's questionnaires (FLEISCHER, STEPPING 2003)

3.2 Questionnaire for Non-managerial Personnel

This lower hierarchical level is used to achieve more detailed information about the rate of KM employment.

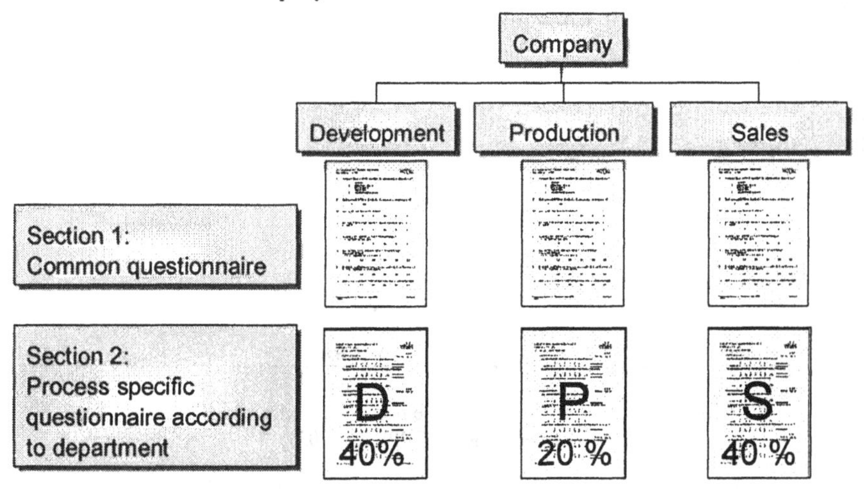

Figure 2. System of non-managerial personnel questionnaire (FLEISCHER, STEPPING 2003)

The questionnaire consists of two sections (Figure 2):

- The first one comprises common questions for all interviewed persons. It tries to highlight the general situation of KM in the company and is a brief version of the department manager's questionnaire in order to find out inconsistencies between managers and non-managers. All interviewed persons of one company assess e.g. the duration to get access to information in general – independently of their work content.

- The second section contains questions regarding the department the interviewed person is working in. This is only feasible since in SME the employees are not split into many working groups but work more or less as a team. They fulfil some steps of the core process chain, which is split into 24 process steps in the methodology of this project (based on preceding works; SPATH, DILL, SCHARER 2001). These steps are spread over the departments research & development, production and sales. Approximately 40 % of the questions treat the subject area of research & development, 20 % of production and another 40 % of sales. For each process step, most typical in- and output items have been specified. In order to achieve a complete overview, all knowledge components according to *Probst* like knowledge identification, acquisition, development, distribution, preservation and use are regarded one by one for each process step (PROBST, RAUB, ROMHARDT 1999).

R&D: capacity planning

Project budgeting relevance: ☐ A ☐ B ☐ C

Does a project budgeting exist? y/n

☐ Yes ☐ No

Show us a budget list for your current project, please. time

Needed time: .... success: ☐ yes ☐ no needed: ☐ yes ☐ no

Up to date:	☐	☐	☐	☐	☐	☐
mark:	1	2	3	4	5	6

Source: .... Media type: ....

Figure 3. Three level structure of the process-specific questionnaire (FLEISCHER, STEPPING 2003)

The process-specific questionnaire has a three level structure (Figure 3) and contains process specific subject areas collecting data about typically needed information. The first level builds the process level. All mentioned 24 process steps are subdivided into several packages of questions. Process steps, packages of questions and the detailed questions themselves are interviewed depending on the work content of the interviewed person. The interviewed person may choose for every package of questions whether it matches with the usual work content to approx. 80 % (relevance A), about 20 % (relevance B) or not at all (relevance C). Depending on these categories, the answers are weighed later in the evaluation phase. If relevance C is chosen, the questions are neither asked nor weighed.

There are several categories of questions in this questionnaire. The exceptional property of this approach is to find out how long certain data gathering and use take. The times are measured at the working place itself. But even if the time is short, the information can be incorrect, out-dated or useless. Therefore, the interviewed person can assess whether the information is up to date and needed.

4. ASSESSMENT

Most questions in the questionnaires are chosen to highlight one particular knowledge component (PROBST, RAUB, ROMHARDT 1999). Therefore it is possible to assign all questions to one of these knowledge components in the database. As a consequence, all companies' knowledge management usage is evaluated based on these knowledge components.

Depending on the question type, either the interviewed person gives his own assessment of the situation in the company on a scale between one and six (one means very good, six very bad) or the answers are linked to a predefined assessment assigned by a working group at wbk based on experience.

All this information is aggregated into one diagram showing the rate of employment of knowledge management instruments and methods. In later stages it will be possible to compare the analysed company with best-practice companies among those analysed earlier by means of the radar diagram.

This aggregation is done in three steps (Figure 4). The first step consists of the aggregation of all process-specific and common answers to one result per department and company. In the second step, all answers of the 3rd section of the department manager questionnaire are evaluated and added to the aggregation per department and company. The last step builds up the fully aggregated radar diagram per company. Each ray represents one of the mentioned knowledge components.

5. RESULTS

5.1 Profile of Interviewed Companies

In summer 2003, several companies belonging to different industrial sectors as shown in Figure 5 were interviewed according to the description above. 56 % of the companies manufacture investment products which can be subdivided into 50 % devices and 50 % machines, which both are mostly built in single-production.

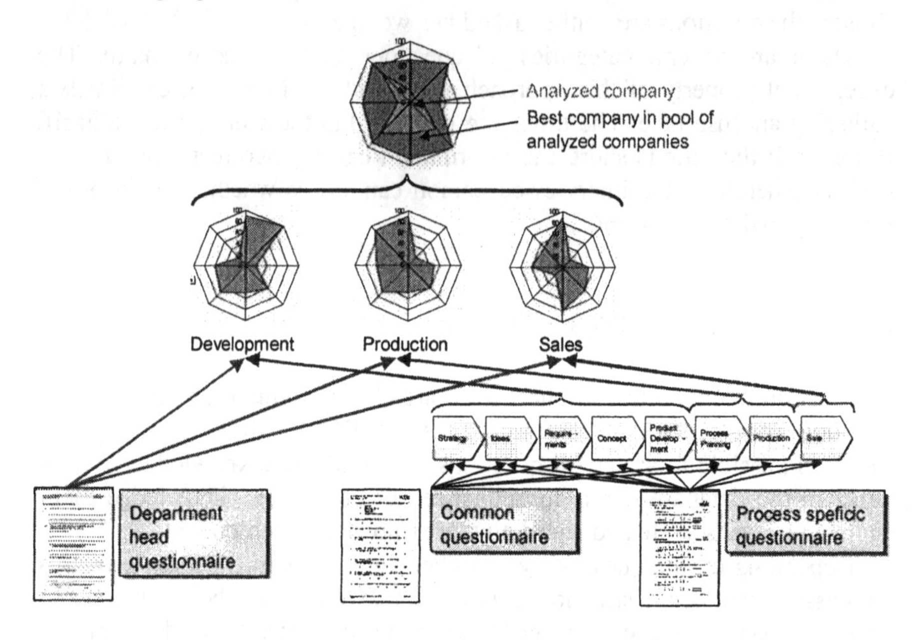

Figure 4. Evaluation system (FLEISCHER, STEPPING 2003)

Not all interviewed companies are small and medium sized enterprises: 37 % of the companies have up to 150 employees, 25 % between 150 and 250 and another 38 % have even more than 1000 employees (as shown in Figure 6).

In total, 52 persons have been interviewed. 79 % of the interviewed non-managerial personnel have professional experience of more than 5 years, 31 % even more than 15 years. The experience in their company is in 46 % of cases less than 5 years but still in 20 % of cases more than 15 years. The time the personnel has been working in the current position is in 63 % of cases less than 5 years, only in 7 % more than 15 years. The group size is in most cases between 10 and 15 persons (44 %). 23 % of groups comprise less than 5 people, another 23 % between 5 and 10 people.

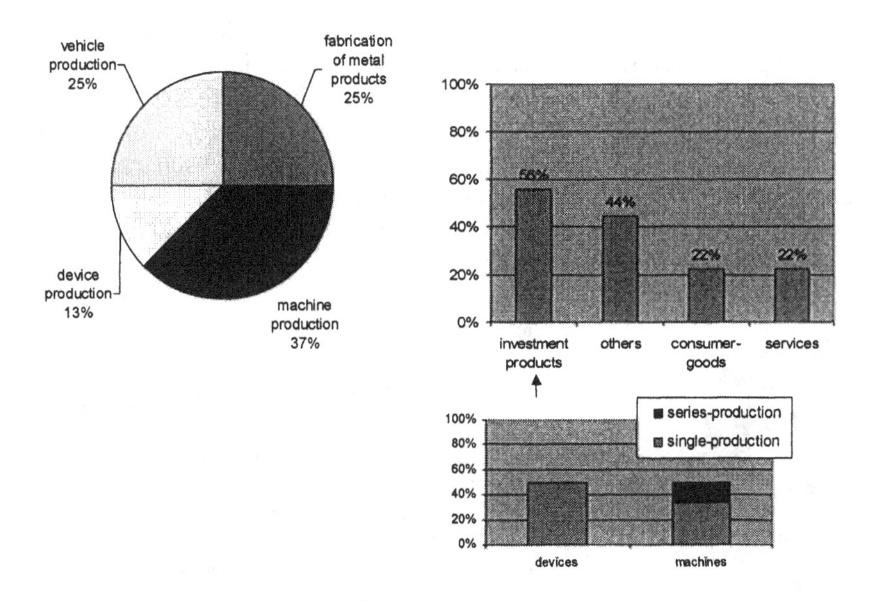

Figure 5. Industrial sectors of interviewed companies

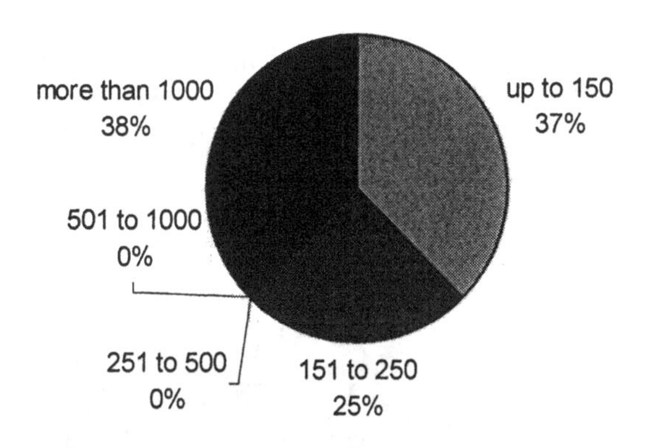

Figure 6. Number of companies' employees

5.2 Aspects of KM Employment

In order to get an idea of how the interviewed person thinks about knowledge and knowledge management, one of the first questions was "What does the term '*knowledge management*' mean to you?" The result is shown in Figure 7: The majority (45 %) thinks of KM as "the way to systematically collect, archive and distribute knowledge" whereas a smaller part

(26 %) has the definition of "a concept for optimal use of knowledge in order to develop new products, processes and business areas" in mind.

It turned out that the most important source of information is the internet followed by business related journals. Very important sources of information are as well the customers and partners like suppliers and consultants. Professional books, seminars and business sector associations are important for more than 7 % of the interviewed staff. Patents and banks are not regarded as important sources for information.

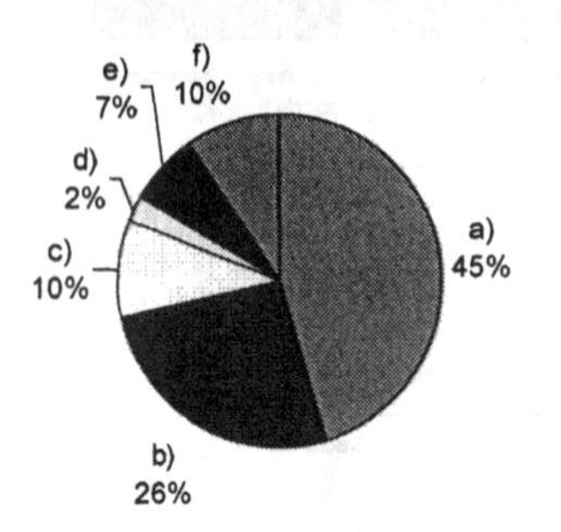

a) ... the way to systematically collect, archive and distribute knowledge

b) ... a concept for optimal use of knowledge in order to develop new products, processes and business areas

c) ... all strategies as a whole in order to use and develop knowledge on all different levels and processes in the company

d) ... less to optimise the knowledge contents than to organise structures, processes and methods

e) ... a way to use knowledge in order to increase productivity

f) ... a possibility to achieve new knowledge

Figure 7. The term „knowledge management" means to me...

The question "What is missing to implement knowledge management methods" showed interesting results: One half is convinced that appropriate software will help and another half says that first of all the budget is missing. Nobody states that a lack of know-how about knowledge management itself exists or that missing interest in knowledge management hinders the implementation of KM.

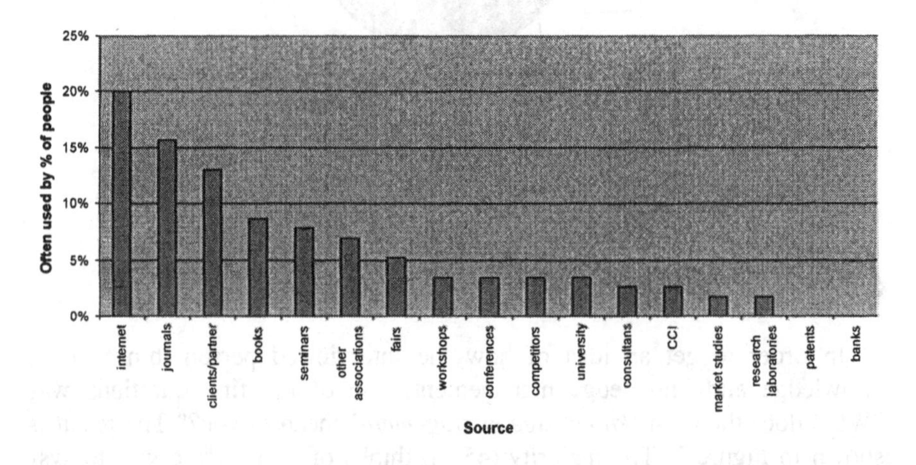

Figure 8. Most used sources for information

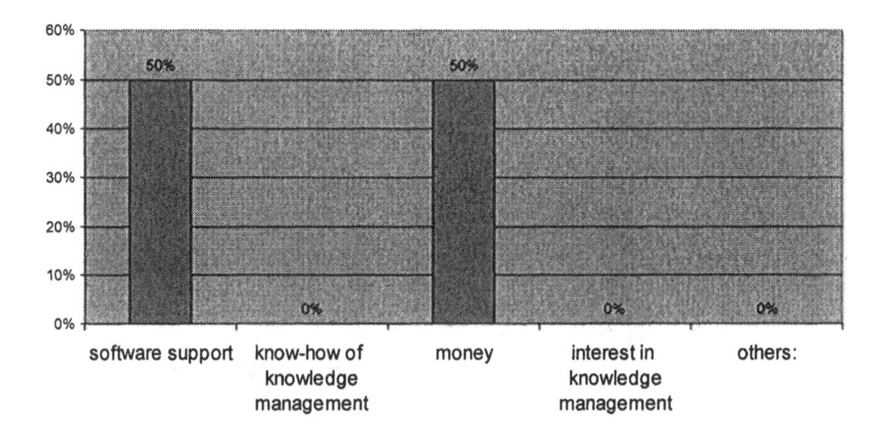

Figure 9. What is missing to implement knowledge management methods?

The purpose of knowledge in the interviewed companies is mainly to optimise the company's products and to achieve advantages in comparison to competitors. Innovation and customer acquisition were less important.

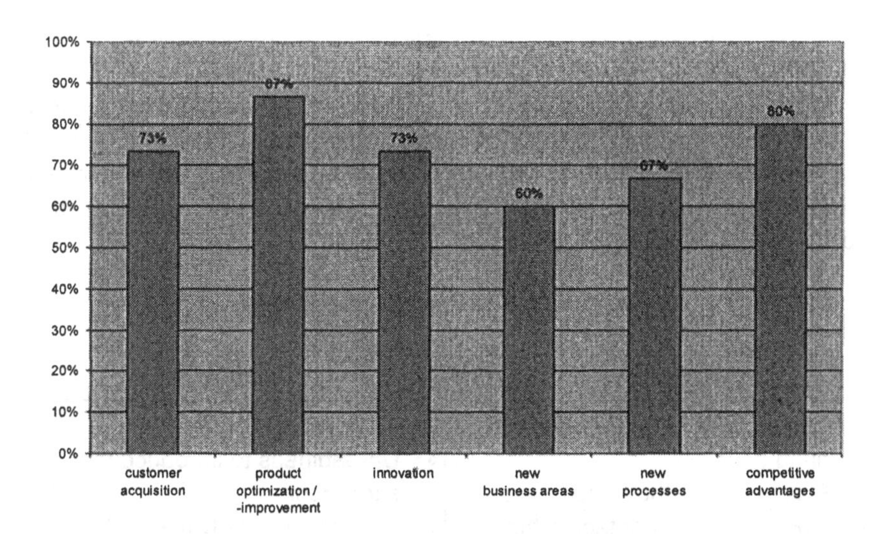

Figure 10. What do you need knowledge for?

Another important aspect of knowledge management is the hand-over of projects and information from a leaving employee to the successor. As shown in Figure 11, most employees see the hand-over on a scale between 1 (very bad) and 6 (very good) only at about 3 if there is a successor and even worse if there is no successor.

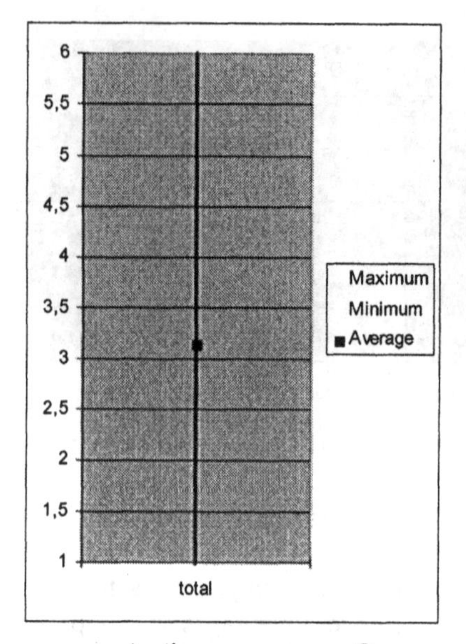

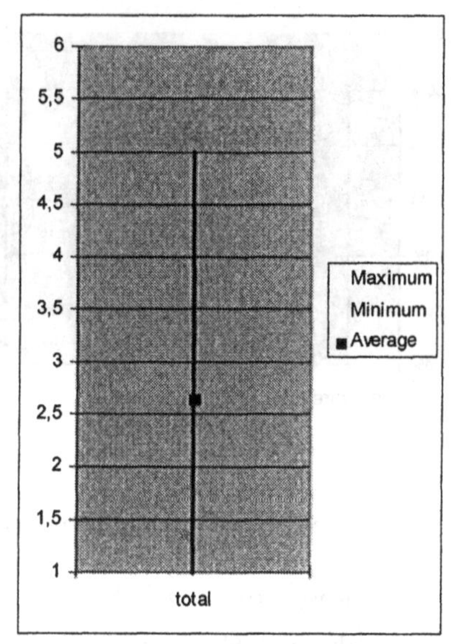

Figure 11. How good is the hand over of knowledge of a leaving employee ...

The key facts why a hand-over appeared to be good or bad are summarised in *Table 3*. For example a defined hand over plan and existing documentation of projects and processes are considered as valuable whereas permanent cancellations of hand over meetings or a lack of time on both sides – leaving person and successor – is seen as bad.

Table 3. What is good/bad for an efficient hand-over?

Good	Bad
• A good atmosphere relieves the hand over • Proceed after the hand over plan • Good personal contact to the successor • Existing documentation • Identification of the employees avoids the "devil my care" mentality • Volunteer switch-over	• Lack of time • Unreadiness to introduce the successor • Social plan (partly the work ends overnight) • Lack of a mentor philosophy • Big gap between the ages • No personal interest in the leaving person • Difficult to assign knowledge in general • Cancellations

An important indicator of employment of existing processes and systems is the existence of so called parallel systems – e. g. using MS Excel® instead of SAP® functionality or using paper instead of software.

The results to the question "Do parallel systems exist?" are shown in Figure 12. In total, only few people see parallel systems in their companies, but this obviously depends on the company and the person itself – and the interviewed person's honesty. The reasons for creating parallel systems are termed as follows:

- Restricted user rights in the "official" system,
- Flexibility, possibility of an individual adaptation,
- Pre-processing often in Excel/Access instead of "official" system,
- No mobile use of "official" system,
- IT systems were introduced without trainings,
- Advantages of one system are not available in others.

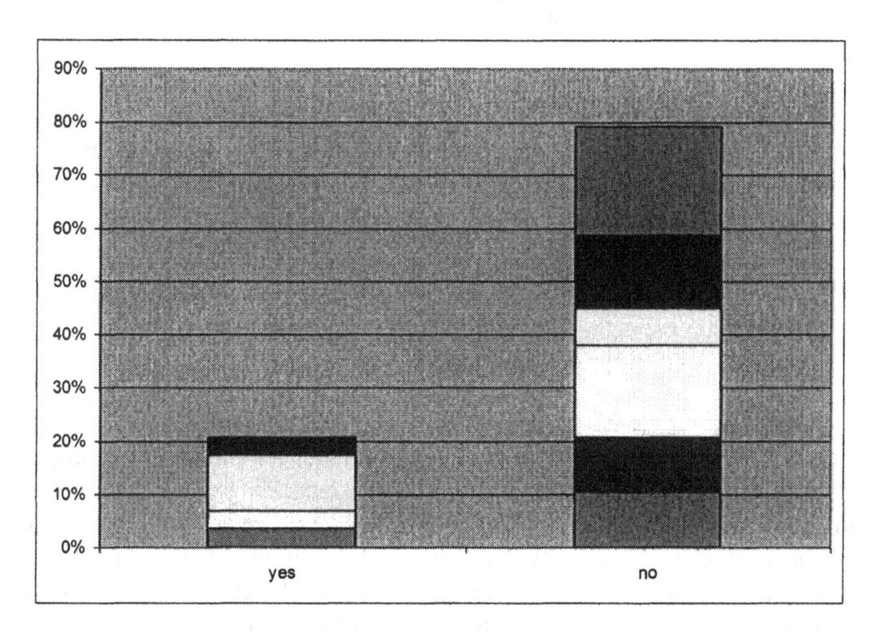

Figure 12. Do parallel systems exist?

Information may be confidential to a specific group of people – not only to external people but as well to different groups of internal persons. Therefore the question "Would you like every person to see every information?" was supposed to be provocative.

Only few people (20 %) say that they like to see and let see everybody every information (Figure 13). All others want to restrict the access to information depending on the topic, the current project, the division/department or the hierarchy.

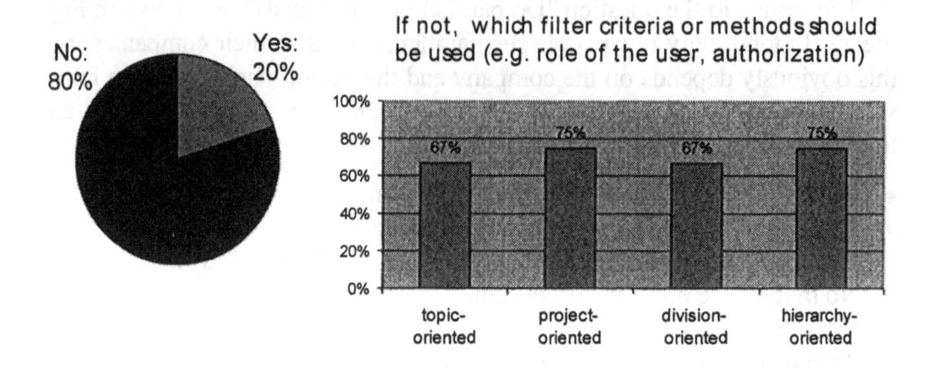

Figure 13. Would you like every person to see every information

Concerning the creation of a common knowledge base to which partners, suppliers and customers have access, the opinion is pretty clear (see Figure 14). The major part of the interviewed persons does not want to design such a knowledge base. It is preferred to distinguish between three groups: suppliers should be able to refer to the knowledge of the company (46 % yes, 47 % partially) whereas customer and external partners should be less integrated (40 % yes).

5.3 Best-Practice Approaches

As already described above, all answers are assessed and stored in a database. The results of all interviews both on manager and on non-managerial level are aggregated per company into one radar diagram. The comparison of all radar diagrams leads to the Figure 15, where for any knowledge component the best company's result gives an impression of how well knowledge management works in general. The fulfilment of only one knowledge component has been met at 100 % by at least one company. All other knowledge components were met at 90 % at highest or in case of knowledge acquisition only about 65 %.

But what are the reasons for these best-practice companies? Why did they succeed in this benchmark? Which solutions were implemented and are used? The following part is divided into four categories: Organisation, Processes, Tools and Culture. For each of these categories some samples of best practice are given.

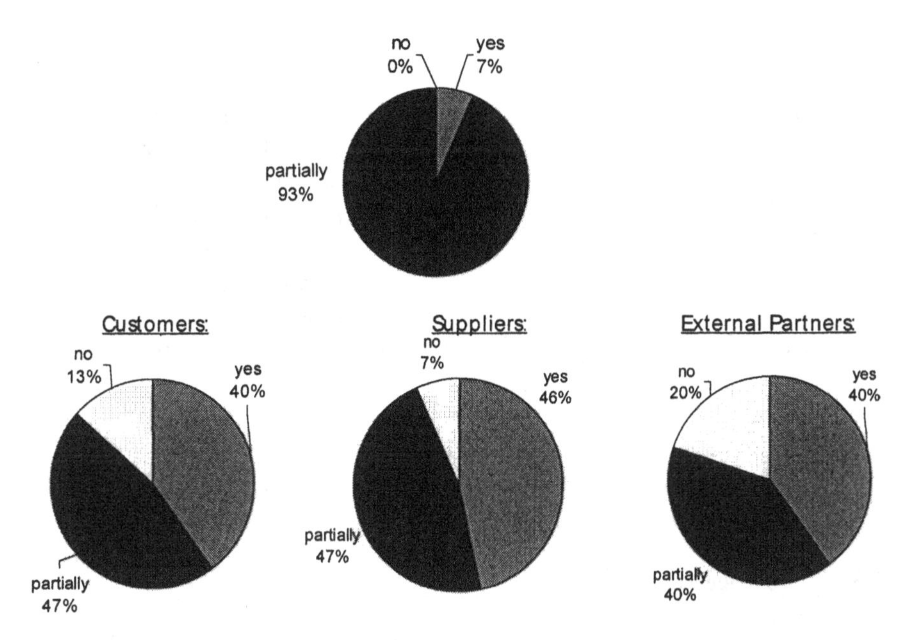

Figure 14. Would you like customers, suppliers and other partners to access the same knowledge base?

5.3.1 Organisation

- For any department particular responsibilities for knowledge management are specified.
- Regular benchmarking of core knowledge is carried out. The results are distributed to all employees.
- The availability of a particular knowledge management budget – even if low – is a signal message to all employees.
- The continuous knowledge progress is supervised by a "knowledge board" that controls target achievement.
- The factor of production „knowledge" is deep-seated in the company's strategy.
- Knowledge objectives are led into the organisation by agreements of objectives.
- Low number of hierarchy levels and short communication support knowledge interchange.

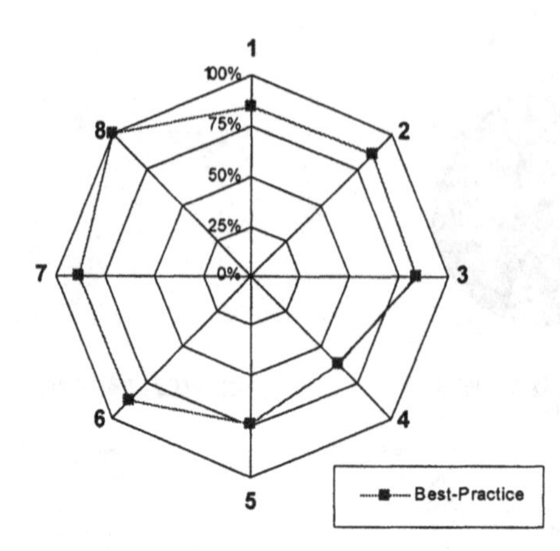

1: knowledge identification
2: knowledge distribution
3: knowledge use
4: knowledge acquisition
5: knowledge development
6: knowledge evaluation
7: knowledge preservation
8: knowledge objective

Figure 15. Best-Practice results per knowledge component

5.3.1 Processes

- Handover-processes are defined, controlled and part of the agreements of objectives – in both cases of existing and non-existing successor.
- The search of internal experts is supported by yellow pages and organisation charts.
- Knowledge-Management intranet-portal designed and developed by the employees to meet their requirements leads to high acceptance and usability, because the key words are deducted from daily work. An additional interface to many company related internet-sites avoids time consuming searches.
- The company formulates long-term vision and mission for new business areas.
- Organisation is based on well-documented processes.
- Knowledge acquisition is a regular process including visits of fairs, contacts to competitors as well as collaborative projects with research institutes.

5.3.2 Tools

- Internal advanced trainings like courses, seminars and workshops are regularly used. Additionally, job rotation und project reviews are used to gain knowledge.
- Giving time for personal development leads to motivated, knowledge sharing staff.
- Research into new subject areas by case-studies.

5.3.3 Culture

- The transmission of knowledge is understood as a part of the job. A positive effect is the reduction of the individual work amount, a possible specialisation and the improvement of the team's output.
- There are a lot of informal communication occasions like whiteboards, bill-boards and coffee-corners.
- The acceptance of mistakes is part of the company's vision.
- Regular "future meetings" are installed where worker, team leader and general management are involved.
- All employees see the company's advantage in sharing information to get or stay better than other companies – and as consequence to keep their jobs.

6. SUMMARY

The herein presented approach presents a survey method and an assessment system to determine the effectiveness of knowledge management employment. The survey method consists of several questionnaires dedicated to different process steps and hierarchical levels of a company. By means of this survey method, it is possible to compare several companies in terms of knowledge management employment and to visualise the differences in simple radar diagrams. Based on the best in class company per ray in the radar diagram, one can deduct the best practices which have led to this good assessment.

REFERENCES

FLEISCHER, J.; STEPPING, A.:
Effectiveness of Knowledge Management - a Process Based Survey Method.
In: Current Trends in Production Management.
Eds.: ZÜLCH, Gert; STOWASSER, Sascha; JAGDEV, Harinder S.
Aachen: Shaker Verlag, 2003, pp. 187-193.
(esim – European Series in Industrial Management, Volume 6)

PROBST, G.; RAUB, S.; ROMHARDT, K. (eds.):
Wissen managen. Wie Unternehmen ihre wertvollste Ressource optimal nutzen.
Wiesbaden: Gabler Verlag, 3rd ed., 1999.

SPATH, D.; DILL, C.; SCHARER, M.:
Vom Markt zum Markt: Produktentstehung als zyklischer Prozess.
Stuttgart: LOG_X, 2001.

Practical Knowledge and Collaboration in Engineering

Päivi Pöyry[1], Markus Mäkelä[1], Jouni Meriluoto[2] and Marju Luoma[3]
1) Helsinki University of Technology, Software Business and Engineering Institute, P.O.Box 9210, FIN-01025 HUT, Finland.
Email: ppoyry@soberit.hut.fi
2) Nokia Research Center, P.O.Box 407, FIN-00045, Nokia Group, Finland.
Email: Jouni.Meriluoto@nokia.com
3) Nokia Networks, P.O.Box 785, FIN-33101 Tampere, Finland.
Email: Marju.Luoma@nokia.com

Abstract: This paper presents the results of a case study conducted in networked product development organisations. In this study, the emphasis is on various dimensions of collaboration and knowledge sharing. The results indicate that knowledge is shared quite fluently in this network of collaborating partners. However, the tacit dimension of knowledge is not yet given enough attention. The information systems and other tools supporting knowledge management are focused on document management and other explicit knowledge, whilst the tactical knowledge is still shared in immediate communication or face-to-face meetings in an unstructured manner. The support for sharing tacit knowledge should be improved for example by encouraging the creation of communities of practice and by providing technological platforms for them.

Keywords: Knowledge management in engineering, Knowledge sharing, Knowledge creation, Collaboration, Distributed product development

1. INTRODUCTION

The conditions of product development engineering work have been changing during the last couple of years. Knowledge Management (KM) has become more important also in engineering. Keywords for today are global product development organisations, requirements for diverse, multiple and fast-changing competencies, strategic partnership networks, virtual teams, and improved organisational memory. The goal of KM in engineering is to facilitate the retain and reuse of information and knowledge created in the product development projects by passing it on to engineers within the project, to other projects and to collaborators. In this study, the information systems supporting knowledge management are expected to store and deliver product-specific information and to support re-using and sharing existing knowledge.

The purpose of this paper is to present the results of a case study conducted as a part of the WISE project. WISE (Web-enabled Information Services for Engineering) is a European Commission funded IST (Information Society Technologies) research project studying knowledge management within engineering environments.

In this study, the emphasis was on various dimensions of collaboration and knowledge sharing within and between organisations and people. We were especially interested in the facilitators and obstacles between strategic partnership companies. In addition, the work practices and work culture affecting the ways of sharing knowledge in the different organisations were investigated.

2. KNOWLEDGE CREATION AND SHARING IN COLLABORATION

The concept of managing knowledge has become increasingly popular both in the practical and in the academic discussion in the fields of engineering and management. Successful management of knowledge-related resources of companies has been recognised as a key basis for acquiring competitive advantage and other organisational success (HUBER 1991), and the acquisition and application of knowledge has even been argued to constitute the focal role of organisations in the society (DYER, NOBEOKA 2000). Virtual communities have significant possibilities of enhancing how this occurs (BIEBER et al. 2002). Today's environment of multi-organisation collaboration in technology alliances calls for improved knowledge sharing and processing capabilities, while information technology has been steadily increasing its importance as a tool of KM (STEIN, ZWASS 1995).

Knowledge is dynamic in nature, because it is created in social interactions among individuals and organisations. Being dependent on time and place (NONAKA, TOYAMA, KONNO 2001), knowledge is context-specific. Knowledge exists in organisations in different forms some of which are tangible in nature while others take a more subtle form. Tangible forms of knowledge have been termed as "articulated" or explicit (HUBER 1991;

HEDLUND 1994), while intangible forms have been labelled "tacit". Explicit knowledge can be expressed in a formal and systematic way, and it can be shared in various forms, such as data, diagrams, and text. Its processing, storing, transmission is relatively easy. Tacit knowledge is harder to formalise because it is local, contextual and closely tied to place. Reconstructing tacit knowledge in a new context is difficult if not impossible. The borderline between tacit and explicit can be viewed as always shifting (HEDLUND, NONAKA 1993; POLANYI 1962).

Tacit knowledge can only be shared through shared experience. Apprenticeship, for instance, is a common way of adopting and sharing tacit knowledge. However, the concept of tacit knowledge does not cover only learnt practical skills, but also intuitions, assumptions and beliefs not expressed publicly (NONAKA, TAKEUCHI 1995). It is an advantage to distinguish explicit but also implicit knowledge from tacit knowledge.

The process of *knowledge creation* is a continuous, self-transcending one in which the boundaries of the old self are crossed and one enters into a new self by acquiring a new context. Knowledge is created in the interactions between individuals and their environments. *Knowledge sharing* refers to the beginning of the knowledge creation process, where tacit knowledge is shared among the individuals through socialisation. Typically this happens in a team with members from various backgrounds (NONAKA, TAKEUCHI 1995). Later in this paper, we have distinct focuses on the creation of knowledge, on its storing and retrieving, on re-using it and on its' sharing. Our analysis is distinct from the majority of prior literature due to our focus on distributed and extended environments.

3. THE CASE STUDY

In this study, we focused on networks formed by strategic partner companies and the engineers collaborating within these networks of global high-tech product development. Fourteen product development engineers and other professionals such as technical writers were interviewed in their working environments. The interviews were carried out in Finland, in the cities of Helsinki, Espoo and Tampere in the premises of the focal company and its partners.

The idea was to apply the research methods traditionally used in ethnography in order to gather rich data and to form a thorough understanding on the subject area, but a full-scale ethnography was not used due to the time limitations and other constraints related to doing research in industrial settings. Thus, a "rapid ethnographic" approach was chosen (MILLEN 2000). The principal research method was semi-structured thematic interview, and these were recorded and transcribed and finally analysed with qualitative content analysis techniques. Originally the interviews were complemented by observations in the workplace, but they proved to have little value in terms of giving new information on the themes studied. Thus, these observations were left out from the data after an initial analysis. In this case study, we investigated these central KM-related issues:

- Creating knowledge
- Storing and retrieving information
- Re-using knowledge from other people and projects
- Sharing knowledge between people and organisations

The interview themes and questions (the interview framework) were prepared and the transcribed interviews were analysed in a group of four consisting of representatives of the focal company and researchers of Helsinki University of Technology. Two members of the analysis group carried out the interviews and participated in transcribing the tapes. The members of the group first read through and analysed the interviews individually. Then a group session was organised so that the individual interpretations could be discussed and a consensus could be reached on the results. There was only little divergence in the interpretations, and the common view was easily reached.

The themes of the interviews supported the analysis, and the interview framework was used as a backbone according to which the interview data was processed. The data was asked the questions defined in the interview framework themes, and the answers found in the data were grouped under the rubrics formed on the basis of the themes. The basic unit of analysis was a set of sentences – including the instance of one sentence – forming a meaningful expression of the subject matter.

4. THE RESULTS OF THE STUDY

4.1 Creating Knowledge

Creating knowledge was understood in various ways depending on the interviewee. For example, to create knowledge referred to inventing some new product or method, to getting a new idea, to producing specifications and other documents, and to coding innovative software. The knowledge creation was seen as an interactive process between individuals and a team; the ideas formed in someone's mind were further refined in one's team, and the results of teamwork and discussions could be further refined as individual work. In most interviews the knowledge creation was located at the focal company; the work between the partner companies and the focal company was divided so that the latter would develop, create and make innovations (product architectures, specifications in some cases innovative, not always) related to the product, and the partners would implement some well-defined part of the product. On the other hand, competence on the environment of the focal company, products, and tools accumulated.

> "*...I'd say that we create knowledge when we create a new product. Then we make something new in a concrete sense. But of course we may invent new ways of doing something. I think that's creating knowledge too*."

> "*Creating knowledge...I think it happens in your head, but you need the rest of the team for it. It's partly inventing something on your own but you have to work it out with the colleagues*."

> "*...we usually implement the knowledge the other company has created. It's a clear division of work. They make the innovation and we do what we are asked to support them. Then of course we may create new ways of making our work in a better way, but that's a different story*."

Most interviewees interpreted knowledge as being tied up to physical products and other artefacts, even as "implementation" of external knowledge. Creation was not seen as an initially iterative generation in discourse as in collaborative engineering. Also the engineering aspect emphasised practical "doing" that was going hand in hand with knowledge creation. That is supporting the new division of tacit knowledge.

4.2 Storing and Retrieving Information

In most cases the information management was handled with a shared storage for project documents to which all partners had an access. In the projects studied, the information or document exchange was highly regulated between the focal company and the partner(s). As a practical detail, version management was found problematic, and there was no way to verify whether the document had been read or not. In addition, the information in the documents as well as the documents themselves were often out-dated. The various responsibilities for up-dating and managing documents caused "detective work" especially for project managers who were trying to find the information needed in the current project. In some cases, project members informed via e-mail when there was a new document or a version containing bigger changes and thus, worth re-reading. In most cases people seemed to be used to asking a certain document is updated or not (usually by e-mail).

"*Yes, we do have a shared place for the project files. But because it depends on the people that upload the documents on the server, it does not always work as it should. The documents and information may be like year old, or even worse. You never know what you'll find there. Or then the most recent document version will not be found there at all, it's just lost in the e-mail with the other trash mail.*"

The most common procedures and means for information exchange included e-mails, mailing lists, shared databases, and circulating regular project reports. Due to the fact that people did not save the documents in the shared system, the documents and information had to be searched for from several places and persons. It was difficult to get people use the information systems. For this reason the newest information was not always available, and the documents could often be out-dated. Clearly, there seemed to be place for improvements in version and change management. Crucial was that either the updated version of a document, or the meta-information about the new version would reach the reader. The user accessing the document management system "never knew what to find" but she had an assumption whether it was valid or not.

4.3 Re-using Knowledge from Others

Most projects reused information from other and previous projects in some form or other (e.g. processes and templates). Many projects were based on the previous projects (e.g. new releases of an existing product), and some projects needed information from other (parallel) projects. According to our results, information from other projects was difficult to find: the interviewees did not know where it was and who could access it. In the opinion of the interviewees information should be stored in one place or at least it should be accessed though one portal. Even more important would be information system and tool integration. An informant of ours claimed that information reuse from systems required personal contacts:

"*...I'd say that I find the information when I look for it. But it requires a lot of detective work. Because you never know where the information is stored and who can access it. And you have to ask from several persons before you find the person that can tell where the information is and who can use it...*"

Information was reused from previous projects if the information was available. In too many cases it was known that a specific document with specific information existed, but it was not known where the document could be found. The old documents could be used as models and they could be modified according to the current project's needs. For example, project plans, templates, and experiences were reused. More information could be reused if it was available. The re-use of information or knowledge still dealt mainly with explicit knowledge and information that could be expressed in writing. However, reusing tacit knowledge was difficult, because it had proved to be almost impossible to capture and to store this kind of highly personal and contextual knowledge. Some parts of tacit knowledge (as practical skills) could be captured with e.g. video devices. Of course this would be just one method to help to share and learn those skills, not capturing and storing the knowledge itself.

Thus, completely virtual work with no face-to-face meetings was not considered feasible, because the current information systems were incapable of supporting the sharing of tacit knowledge. Tacit knowledge is local, contextual and closely tied to place; reconstructing tacit knowledge in a new context is difficult. The sharing of tacit knowledge could be facilitated by supporting communities of practice with the help of information platforms. However, when implicit knowledge is distinguished from tacit knowledge, it is possible to derive explicit knowledge from implicit, in discourse and collaboration.

> "*We usually work so that we move from release to another. So we can use the test cases etc. again in the next release. We can use it as a model and make a new version.*"

> "*...I use the process documents and the slides from other projects. It simply saves so much time when I don't have to make it always from the scratch...*"

Explicit knowledge from history as well as current documents from other people were helping in new challenges. When storing and sharing this kind of knowledge, e-mail and CSCW (Computer Supported Co-operative Work) tools were an advantage. Although implicit knowledge is subjective, and has no firm factual basis, it is *cognitive* and therefore able to be made explicit. That's the case also on organisational level. Information and communication technology helps to move the borderline from implicit towards new explicit knowledge.

4.3 Sharing Knowledge between People and Organisations

The role of information systems was considered central in sharing knowledge. Information systems were used for sharing, storing and exchanging documents that contained the information and knowledge created in the project. However, the systems did not function properly and people did not use them as they were expected to because there was lots of out-dated information and for instance version management was unsatisfactory. Here we can see a chicken and egg situation, typical of many new systems: people do not use or update information in the system because the information in the system is not up to date.

In addition, there were too many systems available, which is confusing; one or two systems/portals would be enough for the interviewees. They also acknowledged that the information systems were not used for sharing informal information and knowledge, such as lessons learned during the project execution. People felt that today there were too many tools, the tools did not function properly and they were not user-friendly, and they were being changed too often. The interviewees wanted a centralised tool or portal through which the information and knowledge could be easily accessed. No more detective work for finding relevant documents or people should be necessary.

> "*At the moment we would not be able to work without the computer systems. They are basic tools for us. But they could be improved... and people should start using them as they are supposed to...*"

> "*...they are rather fragmented. ...and what we hate is that the systems change each year. Just when you have learned to use the system someone tells you it will change. ...There should be some integration and then we build interfaces through which you can move information from one system to another.*"

The biggest *barrier to knowledge sharing* seemed to be the fact that key people were, or at least told they were too busy to share information and knowledge and to answer all the questions asked. This, of course, depended on the project and the project phase. Another problem was that there were too few personal contacts between people working in different organisations. Sometimes company borders, foreign language, and physical distance etc. could be seen as obstacles, but not as severe as the lack of time and contacts. Additionally, the overload of e-mails was seen as problematic: people could not cope with the information overflow and processing all the e-mail messages burdened the interviewees.

The constant lack of time was the major barrier to knowledge sharing in addition to the lacking personal contacts. Differences between cultures could hinder efficient sharing of knowledge; it became increasingly difficult to share knowledge the more "distant" the cultures were. Changes in the economic picture and business situation affected people's willingness to share knowledge; when the market situation was challenging, people wanted to keep the knowledge by themselves, i.e. did not communicate that much in general. Moreover, when extensive organisational changes occurred, the personal networks could fall apart, which complicated the sharing of knowledge.

> *"...it is not about the people being unwilling to help. It's about lack of time. They just don't have the time to answer the questions. The lack of time is the biggest problem."*

> *"A foreign partner is a more complicated case. The cultural difference makes it... A good example is the way in which people from different cultures admit that something does not work properly. ...it is a challenging task to find the problems."*

> *"I feel that the cultural differences are the biggest [barriers]. Language and culture are things that make it more difficult. ...But I don't think that the geographical distance affects; it's the same if you speak on the phone to the neighbour building or to the other side of the earth."*

The factors *facilitating knowledge sharing* included personal contacts that were created in face-to-face meetings. This enhanced the common feeling of trust, or "knowing the colleagues". These contacts between people collaborating in a product development project facilitated the knowledge sharing significantly as they formed a network of resources that could be utilised in different phases of the project. Secondly, good project managers with their own reasonable ways to coordinate the collaborated work were recognised as a factor improving knowledge sharing. Naturally, the agreed procedures between companies with regard to knowledge and information sharing were seen as facilitators.

Moreover, the current organisational culture facilitated knowledge sharing by allowing people to ask questions and to do the work together. However, the distance seemed to affect the success of knowledge sharing: it was easier to share knowledge when people were located geographically close. In addition, it was of course easier to communicate with the native language.

> *"...nice to meet them face-to-face that you know that what kind of a person you are working with. ...it is easier to contact if you have a problem."*

> *"If you know the people, not only from the mail discussions, it helps. Then, if the tools worked right, it would be easier without the extra problems."*

> *"It is easier to communicate in Finnish with the Finnish colleagues. You just don't ask if you have to do it in English."*

4.4 The Effect of the Organisational and Social Structures on Knowledge Sharing

Knowledge was shared in face-to-face situations and through information systems. The role of the information systems was to store the created documents and to enable the reuse and exchange of these knowledge assets. According to the interviewees, knowledge sharing was encouraged by setting common goals and deadlines for project members, but no direct incentives - that would be destined *only* for knowledge sharing - were used. The interviewees were not aware of established ways of sharing knowledge, e.g. the lessons learnt, which were informal knowledge in most cases, seemed to be poorly documented.

There were no significant differences between the focal company and the partners in the ways and culture of working. Due to a long partnership there were common or compatible ways of working together. However, in some cases there were problems with foreign companies due to different cultures, languages, or time zones.

There was no real "knowledge-border" between the different companies as one could have expected, and information and knowledge seemed to flow quite fluently across the company borders. However, one clear bottleneck was identified: Project managers and contact persons controlled the flow of information and knowledge. More direct contacts and communication would be needed between people working in the collaborated projects, since the project managers and contact persons were continuously too busy to answer all the questions directed to them. The partner companies did not always "dare to disturb the busy project manager" and thus suffered from insufficient knowledge. Information sharing was considered difficult due to cultural or geographical distance; meeting people face-to-face in the beginning of the project helped to create an atmosphere of trust that facilitated communication and knowledge sharing via e-mail and phone.

> "*...it's hard to find out about the experiences of the others, because we don't meet these people and nobody writes down the lessons they have learned...*"

> "*...we have worked together for so many years that we know each others' ways almost as well as we were working for the same company. We have kind of grown together.*"

> "*Sometimes it's not nice to stay at work late in the evening just to have a net meeting with the colleagues in the States. The time difference is some kind of an obstacle.*"

"...I personally don't see a knowledge border. There is a gap but it's not that bad. We usually get all knowledge we ask for, if we ask for it."

5. CONCLUSIONS AND DISCUSSION

The results of our interviews can be summarised in short: information and knowledge sharing happens today quite well in the studied network of partners collaborating in product development projects. However, the tacit dimension of knowledge is not given enough attention. The information systems and other tools supporting knowledge management are focused on storing, sharing and searching documents and other explicit knowledge, and the not so easily articulated tacit knowledge is shared in immediate communication or face-to-face meetings in an ad-hoc, unstructured manner. The support for sharing tacit knowledge should be improved for example by encouraging the creation of communities of practice and by providing technological platforms for them.

The organisational differences are not very disturbing between partners that share a history of collaboration. Instead, the processes, ways of working and even organisational cultures are very close to each other. The challenges of collaboration reside in the multi-cultural co-operation: language and different meanings attached to work can be obstacles for efficient project work.

The factors facilitating knowledge sharing include direct personal contacts between the engineers working in partner companies, and face-to-face meetings in the beginning of projects that require extensive virtual communication between distant sites. Information systems also facilitate knowledge sharing, but they need to be improved and integrated in order to support easy access to information and knowledge.

The support for re-using existing information and knowledge is necessary. Most of the information produced in previous or parallel projects is needed in order to be able to plan and run the new project efficiently. This includes also a learning dimension: the lessons learned during the projects may be passed on, and the documents, templates and project plans may be used as models that can be modified according to specific project needs.

When talking about organisational learning it is important to recognise the shifting borderline between implicit and tacit knowledge – non-explicit knowledge is much more difficult to reconstruct when we are talking about tacit, and not only implicit knowledge. ICT can be used to refine implicit knowledge for sharing – "deeper" tacit knowledge is possible to make organisational only by making people to collaborate in practice.

Sharing the knowledge and information created in the product development process is vital for the today's engineering work. Being capable of reusing information and knowledge created in other projects enables the engineers to avoid extra work and to build on the existing knowledge. A mere document management system is insufficient as being unable to support the exchange of informal and often tacit knowledge based on lessons learned, experiences and new practices evolved during a project. Instead, an information system with an integrated approach to knowledge management and engineering work is needed. This system, however, can be a combination of many systems. Furthermore, when implementing such as system, a well-justified KM introduction plan is needed in order not to complicate the engineers' work with another new information system. System development should be based on existing communities. In addition to creating viable systems supporting collaborative work, more emphasis must be laid on formalising, training and creating guidelines and coherent practices on how to use KM systems and portals, because human adaptation requires time, change of mindset etc.

REFERENCES

BIEBER, M. et al.:
Towards knowledge-sharing and learning in virtual professional communities.
In: The Proceedings of the 35th Annual Hawaii International Conference on System Sciences.
Los Alamitos, CA et al.: IEEE Computer Society, 2002.

DYER, J. H.; NOBEOKA, K.:
Creating and managing a high-performance knowledge-sharing network: The Toyota case.
In: Strategic Management Journal,
Chichester, 21(2000)3, pp. 345-367.

HEDLUND, G.:
A model of knowledge management and the N-form corporation.
In: Strategic Management Journal,
Chichester, 15(1994)SUMME/2287, pp. 73-90.

HEDLUND, G.; NONAKA, I.:
Models of knowledge management in the west and Japan.
In: Implementing Strategic Processes, Change, Learning, and Cooperation.
Eds.: LORANGE, P. et al.
London: Basil Blackwell, 1993, p. 117-144.

HUBER, G. P.:
Organizational learning: The contributing processes and the literature.
In: Organization Science,
Providence, RI, 2(1991)1, pp. 88-115.

MILLEN, D. R.:
Rapid Ethnography: Time Deepening Strategies for HCI Field Research.
In: Interactive systems: processes, practices, methods, and techniques.
Ed.: BOYARSKI, David.
New York, NY: Association for Computing Machinery, 2000, pp. 280-286.
(Proceedings of Designing Interactive Systems DIS'00.)
NONAKA, I.; TAKEUCHI, H.:
Knowledge Creating Company.
New York, NY et al.: Oxford University Press, 1995.
NONAKA, I.; TOYAMA, R.; KONNO, N.:
SECI, Ba and leadership: A unified model of dynamic knowledge creation.
In: Managing Industrial Knowledge: Creation, Transfer and Utilization.
Eds.: NONAKA, I.; TEECE, D.
London: Sage, 2001, pp. 13-43.
POLANYI, M.:
Personal Knowledge: Toward a Post-critical Philosophy.
New York, NY: Harper, 1962.
STEIN, E. W.; ZWASS, V.:
Actualizing organizational memory with information systems.
In: Information Systems Research,
Providence, RI, 6(1995)2, p. 85-117.

Work Process Knowledge

A Keyword of Modern Competence Development Focussed on the Chemical Industry

Thomas Scheib
University of Dortmund, Lehrstuhl für Technik und ihre Didaktik I, Baroper Strasse 301, D-44227 Dortmund, Germany.
Email: scheib@ltd1.mb.uni-dortmund.de

Abstract: Modern production concepts like Total Quality Management, Lean production, Small factory unit or Business reengineering equate the human resource with the organisation and the technology. Therefore the knowledge of the workers about the execution and the optimisation of their working processes get an overriding importance. The so-called "Work process knowledge" becomes a new keyword in competence development. The following essay shows the difference as well as the relation between Work experience, Key competences and Work process knowledge. Based on that definition a model for the development of work process knowledge will be described. Finally difficulties and possible ways for the acquisition of working process knowledge are shown. These aspects are exemplarily applied to the work in the chemical industry.

Key words: Work process knowledge, Work experience, Expert groups, Cognitive apprenticeship

1. INTRODUCTION

Nowadays process optimisations are more likely to cause growth potential than technical developments. As a general rule, process optimisation is realised by the new organisation or reorganisation of working and business processes. Especially for the analysis and optimisation of the working processes, the experience and knowledge of the employees is of increasing importance.

With their work process knowledge, including experience and the knowledge of interactions and possible optimisations in the working process, employees make a substantial contribution to process and product innovations. The number of suggestions for improvement of employees can prove this. Therefore, the development of organisational competence and the promotion of the motivation to contribute it could become a decisive entrepreneurial aim in the next years. This also includes the creation of organisational options (knowledge, will, allowance).

2. THE STRUCTURE OF WORK PROCESS KNOWLEDGE

Work process knowledge is one of the keywords of modern competence development, but as well as the key competences this expression is not clearly defined. Literally, the work process knowledge is the essential knowledge to carry out a working process. In respect of integral acting, work process knowledge includes the knowledge of target setting, planning, execution, control and valuation of the working process. If the working process is described as a limited number of operations, which need to be carried out, the definition will be too narrow.

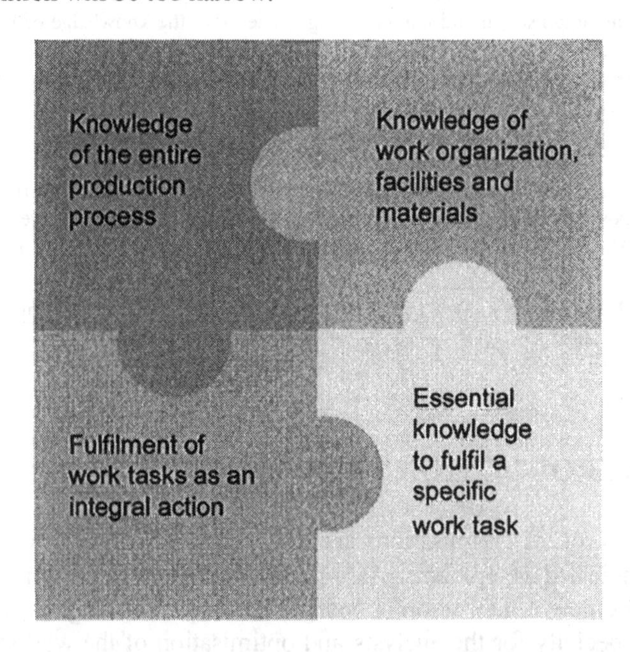

Figure 1. Definition of work process knowledge

Work process knowledge includes also the knowledge of work organisation, working capital and materials. It is especially important to be familiar with the equipment, its "snags", potential faults and the appropriate measures to solve the problems.

In chemical production processes it is also necessary to know the plant structure as well as the function of constituent elements. In every company this knowledge is different. Following the work process knowledge is rather depending on the company than on the profession.

Process orientation of modern production and management concepts like Total Quality Management, Lean Production or Business Reengineering further enhances work process knowledge.

The work process cannot be seen separately from previous and subsequent processes or the entire process. Work process knowledge embraces the knowledge of the connections of the production process itself and the embedding of the work process in the entire production process. Additionally, it includes the knowledge of the organisational structure and the processes prior or subsequent to each work or production process and how to implement and improve these.

To acquire work process knowledge, transparency of operational performance is absolutely essential. Transparency of the process structure refers to the separate phases of the process and their connections. Transparency of performance in the area of customer satisfaction, quality, time and costs is important to focus on weaknesses and optimisation potential in the process (KLING 2000).

Work process knowledge is of special importance, if the work process itself does not function properly. Besides the usual work routine, the question arises how malfunction and problems can be solved. Therefore work process knowledge includes the knowledge of how to prevent, realise and rectify malfunctions.

This aspect is especially considered in the continuous production processes of the chemical industry. The normal work there is increasingly characterised by data collection. Process control itself requires relatively less knowledge as long as the process works properly. Here the understanding of the production process and the facilities structure is just needed to a small extent.

In the case of malfunctions the required knowledge rises immensely to be able to minimise and to eliminate the interferences. The acquisition of knowledge of the plant structure becomes even more difficult due to the increasing level of abstraction of the process control (CAVESTRO 1989).

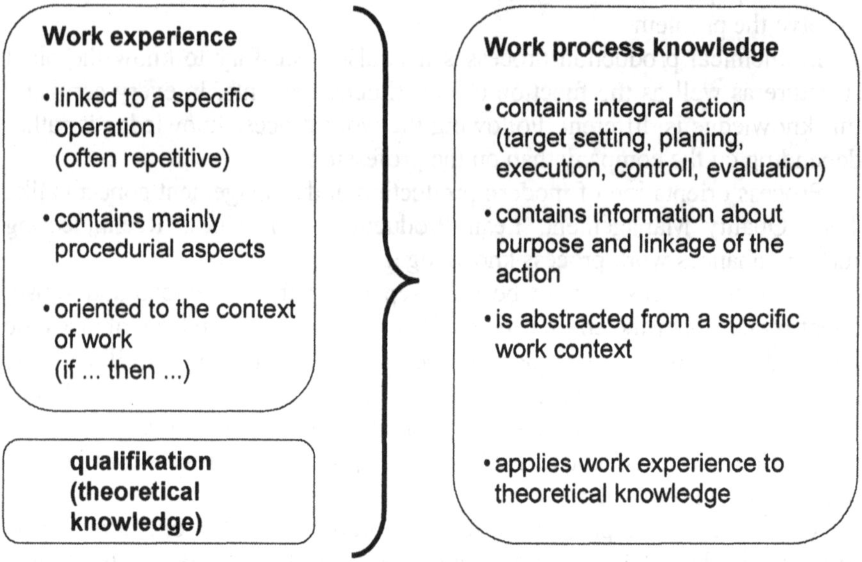

Work experience

- linked to a specific operation (often repetitive)
- contains mainly procedurial aspects
- oriented to the context of work (if ... then ...)

qualifikation (theoretical knowledge)

Work process knowledge

- contains integral action (target setting, planing, execution, controll, evaluation)
- contains information about purpose and linkage of the action
- is abstracted from a specific work context
- applies work experience to theoretical knowledge

Figure 2. Work process knowledge and work experience

Work process knowledge can be described as a fusion of work experience and theoretical knowledge. Work experience always refers to a specific context that means a concrete work situation. Theoretical knowledge, which is mostly acquired through formal learning, is generally separated from concrete work situations and therefore hard to apply in such a work situation. So, on the one hand work process knowledge is the work experience, which can be reflected through theoretical knowledge. On the other hand, work process knowledge can be understood as specialised knowledge which is related to concrete work situations and enriched with appropriate knowledge.

The work in the process control centre in the chemical production is a good example to make it clear. A chemical worker in his final year of apprenticeship is not capable of running the production facilities yet, because work experience including special knowledge about the facilities is missing. The other way round the experienced mechanic in the maintenance, who knows the facilities in detail, lacks the specific chemical knowledge for the work in the process control centre. Theoretical knowledge and work experience are both necessary to control such a process.

The expression "Work process knowledge" could imply that it just deals with declarative knowledge. On one hand it is not only one-dimensionally limited to technical aspects. In the case of interferences fast material requirements planning and strategies for the solution of the problems have to be developed. In this way shortfall in production shall be minimised. This requires planning knowledge, which goes beyond the pure technical and has procedural parts. On the other hand declarative knowledge is not sufficient to eliminate malfunctions (solve problems). The employee often resorts to heuristic procedures of trial and error, hypothesis-forming and problem structuring and solving (CAVESTRO 1989), which contain also procedural knowledge. Further more work process knowledge generated out of work experience also includes contextual elements.

A correlation of work process knowledge only with problem solving holds the risk to include also methodical and social-communicative competences. In that way expanded, the expression "work process knowledge" should rather be replaced by the expression "work process competence" (SCHWERES 1998). But you have to take into account that it becomes difficult to distinguish this expression from the term "key competences". Therefore, the interpretation of "work process knowledge" just includes the knowledge itself. The knowledge certainly goes beyond declarative elements, but it separates clearly from competences to apply and acquire the knowledge (ability) and the willingness to use it.

So far, this text shows that work process knowledge includes a variety of aspects, e.g. operational or system-depending aspects. These aspects depend on the applied technical and the existing organisational structure. A change in these areas always involves a change in the needed work process knowledge and therefore requires an alteration of the definition of its content.

3. THE ACQUISITION OF WORK PROCESS KNOWLEDGE

Work process knowledge is largely gained in the working process itself. The reflection of practical experience and its connection with existing know-how transforms it to work process knowledge (DERBOVEN 2002). In this way, work process knowledge develops from learning out of experience and from intentional learning. From knowledge out of experience and theoretical knowledge, knowledge of action arises. The difficulties of this connection can be found in the incoherent structure of the two learning methods. Generally, theoretical knowledge is gained outside a concrete working context.

In contradiction to that experience emerges in the concrete working process, which often crosses scientific fields. Reflection of experience through theoretical knowledge means the interaction of taking things out of context (de-contextualisation) of concrete working experience and the contextualisation of existing theoretical knowledge.

The acquisition of "new" work process knowledge is linked to new working situations. They develop from technical or organisational changes or out of problematical situations, e.g. interferences in the working processes.

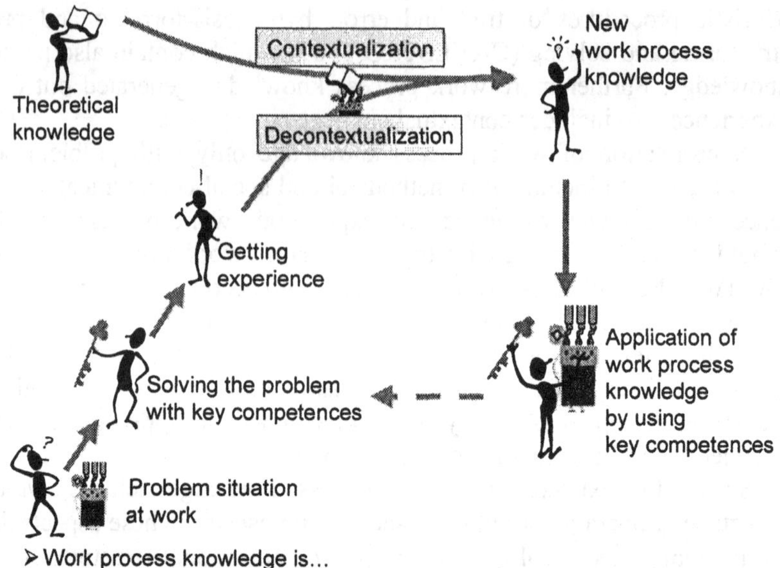

Figure 3. Development of work process knowledge

To cope with these situations work process knowledge and key competences like problem solving methods, the acquisition of information and the absorption of external experience (through cooperation and communication) have to be utilised. The process of solving problems creates work experience. To develop work process knowledge on this basis, work experience has to be substantiated with the help of theoretical knowledge (Decontextualisation). Simultaneously, theoretical knowledge has to be related to the concrete work situation (Contextualisation). Following, in respect to the application, theoretical knowledge is enriched and new structured. Work experience, resulting from the concrete work situation, is abstracted. To apply and to pass on the new acquired knowledge to similar work situations, key competences like the ability to communicate are necessary.

Enlarged work process knowledge is not only helpful in comparable work situations, but also changes and supports the process of solving problems in different situations.

But the problem of the acquisition of work process knowledge lies in new or problematic work situations. In problematic situations, especially when the process is interfered, there is mostly no time for reflection and for the learning process to gain work process knowledge. But the knowledge, which is necessary to solve the problems, can only be developed in such situations. In the extensive automatic production of chemical products this phenomenon becomes very clear. If the production works properly, only less knowledge is necessary to handle the control room. The computer-aided production and schematic presentation reduce the required knowledge to a great extent. The complexity of the production process remains a secret to the operator. Exactly this aspect intensifies the problems, if there is interference in the process. The complexity of modern control systems, which normally makes work easier, causes additional difficulties here. On the one hand the operator has to be acquainted with the "additional" system and on the other hand he loses the direct contact to the facilities (vibrations, noise) and to the process. However, this contact is necessary to develop work process knowledge.

Consequently, computer-aided, automated, continuous production in the chemical industry complicates the acquisition of work process knowledge in three ways. Firstly, automation requires consolidated knowledge. Secondly, the automation "separates" the employee in the control room from the facilities and therefore takes away the possibility to develop an understanding for the system. Furthermore, the control system is capable of eliminating interferences on its own or has at least mechanisms to reduce interferences. Consequently, the control systems prevent some possible learning situations.

So there is an unresolved contradiction of the increased formalisation of work and the existence of breakdowns, unforeseen incidents and errors (CAVESTRO 1989). Remaining interferences are all the more complex and grave and therefore difficult to deal with.

4. ADVANCEMENT OF WORK PROCESS KNOWLEDGE

The advancement of work process knowledge is often associated with the key phrase "Learning in the working process". For example participation of the plant operators in the construction and the run-in of the plant is an ideal precondition to develop know-how of the plant structure (JÜRGENS 1990). Mostly, this ideal situation does not appear. Therefore experience has to be gathered through different ways. The lacking opportunities given by control units to perceive the process, make it more difficult to get this experience and to acquire work process knowledge. Beyond the tasks "Learning in the working process" is influenced by job design, job organisation, self-image and learning culture of the company as well as qualifications and motivation of the employees.

In modern production concepts, responsibility is delegated and self-control increased to enlarge the scope of the operating units to give them the opportunity to acquire knowledge on their own. Employees even demand this scope. An integral action, which is a precondition of the learning process, is also considered in the concepts (HACKER 1986). In addition, the social organisation of the working process (cooperation and coordination) has a supporting function for the acquisition of competences.

For the acquisition of work process knowledge the cooperation of the employees is relevant. Here the organisational form of working groups supports the acquisition of work process knowledge. Differences can be realised in the constellation of the groups. A group of "specialists" enlarges the spectrum of knowledge available in the group, but makes the exchange of knowledge more difficult, as the perspectives of the group members differ a lot.

The work structure principle of job rotation supports the development of work process knowledge in the group. With this principle special requirements are linked which go beyond the formally demanded competences in groups (JÜRGENS 1990).

For the development of new work process knowledge, the reflection and de-contextualisation of work experience as well as the possibility to relate theoretical knowledge to specific tasks is important.

Besides the above-mentioned prevailing conditions, employees have to be able to initiate self-learning processes. This requires the development of organisational and structural aids. Following, two examples will be described which support the acquisition of work process knowledge in an early phase of working life that is characterised by less work experience.

4.1 Experience Circles or Cross-Training

Experience circles or groups deliver a possibility to change knowledge to experience. Experience groups are based on the concept of quality circles. Quality circles are/were often used to discuss group-related, global working problems. Experience groups do not concentrate on the problem layer, but access to concrete solutions and working experience. In these groups, individual know-how and skills referring to a specific situation are transferred to colleagues. Experience of the group members overlaps and therefore broadens. It is also a step to transfer the specific experience to an abstract level. The other way round, it is also possible to relate theoretical knowledge to the experience. The essential work process knowledge is not only linked to an individual person, but to the whole work group.

DAVENPORT (1993) pursuits a similar approach, which he describes as "Cross-Training". This means that colleagues teach each other according to the motto: "You teach me, and I teach you" (DAVENPORT 1993, p. 107). This principle can be transferred to the apprenticeship and therefore takes up the old approaches in the middle Ages. Young and inexperienced colleagues who have got a fundamental theoretical knowledge work together with experienced colleagues. Following, the experienced colleagues can support the young and the young colleagues can help the older ones to reflect upon their experience. Old-fashioned theoretical knowledge of older colleagues can be replenished and updated with the current specific knowledge of the younger colleagues. This concept can only be successful, if the relationship of the work group members is cooperative, free of prevailing prejudices and based on mutual confidence.

4.2 Working and Business Process Orientated Apprenticeship

Orientation at working and business processes in the apprenticeship aim at an enforced use of theoretical knowledge and skills in practical experience. The learning process itself focuses on the realisation of operational tasks. Here, the acquisition of theoretical knowledge is linked with precise working experience. Methodical and social competences become realistic. With the solution of these operational tasks professional competences (e.g. knowledge and skills) should be increased.

In English speaking countries a comparable and wide spread approach is the "Cognitive Apprenticeship". This approach follows the constructivism and includes four basic principles:

- authenticity of the task and the learning situation,
- practical and situated application,
- multiple contexts,
- social context.

A possible implementation of these approaches can be found in tangible tasks of operational projects for the trainee. Through these tasks, the learning process is orientated towards technical, organisational, and social conditions in modern work systems. Additionally, the trainee gets the opportunity to gain work experience – already during the apprenticeship. An example in the apprenticeship as a chemical worker is the immediate experience of the high complex facilities, which cannot be replaced by representations (for example by schematic representations in a simulation). Furthermore the integration of the work process in the total operational process and its organisation was of great interest for the trainees.

Limits of working and business orientation can be found in the acquisition of work process knowledge in tangible problem situations. An interference cannot be planned. Problem situations are a challenge themselves, which make it difficult to involve trainees, but they can be trained with computer-based simulation programs. The concrete senso-motoric experience as described above, which is of specific importance in real cases, can be gained only partly and via media.

Further difficulties of application-orientated imparting principles can be found in the externalisation of cognitive processes and the accessibility of knowledge. Many aspects are kept on the level of work experience and cannot be transferred to similar work situations. All the more it is difficult to transfer these aspects to colleagues.

5. CONCLUSION

In my opinion, there is not the silver bullet to advance work process knowledge. For the advancement of work process knowledge the interaction of many factors is important. These factors relate on the one hand to the task and the work place. This includes for example complete action, sufficient Leeway for decision and design, permanent and appropriate challenges and problems as well as group-orientated organisational structures.

On the other hand these are factors, which relate to the employee. Besides the de-contextualisation of working experience in special situations and the contextualisation of theoretical knowledge, it is also of special interest to advance and motivate self-learning aptitude.

REFERENCES

CAVESTRO, William:
Automation, new technology and work content.
In: The transformation of work.
Eds.: WOOD, Stephen.
London, New York: Routledge, 1992.
DAVENPORT, Thomas H.:
Process innovation: Reengineering work through information technology.
Boston: Harvard Business School Press, 1993.
DERBOVEN, Wibke; DICK, Michael; WEHNER, Theo:
Die Transformation von Erfahrung und Wissen in Zirkeln.
In: Lernfeld: Arbeitsprozess.
Eds.: FISCHER, Martin; RAUNER, Felix.
Baden-Baden: Nomos Verlagsgesellschaft, 2002, pp. 369-392.

DYBOWSKI, Gisela:
Berufliches Arbeitsprozesswissen.
In: Lernfeld: Arbeitsprozess.
Eds.: FISCHER, Martin; RAUNER, Felix.
Baden-Baden: Nomos Verlagsgesellschaft, 2002, pp. 355-368.
FISCHER, Martin:
Von der Arbeitserfahrung zum Arbeitsprozesswissen.
Opladen: Leske + Budrich, 2000.
HACKER, Winfried:
Arbeitspsychologie.
Bern, Stuttgart, Toronto: Huber, 1986.
JÜRGENS, Uwe:
Aktueller Stand von Produktionssystemen – ein globaler Überblick.
In: angewandte Arbeitswissenschaft,
Köln, (2003)176, pp. 25-36.
KLING, Jens:
Geschäftsprozessorientierte Personalentwicklung.
Wiesbaden: Gabler, 2000.
SCHWERES, Manfred:
Arbeitssystemwissen oder Arbeitsprozesswissen in der Berufsausbildung?
In: Die berufsbildende Schule,
Berlin, 50(1998)5, pp. 159-164.

Knowledge Management Issues For Maintenance of Automated Production Systems

Jacek Reiner[1], Jan Koch[1], Irene Krebs[2], Stefan Schnabel[1] and Thomas Siech[1]

1) Wroclaw University of Technology, Centre for Advanced Manufacturing Technologies, Lukasiewicza 3/5, PL-50-371 Wroclaw, Poland.
Email: {j.reiner, j.koch, s.schnabel, t.siech}@camt.pl

2) Brandenburg Technical University of Cottbus, Faculty of Mechanical, Electrical and Industrial Engineering Chair of Information Systems, Universitätsplatz 3-4 D, D-03044 Cottbus, Germany.
Email: krebs@iit.tu-cottbus.de

Abstract: Today's maintenance approach is functional oriented and mostly it is understood as hardware preventative actions. The changing share of mechanics, electronics and software cost of automated production systems call for new competences and responsibility for efficient maintenance. The paper presents automated production system which is considered as socio-technical system whose effective maintenance is based on knowledge management. In this paper Object-Oriented Modelling and UML were pointed to, as enhancing communication and documentation effectiveness in interdisciplinary teams responsible for maintenance.

Keywords: Maintenance, Automation, Knowledge management, Human-machine, Object-oriented, UML

1. INTRODUCTION

Looking for high product quality, production flexibility and cost cutting, the automation level of production systems is growing. The PCs, Internet services and even multimedia (vision systems) are migrating from the office world into industrial environment. Thanks to the increasing role of software in automated production systems, which is comparable with mechanics in

40 % share (PRITSHOW et al. 2000), the system functionality is developing, while their usage should become more and more user-friendly.

Do such superlatives characterise the maintenance of the above systems? It is visible, that the advance, scale, complication and complexity of the above systems is growing rapidly. Particularly, the complexity is inseparably linked with human-being, and its reduction is not achievable by decomposition, but only by systematic approach. The systematic approach is essential for new development, but as well as for service, modernisation and upgrade. The two second activities are essential for dynamic changes of products and production technologies within the life cycle of automated production system.

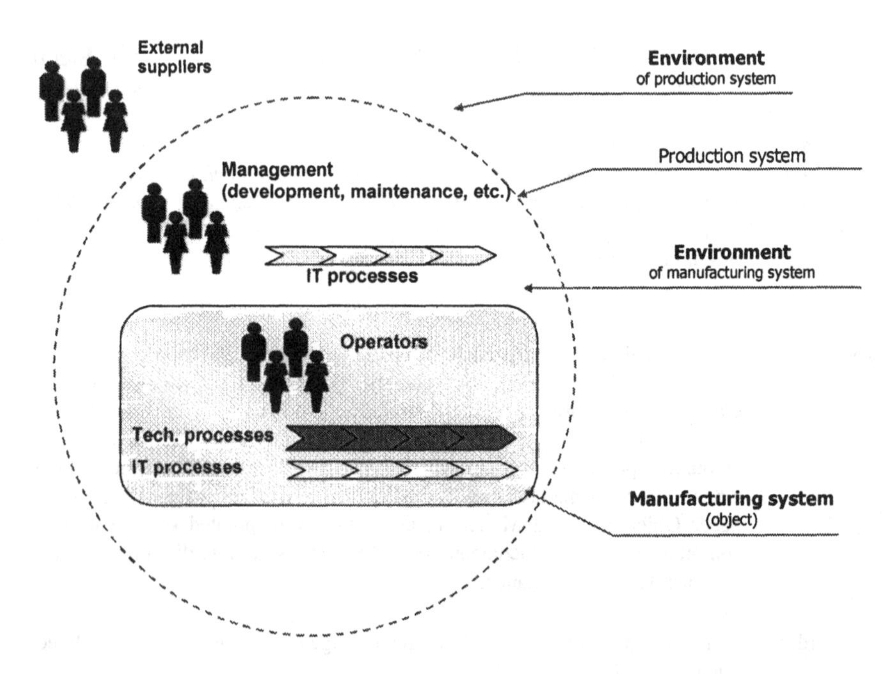

Figure 1. Socio-technical system

Therefore, a holistic approach to automated manufacturing systems, which integrates people, technological and IT processes into socio-technical systems for maintenance, is essential (Figure 1). The attention should be focused even more on strategy level of the "human–technology", reaching its philosophical roots ("is the robot the extension of human's arm, or is the human the extension of robot's arm?") (EVERSHEIM et al. 2002). Such holistic approach will appreciate maintenance process – the life cycle of automated production systems.

2. MAINTENANCE OF SOCIO-TECHNICAL SYSTEMS

"Assets maintenance" is the most capacious term regarding maintenance, which includes: buildings facilities, transportation means, production systems, etc. The maintenance of automated production systems through socio-technical perspective distinguishes three main domains: hardware, software, and human-being.

1. Hardware maintenance is a well developed division which supports production equipment. The service role (breakdowns reparation), is minimised by preventive approach. The domain has developed variety of maturity integrated methods, techniques and tools e.g. TPM (Total Productive Maintenance). A lot of services is outsourced. The relations between equipment users (production) and its carers (maintainers) is differently organised. Mainly the supervision, e.g. for new developments and modifications on the plant, is restricted to the planning division.
2. Software maintenance consists of four types of actions: corrective, adaptive, perfective, preventative. Activity share shows that more than 75 % of them concerns new functionality and safety upgrade or development. Because the software is not so open and more extensive, maintenance requires deep knowledge about internal SW structure and behaviour, as well as development environment. Therefore, the user commitment for maintenance is limited. On the other side, the software can offer very extensive configuration. The role responsible for SW maintenance is an "administrator", whose responsibility grows for Internet connected systems.
3. "Human maintenance" – as an expression is not used. Staff development is managed by Human Resources (HR) or Human Capital Management divisions. Through technical and social competences profiles, the people's carrier development paths are planned. The motivation and engagement are stimulated with difficulty (GRAF, HENCKEL 2003) while effective team working pays important role.

The above analysis indicates that each area has developed not only its own tools and methodology but also language and even its sub-culture. Despite of process oriented approach for analysis of value chains, the maintenance is not treated as integration of the above components. The maintenance of socio-technical system is complex, therefore, the problem decomposition hides the inter-domain problems.

1. Classical maintenance and planning divisions require different competences than software maintenance. The hardware and software life cycles depend on each other, and therefore, today's planning division should be

integrated more efficiently with software maintenance and development teams. The integration will be especially difficult if the software maintenance is supported by external companies (outsourcing).

2. Automation leads to replacement of humans manual or cognitive abilities. Interfaces are becoming more user-friendly, but it paradoxically appears that growing "machines intelligence" requires more knowledge from the user. It does not concern a typical operation but emergency, errors, warnings, configuration or manual modes. The variety of configuration possibilities makes the system more flexible, but simultaneously weakens the system safety and reliability. Therefore, the user needs more training and "playing" with machine configuration for refreshment of system functionality, which is not frequently used.
3. The high specialisation, distinct jargon between domain experts requires better communication between them. The understanding can be only expanded top-down, which means from the strategic and logical issues which should truly integrate all of the team members.

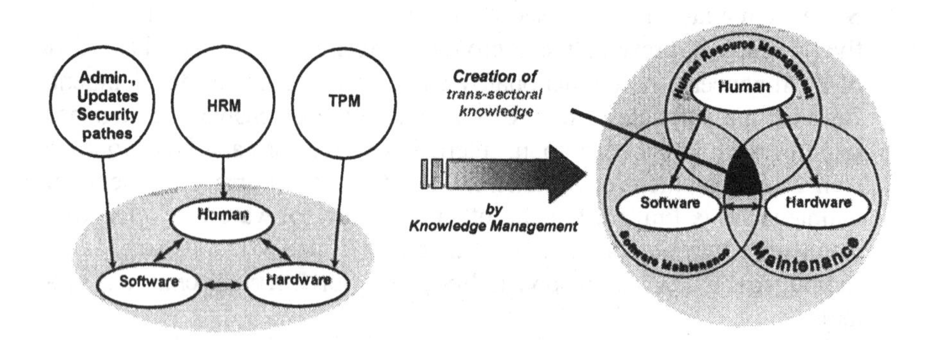

Figure 2. Maintenance of Socio-Technical System
(Source: KOCH et al. 2003, p. 178; modified)

Researching for strategic approach referring to the above identified problems directs to better information management, team working and knowledge transfer.

3. KNOWLEDGE MANAGEMENT

Presently, knowledge management (KM) provokes lively discussions in the scientific and industrial world. Some believe that KM is the panacea for today's company's difficulties, the others sceptically call it just "fashion" or even "new religion".

The discussion entry point should be unequivocal defining of the term "knowledge" – which seams to be impossible. If we accept that knowledge is only in human heads, then we have to radically distance from information technologies, which can only support information storage and communication. In this case the human is irreplaceable for knowledge creating and usage. Knowledge communication and processing cannot be efficiently realised based on information technology.

Is knowledge measurable in that case? There are several approaches how to measure human knowledge using competences tests. Analogous to intelligence testing (IQ, emotional intelligence) the question of reliability and how to keep the captures updated is still open.

There are two main approaches to knowledge processes supporting. The first one is "knowledge management" which tends towards control by mapping, measurement, evaluation with IT support. The second approach is called "knowledge creation" in order to stress that the rational management pays insignificant role. The creation is based on "soft factors" like emotions, understanding and community. The above approaches focus human role differently (object or subject), which derive from two philosophical roots of Western or Japanese culture (TAKEUCHI 1998). Therefore, the knowledge support implementation has to respect environmental culture.

There are several different approaches for knowledge management, which start with problem domain analysis – see Table 1.

Table 1. Different approaches for KM implementation

Dimensions	Approach	Author
	episteme, techne, phronesis	Arystoteles
Type of knowledge	know-what, know-why, know-how, know-who	Lundval, Johnson
Type of knowledge	implicit, explicit	Nonaka, Takeuchi
System components	human, technology, organisation	Bullinger, et al.
Supply chain	internal – external	Ruemler, Reinhard
Core activities	identification, creation, storage, allocation, use, oblivion, evaluation	Probst, et al

The implementable solutions supporting knowledge management are called KM tools. They are represented by variety of classes: information exchange, information mining, creativity, etc.

4. KNOWLEDGE MANAGEMENT TOOL FOR MAINTENANCE

The analysis of the complex automated production systems indicates that maintenance effectiveness will depend more and more on knowledge of the system. The new automated production systems include hardware, software, documentation, training and support. The inconsistency between each of them can occur if understanding or resources are not sufficient.

Multidisciplinary competences and distributed responsibility among the users of automated production systems pay attention to need and support for knowledge sharing and archiving.

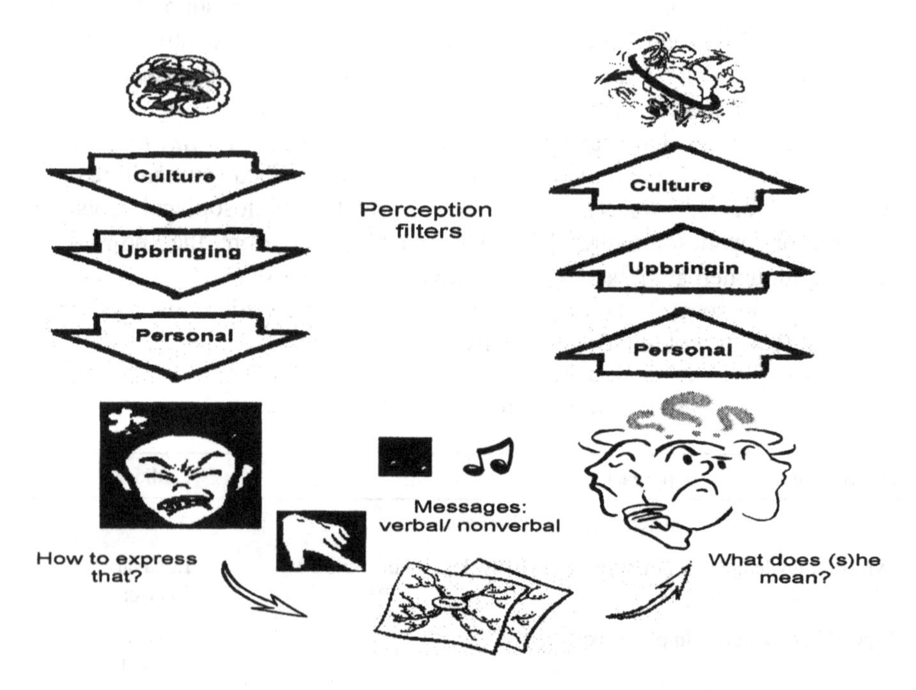

Figure 3. Interpersonal communication [JR]

At CAMT (REINER 2003), "Object-Oriented Modelling and UML for Manufacturing Automation" was proposed, as a knowledge management tool. The tool can be also implemented to support automation maintenance processes.

The knowledge management tool is based on interpersonal communication analysis. The communication chain (Figure 3) identifies perception filters formed by culture, upbringing and personal abilities. The filters determine paradigm – the way of thinking and building mental models. Diagrammatic which deals with graphical communication languages, offers tools for projection of the mental models. Languages with defined syntax and semantic are means for human communication and produce tangible expressions. The proposed knowledge tool uses object-oriented paradigm and Unified Modelling Language (UML). The solution is formulated as layered competences model. Above the lowest layer of Manufacturing Automation, the Modelling, Objectory and UML are defined (Figure 4). The model forms a platform for four modelling competences: business automation processes, logical process structure, use cases and systems dynamics. The efficient use of proposed methodology requires appropriate implementation. The implementation process consists of four stages: training, demonstrator building, expansion and usage. The demonstrator delivers evaluation for introduced knowledge management projects.

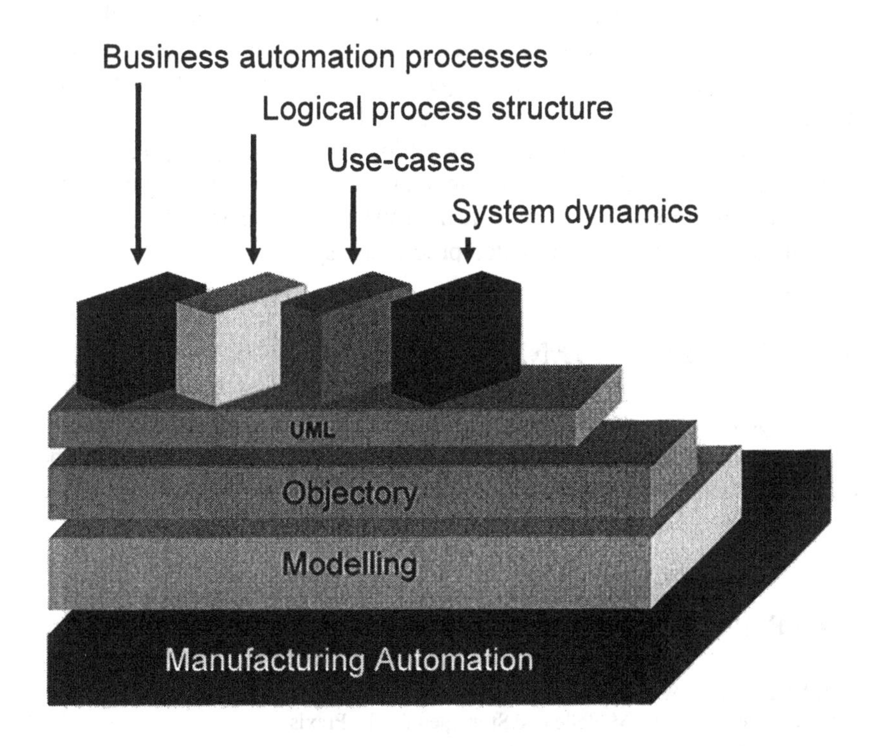

Figure 4. Communication tool OOMMA [JR]

The proposed methodology was implemented in an interdisciplinary team responsible for operation, maintenance and development of automated production system. The evaluation showed that objectory and UML can be easily learned and used by non programmers for modelling real world and communication. The methodology was used as a communication platform for better knowledge exchange between experts and to support knowledge consistency.

5. CONCLUSION

In the above article the reader's attention was turned to the influence of the introduction of the PC's, the Internet services and multimedia from the office world, on the understanding of maintenance of automated production systems. The need and profits resulting from integrated process analysis of socio-technical system (human, technology, information) was illustrated.

Critical discussion of the knowledge management issues proves that despite incoherence and misunderstandings, its elements may and should support maintenance processes in a company. For their success it is important to consider human-being and organisational culture in the foreground, and information technology in the background.

In this article Object-Oriented Modelling and UML were pointed to, as enhancing communication and documentation knowledge management tool, which can improve effectiveness in interdisciplinary teams responsible for maintenance of advanced automated production systems.

ACKNOWLEDGEMENT

The above research has been performed in an international and interdisciplinary team (CAMT-PWr and BTU-Cottbus) thanks to financial support of the European Commission within the project Centre of Excellence (ICA1-2000-70011).

REFERENCES

BULLINGER, H. J.; WÖRNER, K.; PRIETO, J.:
Wissensmanagement – Modelle und Strategien für die Praxis.
In: Wissensmanagement – Schritte zum intelligenten Unternehmen.
Ed.: Bürgel, H. D.
Berlin, Springer 1998.

EVERSHEIM, W. et al.:
Mit e-Engineering zum i3- Engineering.
In: Wettbewerbsfaktor Produktionstechnik: Aachener Perspektiven.
Edts.: KLOCKE, Fritz; EVERSHEIM, Walter; SCHUH, Günther; PFEIFER, Tilo; WECK, Manfred.
Aachen: Shaker Verlag, 2002.
GRAF HENCKEL VON DONNERSMARCK, H.:
Überwindung von Grenzen als Herausforderung der Erfahrung eigener Begrenztheit.
Münchener Kolloquium 2003
KOCH, Jan; KREBS, Irene; REINER, Jacek; SCHNABEL, Stefan; SIECH, Thomas:
Knowledge Management Issues for Maintenance of Automated Production Systems.
In: Current Trends in Production Management.
Eds.: ZÜLCH, Gert; STOWASSER, Sascha; JAGDEV, Harinder S.
Aachen: Shaker Verlag, 2003, pp. 174-180.
(esim – European Series in Industrial Management, Volume 6)
LUNDVALL, B. A.; JOHNSON, B.:
The learning economy.
In: Journal of Industry Studies,
Kensington, 1(1994)2, pp. 23-42.
NONAKA, I.; TAKEUCHI, H.:
The Knowledge Creating Company.
New York, NY: Oxford University Press, 1995.
PRITSHOW, G.; WELDE, K.:
Übersicht über Prozessschnittstellen - Zukunftsperspektiven.
IuK - Schnittstellen in der Produktionstechnik.
Düsseldorf: VDI-Verlag, 2002.
(Fortschritt-Berichte VDI, Nr. 593)
PROBST, G.; RAUB, S.; ROMHARDT, K. (edts.):
Wissen managen. Wie Unternehmen ihre wertvollste Ressource optimal nutzen.
Frankfurt/Main: FAZ, 3rd ed., 1999.
REINER, J.:
Object Oriented Modelling for Manufacturing Automation.
Wroclaw, University of Technology, PhD Thesis, 2003.
RÜMLER, Reinhard:
Wissensbarrieren behindern effektives Wissensmanagement.
In: Wissensmanagement,
Mindelheim, (2001)5, pp. 24-27.
TAKEUCHI, H.:
Beyond Knowledge Management: Lessons from Japan.
Edt.: TAKEUCHI, H.
June 1998.
http://www.sveiby.com/articles/LessonsJapan.htm, 25.07.2003.

Factory Planning Modules for Knowledge Sharing among Different Locations

Michael F. Zaeh and Wolfgang Wagner
Technische Universitaet Muenchen, Institute for Machine Tools and Industrial Management (iwb), Boltzmannstrasse 15, D-85748 Garching, Germany.
Email: {michael.zaeh, wolfgang.wagner}@iwb.tum.de

Abstract: In order to succeed in close-to-saturated markets, companies must offer services that are tailored to suit their customers' individual needs. This necessitates intensive interaction between the supplier and the customer. In order to produce increasingly individualised products close to customer markets, in comparison with today's production structures, a much larger number of factories will be necessary. The factories are distributed locally and directly in the actual markets and therefore have to be adapted to market-specific conditions. Decentralised knowledge needs to be utilised in order to exploit location-specific potentials, and the experience and knowledge about improvements that have been realised in production processes has to be shared among different locations. The use of preconfigured planning modules makes it possible to plan and re-plan factories relatively quickly on the one hand and facilitates reliable planning on the other hand. Digital tools and interactive input media support participatory planning and promote the documentation of knowledge as well as the transfer of knowledge among different locations.

Keywords: Individualised products, Close-to-Market production, Factory planning modules, Media for interactive planning

1. INTRODUCTION

The current situation of most production companies is characterised by increasing pressure from global competition. Today's customers are better informed and demanding as well as price-conscious. Along with this, the fast technological progress, due to the development of electronics, for example, makes the innovation cycles of mechatronic products shorter. The ability to react quickly to the markets' needs is becoming more and more important for production companies (NOFEN et al. 2003). Due to the globalisation of the markets, companies influence the development, production and sales in worldwide networks. In fast-moving buyers' markets, companies must offer innovative products and services that are tailored to their customers needs, together with short delivery times and moderate prices. Companies frequently react to this situation by designing products that are composed of prefabricated components that can be combined to form different product variants. In order to improve customer orientation, companies that are active worldwide perform parts of the product design and development work close to important selling markets of the products. For example the Japanese car manufacturer NISSAN (2004) developed the design for a new car for the European market near Munich in Germany. In some cases, not only the development but also the production operates according to market demands. Simplifying logistics or realising short delivery times by reducing transport times and waiting times are the major motivations here. Thus, in addition to improved interaction, the customer benefits from shorter delivery times and an improved observance of schedules (MILBERG 1996) due to production close to the market. When the added value is produced close to the market, market entry barriers such as local content regulations or sales taxes for imported products can be avoided. Considering that duties and import taxes can be up to 245 per cent of the price, of imported cars for example (RAUCH 1997), production can be economical if it is close to the market.

However, the efforts of companies to improve customer loyalty by offering more product variants and increasing customer orientation can bring about disadvantages as well. It is often difficult for a customer to overlook the enormous number of product variants available on the market. As the products have been developed for an average demand of a specific market, they do not necessarily correspond to a customer's needs. Due to the complexity of technical products, customers are frequently forced to buy features they do not need. On the other hand, production companies risk developing products and product features which are not well received in the customer markets while the production of numerous product variants makes production workflows and logistics complex at the same time.

An effective competitive advantage can be achieved if it is possible to offer specific products to customers. In the case of sophisticated technical products, there are many degrees of freedom by which customer requirements may differ. With today's product development methods, production processes and distribution channels, the augmented customer orientation is hardly realisable.

Replacing standard products or standardised product variants with individualised technical goods and individualised services has different effects on the product creation process. Among other things, it requires new strategies for interaction with the customer, methods for product adaptation, procedures for factory and production planning and new production techniques for manufacturing materials such as metals or plastics to individual customers' wishes in an economical way.

The Collaborative Research Centre 582 "Production of Individualized Products Close to the Market" is a cross-faculty project in which researchers from different disciplines such as product development, factory planning, and development of production techniques work together to establish innovative approaches, so that customers can obtain individualised products under comparable conditions of series production. The intention to offer products that can be individualised to customer wishes and to produce these goods directly in the markets influences factory planning to a considerable degree.

2. INDIVIDUALISATION AND CLOSE-TO-MARKET PRODUCTION

2.1 Technological Preconditions

Preliminary production of anything other than basic technical components is almost unthinkable in the case of far-reaching individualisation, since the customer's requirements cannot be reliably predicted. The production of personalised product components, where the customer can designate the form, functions and material, can only begin following product adaptation in line with customer wishes (Figure 1; KRESS 2000). This is not possible in an economical way using conventional production processes, because the cost of developing and building tools or the effort to set up machines would only affect one part. Classical economies of scale can not be achieved, so other potentials of cost reduction have to be made accessible. Therefore production processes are required that specially focus on production with "batch size one".

Processes such as sheet metal forming by computer-controlled drifting or rapid manufacturing of individual components, which allow the manufacturing of metal or plastic parts using tools that are not attached to the specific geometry of a part, are developed in different sub-projects of the Collaborative Research Centre 582. As the economies of scale are reduced, other potentials of cost reduction, such as economies of learning, are required (PILLER 2001), that depend on the number of executions of production processes and the complexity of the process rather than the number of identical parts produced. Economies of scope are based upon the use of production resources for manufacturing more than one product type. These economies offer potential especially when the new production processes are applied: different parts can be manufactured, but they do not require the typical setup efforts. This contributes to benefiting from capacities of the resources which might eventually be unused when focusing on one-product-manufacturing. Economies of integration characterise the combination of economies of scope and economies of scale. Both of them aim to cover the cost of resources by augmented production output.

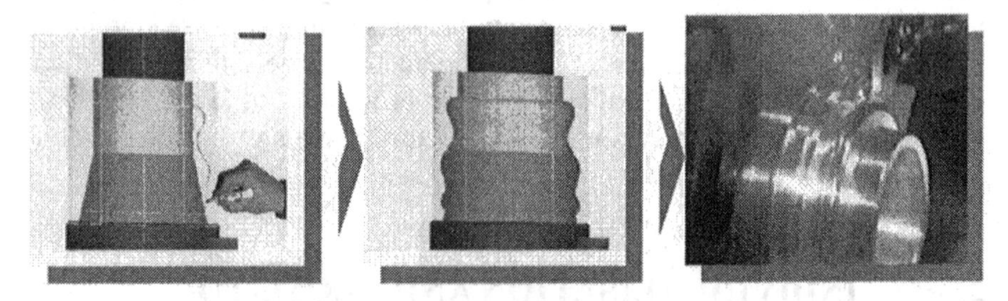

Figure 1: Manufacture of individualised parts

The manufacture of individualised products therefore requires standardised processes that can be used in the same form independently of the distinctness of the individual product.

2.2 Closeness to Customer Markets

Product individualisation, which can satisfy customers, is only possible if the supplier can accurately query the customer's wishes. This requires intensive interaction between the supplier and the customer. For this purpose, companies servicing world markets must go to the customer.

3. EFFECTS ON FACTORY PLANNING

Closeness to the customer and close-to-market production are vital factors for successfully manufacturing personalised products. Compared with today's production structures, a much larger number of close-to-market factories will be necessary in order to meet these requirements. These so called mini-factories can be described as units comprising sales, design and production (REINHART et al. 2000). Mini-factories develop and generate customised goods close to customer markets. Customers have the possibility to individualise their products. The adaptation of individual products to customers' wishes takes place starting from a pre-developed product spectrum. In the mini-factories, customer-specific parts are manufactured and assembled together with order-neutral parts to form a customer-specific technical product.

Close-to-market production of individualised products opens up new potential for companies, but also makes them face new factory planning tasks which have to be dealt with using new methods.

3.1 Dispersion of the Total Number of Pieces

As a result of a large number of locally distributed close-to-market mini-factories, coupled with shorter innovation cycles for products and production technology, a great effort for frequent production planning and re-planning is to be expected. Another important fact is that mini-factories are distributed locally and directly in the actual markets. Individual factories must therefore be adapted to market-specific conditions. Thus, although mini-factories are similar they also have a distinct location-specific character. However, these individual mini-factories have comparable processes, which is a prerequisite for using knowledge in multiple locations.

Decentralised knowledge must be utilised in order to exploit location-specific potential. The total number of pieces produced is distributed to different mini-factories. Nevertheless, if production at different locations does not start simultaneously, it should be possible to tap the experience effects of the overall group in the best possible way. Therefore it is necessary to share the experience and the knowledge about improvements that have been realised in production processes among different locations. An attempt must be made to consider as many aspects as possible for the factory to be planned and to optimise these in the planning stages.

3.2 Factory Modules

The goal conflict between planning many different variants of mini-factories and reliable, fast and comprehensive planning can be solved by creating groups of subunits. Preconfigured modules with defined interfaces are adapted and can be combined according to the location. This also allows new versions to be developed and integrated. Another advantage of a modular production structure is the possibility to identify affected modules and to isolate adaptations when changes in production become necessary (SCHUH et al. 2003). Functioning subunits should be employed with a view to avoiding optimisation loops, and the optimisation of these subunits should make it possible to use and disseminate proven basic knowledge within the company.

Under these conditions, strategies based on construction systems must be adapted to factory planning tasks (REINHART et al. 2000; EVERSHEIM 2001). In factory planning tasks, equal platforms that are suitable for different cases are hardly available. On the one hand, new manufacturing facilities or changes have to be incorporated into existing structures. On the other hand, when planning new projects at different locations in different countries, repetition effects can hardly be achieved: in general, conditions such as legal regulations or the connection to the infrastructure are different. Therefore, having possibilities to effect adaptations is necessary.

When using modular construction systems in the field of product development, for example, a distinction is made between planning construction sets that contain modules that have an equal or different hierarchy (Borowski, 1961). In addition, modules can differ (KOHLHASE 1997) in whether

- their application is required or optional
- they are abstract or concrete (substantial)
- they are elementary or configured modules.

The so-called configured modules are composed of subordinate modules. At the lowest level, there are elementary modules that cannot be configured further. There are different kinds of modularisation that depend on the degrees of freedom for the combination of different modules (PILLER 2001). Planning mini-factory structures requires the so-called individual modularisation to be adapted. The basic principle for this purpose is a common basic architecture (abstract), complemented with a determined or variable number of modules (ULRICH et al. 1991). In the field of process design, modular design integrates operations for processing modules with defined interfaces and associated input and output parameters (AURICH et al. 2003). Process modularisation and concurrent integration of the corresponding information helps to abolish the dissociation of tasks that belong together from the contents (REINHART, GLANDER et al. 2000). With this approach, separate modules and the associated information are planned in advance, although their composition is not.

Single factory modules thus are units with defined interfaces that are able to execute one or several machining steps for the product (Figure 2). They have input and output parameters for the material flow. These input and output parameters are created and required by upstream and downstream modules. The structure of factory modules corresponds to the form of configured modules and allows location-specific adaptations. These configured modules consist of elementary modules that can not be modified further. The elementary modules contain the components of the manufacturing facilities that ensure the quality of the production.

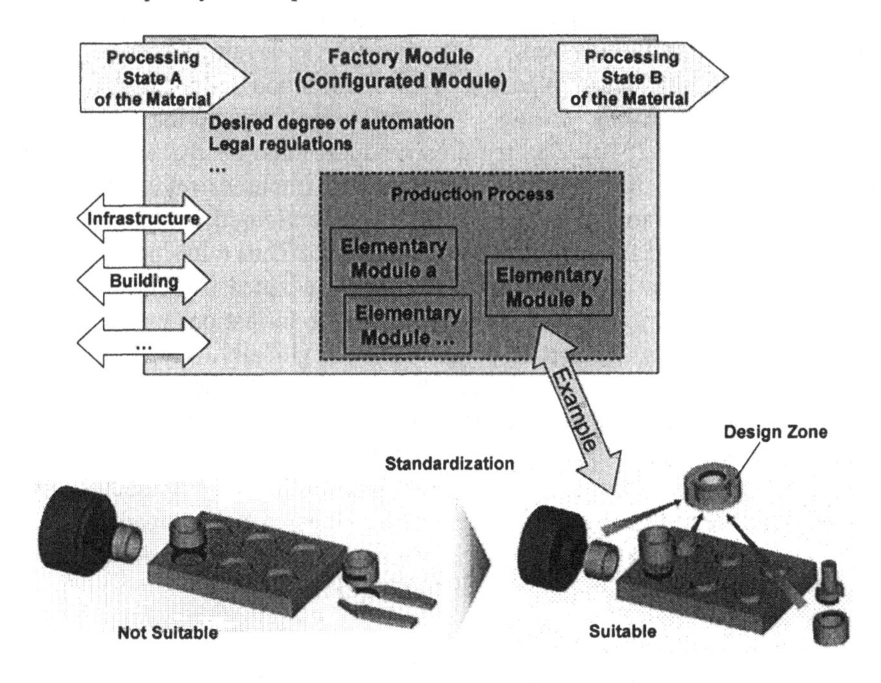

Figure 2: Configured and elementary modules

Figure 2 shows an example of standardised production utilities: due to appropriate design, the devices for manufacturing, transporting and handling individual product components are constructed in a way so that they fit product areas in which the customer is not interested. They are independent of the design zone that can be adapted to individual customer wishes, for example the exterior shell (KRESS 2000). As these devices can be applied without major adaptations, not every new alignment of the production will cause efforts, for example, to develop and build the production utilities. Planning, manufacturing and setting up appropriate production utilities take a large share of time in projects (RONG et al. 2003). Components that need not be developed for specific use cases, for example reusable, product-neutral apparatuses, are cost-saving and quickly available (COMSTOCK et al. 2000). This is also a precondition for manufacturing large numbers of unique items using industrial processes and mastering production time and costs at the same time.

3.3 Factory Planning Modules

When aiming to develop many mini-factories that are comparable but can be adapted to different locations, it is worthwhile to develop structures for a construction set for factory planning. This construction set contains configured modules. These include preconfigured structuring solutions (REINHART et al. 2000). The required modules must be characterised for the different kinds of production processes and resources required by the company. When planning a factory, these preconfigured components are selected and adapted for a specific planning project. Thus mini-factories possess a common basic structure with specifically adapted sub-units. Due to the modular structure of the factories, comparable formations can be found at different locations. Production processes in the globally located mini-factories correspond to the same standards. This supports the application of experience acquired at other locations. Optimisation procedures occurring during operation must therefore be incorporated in the planning to ensure that the building block system is enriched with information. In such cases, the effects of learning must be realised by the repeated and comparable execution of processes. The knowledge about processes has to be documented and reused like a "platform" (EVERSHEIM, SCHUH 2003).

4. IMPLEMENTATION AND TOOLS

Adaptations of factories must be planned as accurately as possible in order to allow the factories to operate at the most economical operating point. The large number of factories requires planning scenarios to be generated promptly, i.e. short-term fluctuations in demand must permit fast re-planning and adaptation that cost little time and effort. These updates are frequently unsuccessful due to the heterogeneity of planning and consequently give rise to redundant data. In planning tasks only a minor part of the time is needed for decisions, whereas a major part is needed to gather, edit and document information (YOUNIS et al. 1997). Planning support should be provided for this reason.

Concerning factory planning, important requirements are the ability to change planning scenarios quickly, the possibility for distribution and dissemination of information among different departments and locations, and the ability to handle large quantities of information. High-performance digital planning tools are a prerequisite for this. The introduction of tools for the so-called "digital factory" or "virtual production" (REINHART et al. 2003) is expected to reduce planning times by up to 40 %. Around 80 % of this progress can be traced back to the development of appropriate methods and new work procedures (REINFELDER et al. 2003; HALLER et al. 2002). The objective is to plan as thoroughly as possible and to facilitate planning procedures using the concept of factory modularisation and planning modularisation. For the purposes of the application of close-to-market mini-factories, tools must be specifically configured so that the planning method with preconfigured modules can be realised using the tools of the digital factory. The relations between planning objects and their properties such as occur in the integrated planning of factory structures are complex. They generally change during the planning and development phase and must be updated during operation. A primary goal is therefore to design the planning tools so that integrated planning with intuitive media is supported, which facilitates module handling. Planning with interactive input fields and simultaneous 3D visualisation has proven promising for planning factory layouts (WESTKÄMPER et al. 2001; WIRTH et al. 2001). The basic idea of these input media is for a group of planners and skilled workers to discuss layout changes together and enter these via touch-sensitive or imaging screens. Explicit as well as implicit knowledge (NONAKA et al. 1995) can enter into this process and is documented. Similar input media should be used to handle complex planning modules (ZAEH et al. 2003). Adaptations that are incorporated in the layout are assessed in a 3D display. The tool for layout planning is coupled to a process planning tool. Therefore changes can be updated immediately in a process planning database which is the basis for data-integration in process planning (Figure 3).

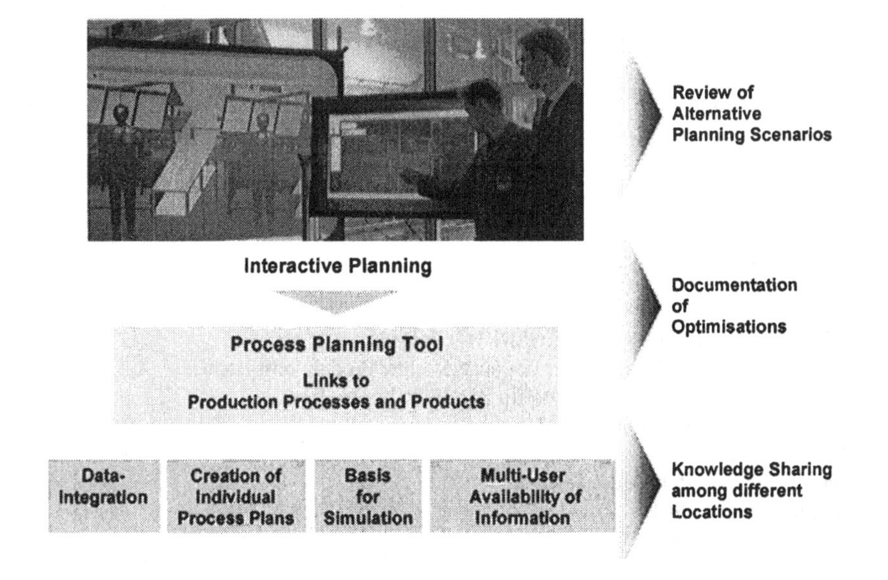

Figure 3: Media for interactive planning

5. BENEFIT

The application of preconfigured factory planning modules and the usage of standardised production resources causes mini-factories to have a common basic structure with sub-units that are adapted to specific locations. This facilitates the transfer of optimisations. Further improvements, at other locations for example, need to be acquired from the actual situation (KARLSSON et al. 2003). This is only possible if the knowledge of optimisations is available. Based on the demonstrated planning procedures that are backed up by interactive media, optimisations are documented and can be shared among different locations. The production processes in the globally distributed mini-factories correspond to the same standards, and it therefore becomes possible not only to exchange experience but also tools or production facilities. For a comparable global process standardisation, the Honda company has calculated that the investments in factories for production starts can be halved in the future (SHIPLETT 2003).

The use of modularised planning makes it possible to reduce the efforts for factory planning and accelerates the planning process. Optimisations that flow back into the modules allow experiences gained during production to be incorporated in the knowledge base, thus helping companies to manufacture the products requested by the customer quickly and cost-effectively. This permits fast and economical planning of close-to-market factories.

ACKNOWLEDGEMENTS

The Collaborative Research Centre SFB 582 "Production of Individualised Products Close to the Market" is funded by the German Research Foundation (Deutsche Forschungsgemeinschaft DFG).

REFERENCES

AURICH, J. C.; BARBIAN, P.; WAGENKNECHT, C.:
Prozessmodule zur Gestaltung flexibilitätsgerechter Produktionssysteme.
In: ZWF Zeitschrift für wirtschaftlichen Fabrikbetrieb,
München, 98(2003)5, pp. 214-218.

BOROWSKI, K.-H.:
Das Baukastensystem in der Technik.
Berlin: Springer 1961.

COMSTOCK, M.; OSSBAHR, G.:
Hyper-Flexibility – A Concept for a New Dimension in System Variability.
In: The manufacturing system in its human context - a tool to extend the global welfare.
Stockholm: CIRP College International pour l'Etude Scientifique des Techniques de Production Mecanique, 2000, pp. 372-377.
(Proceedings of 33rd CIRP International Seminar on Manufacturing Systems)
EVERSHEIM, W.:
Die Einmalaufgabe mausert sich zum Dauerprozess.
In: Industrieanzeiger
Leinfelden-Echterdingen et al., 123(2001)40, pp. 76.
EVERSHEIM, W.; SCHUH, G.:
Standard, individualisiert – individuell.
In: Marktchance Individualisierung.
Eds.: REINHART, G.; ZÄH, M. F.
Berlin: Springer, 2003, pp. 71-77.
HALLER, Eberhard; SCHILLER, Emmerich F.:
Digitale Fabrik bei DaimlerChrysler – Herausforderungen und Chancen.
In: Automobil Forum Ludwigsburg,
Landsberg, 2002, pp. 2/1 – 2/9.
KARLSSON, T.; OSCARSON, J.:
Information Transfer for Virtual Manufacturing Systems.
In: Progress in virtual manufacturing systems.
Ed.: BLEY, Helmut.
Saarbrücken: CIRP College International pour l'Etude Scientifique des Techniques de Production Mecanique, 2003, pp. 131-136.
(Proceedings of 36rd CIRP International Seminar on Manufacturing Systems)
KOHLHASE, N.:
Strukturieren und Beurteilen von Baukastensystemen. Strategien, Methoden, Instrumente.
Süsseldorf: Verein Deutscher Ingenieure, 1997.
KRESS, M.:
Maßgeschneidert auf virtueller Basis.
In: Die Neue Fabrik 2000, Produktion im Wandel.
Landsberg: Moderne Industrie, 2000, pp. 28-29.
MILBERG, J.:
Montieren wo die Märkte sind.
13. Deutscher Montagekongreß.
Landsberg: Moderne Industrie, 1996.
NISSAN:
http://www.nissan.co.at/AKTUELL/NEWS/2001/24_Sunderland.html, 09.02.2004
NOFEN, D.; KLUSSMANN, J.; LOELLMANN, F.:
Transformability by Modular Facility Structures.
In: Proceedings of the CIRP 2nd International Conference on Reconfigurable Manufacturing.
Ann Arbor, MI: University of Michigan College of Engineering, 2003.
NONAKA, I.; TAKEUCHI, H.:
The Knowledge Creating Company.
New York, NY: Oxford University Press, 1995.

PILLER, Frank T.:
Mass Customization.
Wiesbaden: Gabler Verlag, 2001.
RAUCH, M.:
Aufschwung am Nil,
In: BMW Magazin,
München, (1997)4, pp. 84-85.
REINFELDER, Anton; KOTZ, Thomas:
40 Prozent weniger Planungszeit.
In: Automobil Produktion,
Landsberg, 16(2002)5, pp. 34-35.
REINHART, Gunther; EFFERT, C.; GRUNWALD, S.; PILLER, F.; WAGNER, W.:
Minifabriken für die marktnahe Produktion.
In: Zeitschrift für wirtschaftlichen Fabrikbetrieb,
München, 95(2000)12, pp. 597-600.
REINHART, G.; HAAG, M.; FUSCH, T.; WAGNER, W.:
Mit der Digitalen Fabrik zur Virtuellen Produktion.
In: Proceedings of Münchener Kolloquium: Grenzen überwinden – Wachstum der neuen Art.
München: Utz, 2003, pp. 137-158.
RONG, Y.; HAN, X.:
Computer-aided Reconfigurable Fixture Design.
In: Proceedings of the CIRP 2nd International Conference on Reconfigurable Manufacturing,
Ann Arbor, MI: University of Michigan College of Engineering, 2003.
SCHUH, G.; VAN BRUSSEL, H.; BOËR, C.; VALCKENAERS, P.; SACCO, M.; BERGHOLZ, M.; HARRE, J.:
A Model-Based Approach to Design Modular Plant Architectures.
In: Progress in virtual manufacturing systems.
Ed.: BLEY, Helmut.
Saarbrücken: CIRP College International pour l'Etude Scientifique des Techniques de Production Mecanique, 2003, pp. 369-373.
(Proceedings of 36rd CIRP International Seminar on Manufacturing Systems)
SHIPLETT, R.:
Jellyfish statt Fischgräten.
In: Automobil-Produktion,
Landsberg, (2003)4, pp. 72-73.
ULRICH, K. T.; TUNG, K.:
Fundamentals of product modularity.
In: Issues in design-manufacture-integration.
Ed.: ANDIE, S.
New York, NY: Amercian society of mechanics engineers (ASME), 1991, pp. 73-79.
WESTKÄMPER, Engelbert; BRIEL, Ralf von :
Continuous improvement and participative factory planning by computer systems.
In: Annals of the CIRP,
Berne, 50(2001)1, pp. 347-356.
WIRTH, Siegfried; GÄSE, Thomas; GÜNTHER, Uwe:
Partizipative simulationsgestützte Layoutplanung.
In: wt Werkstattstechnik,
Düsseldorf, 91(2001)6, pp. 328-332.

YOUNIS, M. A.; WAHAB, A. M.:
A CAPP Expert System for rotational components.
In: Computers and Industrial Engineering,
Amsterdam, 33(1997)3-4, pp. 509-512.
ZAEH, M. F.; WAGNER, W.:
Planning Minifactory Structures for the Close-to-Market Manufacture of Individualized Products.
In: Proceedings of the MCPC 03: 2nd Interdisciplinary World Congress on Mass Customization and Personalization 03.
Eds.: REICHWALD, R.; PILLER, F.; TSENG, M.
München: TU München, 2003.

A Competence Approach in the Experience Feedback Process

Jorge Hermosillo Worley[1], Holitiana Rakoto[2], Bernard Grabot[1] and Laurent Geneste[1]

1) *Ecole Nationale d'Ingénieurs de Tarbes, Laboratoire Gestion de Production (LGP-ENIT), 47, Avenue d'Azereix - BP 1629, F-65016 Tarbes Cedex, France. Email: {hermosi, bernard, laurent}@enit.fr*
2) *ALSTOM Transport, Rue du Docteur Guinier – BP 4, F-65600 Séméac, France. Email: holitiana.rakoto@transport.alstom.com*

Abstract: The capitalisation of the know-how and experiences becomes a major issue of the industrial world, especially in large companies. Lesson learned techniques and experience capitalisation are possible methods for allowing the companies to increase their knowledge on their internal processes. This paper aims at presenting a study carried out with Alstom Transport on the "Experience Feedback" and "Lesson Learned" problems. We show how an Experience Feedback (EF) process, mainly aiming at transforming data into information, then information into knowledge, can benefit from an explicit modelling of concepts like *role*, *competence* and *knowledge* of the actors. We also show how these concepts may help to better identify the needs and potentialities of the actors, with a twofold goal: increasing the efficiency and acceptability of the EF system to be implemented on one hand, and improving the implication of the human resources in the technical processes on the other hand.

Key words: Experience feed-back, Role, Competence, Knowledge

1. INTRODUCTION

In order to face the quick variations of their environment, the dynamics of industrial companies have considerably increased, both in terms of product evolution, organisational changes and people mobility. Therefore, the capitalisation of the know-how and experiences becomes a major issue of the industrial world, especially in large companies which may be subject to an important turn over, either internal or external. Lesson learned techniques and experience capitalisation are possible methods for allowing the companies to increase their knowledge on their internal processes (see for instance AHA 1999; BICKFORD 2000; DELAHAYE 1996). Therefore, being able to re-inject the lesson learned into operational industrial processes becomes a strategic issue of nowadays companies.

This paper aims at presenting a study carried out with Alstom Transport on the "Experience Feedback" and "Lesson Learned" problems. This company designs, industrialises and ensures the maintenance of high technology devices such as power modules or command platforms of the traction part of trains, metros or tramways.

An essential factor for customer satisfaction is the high level of reliability of the products, which must be taken into account form the design phase on (Design For Reliability) and during the whole life cycle of the products. Until a recent period, the management of the reliability aspects (including diagnosis of problems and solutions) was completely dependent on the deep technical expertise of specialists whose intellectual assets clearly constitute a technical patrimony of the company. In that context, a project aiming at formalising a lesson learned process dealing with the expertise on defaults on large components has been launched in 2001, in order to structure and capitalize explicit knowledge, but also to optimise the involvement of the experts, which constitute a scarce resource in the diagnosis and solution search processes.

We show in this paper how an Experience Feedback (EF) process, mainly aiming at transforming data into information, then information into knowledge, can benefit from an explicit modelling of concepts like *role*, *competence* and *knowledge* of the actors. We also show how these concepts may help to better identify the needs and potentialities of the actors, with a two-fold goal: increasing the efficiency and acceptability of the EF system to be implemented on one hand, and improving the implication of the human resources in the technical processes on the other hand.

The structure of the article is as follows: the experience feedback experiment launched in Alstom is described with more details in the second section. A modelling framework aiming at allowing a better description of the way the human resources are involved in the technical processes is suggested in section 3, whereas the application of this framework to the feedback process is described in section 4. Conclusions and perspectives of this work are given in the last section.

2. EXPERIENCE FEEDBACK IN ALSTOM

The input of the EF process in Alstom is the occurrence of an unexpected situation (an *event*) during the life cycle of a device sold to a customer. Since the event expresses an unexpected failure, a solving process is set up as soon as it is detected. This process corresponds to a sequence of activities defined according to a workflow that will lead, in the best situation, to the resolution of the problem that resulted in the event (see the upper arrow of Figure 1).

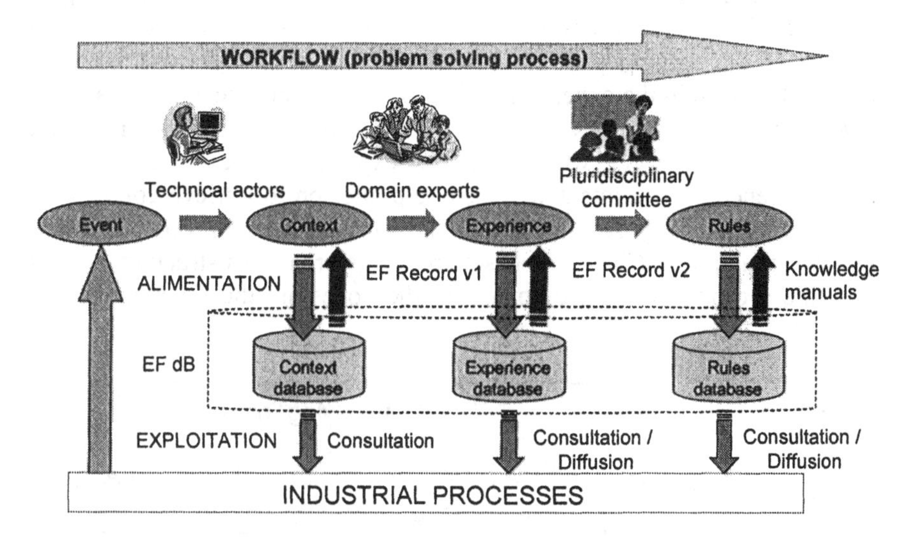

Figure 1. Global process to capture experience feedback from an event

Several activities are conducted by different groups of actors during this expertise, and will be developed later on, namely:

- "*Define*", where the problem is carefully stated. This activity is performed by a group of technical actors who first enhance the event description with information about the context in which it occurred (situation, working environment, first symptoms, etc.),
- "*Plan*", where the schedule of the expertise tasks is defined.
- "*Analyse*", where the identification of the problem is made and solutions are suggested. A group of experts is built for that purpose, led by the product manager, and consisting mainly of experts from the different technical areas involved in the component (electronics, electrical engineering, mechanical engineering, etc.), plus an expert on reliability.
- "*Verify*", where tests are performed on the products in order to check that the solutions are relevant, i.e. that the failure can not occur anymore.

- And when possible "*Generate rules*", aiming at making a generalisation on several past analyses through rules which may concern the diagnosis phase, the manufacturing of the component, or even its design.

In that context, and beyond the adoption of a good solution for the problem which occurred, the aim of the experience feedback process is to capitalise the results of the expertise and to learn from it.

All along the process, the experts are chosen according to the nature of the problem to solve, determined by the type of occurring event. In that purpose, a grid is available describing the various skills of each expert. Each activity of the solving problem process feeds the EF database with EF records that contain the expert analysis produced during the activity (see the lower part of Figure 1). As stated above, the technical actors describe first the *context* of the event, which helps the experts to understand what may have happened. This context will also help later on to retrieve comparable problems in the database. The *tests* which are performed in order to check the first assumptions (*"Verify"* phase) are then stored. This structure enables to define a product/expertise net where the nodes are the EF records (see Figure 2).

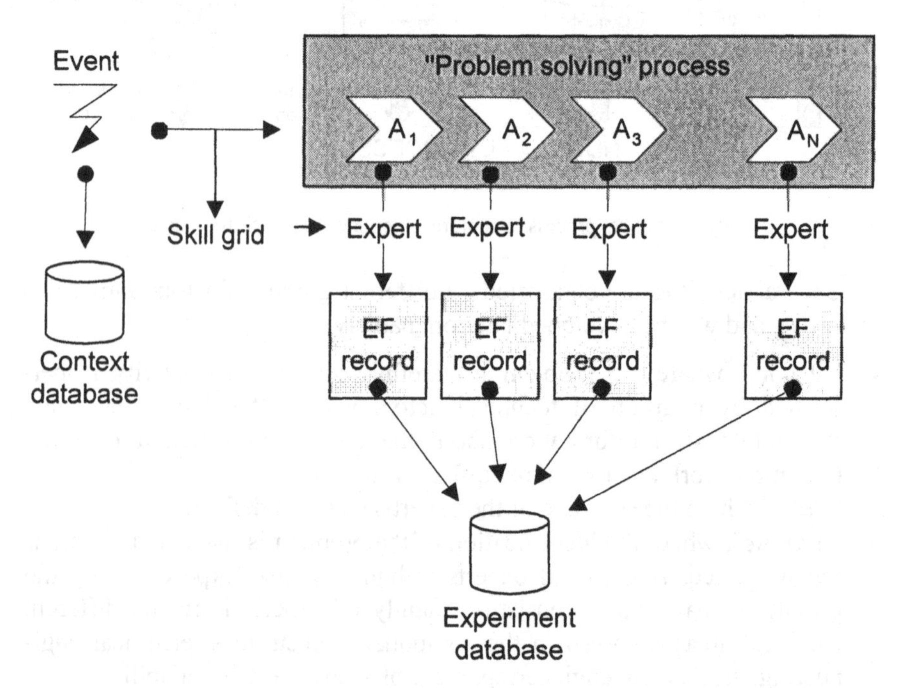

Figure 2. Experiments capitalisation

When the solving process is achieved, it may be judicious to build rules that will be systematically applied in future developments of similar product. These rules generalise and reinforce a set of previous experiences; the elaboration of such rules involves one or several experts who will propose solutions aiming at avoiding future problems according to a set of previous experiments. Decisions are translated into rules which, once incorporated into operational industrial processes, should prevent the failure to occur on the same or on a similar product.

In all companies, high level experts are a scarce resource. Therefore, different categories describing levels of expertise have been identified at Alstom in order to prepare a better assignment of these resources to tasks. These categories have been defined according to the years of experience, recognised skills, or types of problems already solved. They are: engineer, specialist, junior expert, senior expert and master expert. Of course, experts of higher levels are more seldom and costly. It is consequently very important to optimise their allocation to the listed activities, but also to prepare the access of young experts to higher levels of expertise. For that purpose, a study has been launched in order to apply a framework developed in the research laboratory of *Ecole Nationale d'Ingénieurs de Tarbes* (ENIT) aiming at better identifying the characteristics of the people (operators or decision makers) involved in business processes.

3. A MODELLING FRAMEWORK FOR HUMAN RESOURCE-BASED BUSINESS PROCESSES

Many different sources of improvements are currently implemented in nowadays companies but although the involvement of the human resources is always considered as the key of success, it is interesting to notice that the actors are seldom explicitly described in the "as-is" or "to-be" processes. Following this observation, a modelling framework has been suggested in by HERMOSILLO et al. (2002) aiming at better integrating the human resource aspect into business processes. This framework is described here after a short panorama on the domain, and its relevance to the experience feedback process is emphasized.

3.1 Competences and Roles in an Industrial Context

The models allowing to represent the characteristics of the human workers have considerably evolved since the emergence of the two main ones:

- the *trade* model, coming from the guilds in the Middle Ages,

- the *job*, *employment* or *position* model of the XVIIIth century, greatly promoted by the Taylorism at the beginning of the industrialization process.

In that context, the *qualification* model, still in use in most of the companies, gives a "Fordian" view on industrial manufacturing, mainly in order to help the definition of minimum salaries (PARADEISE, LICHTENBERGER et al. 2001). In response to the necessity to promote continuous improvement and flexibility, a new model has emerged in the 80's: the *competence* model. Instead of assessing a worker by comparison between pre-defined activities related to a workstation and the ability of a worker to perform these activities, it mainly consists in directly qualifying the person on the base of the competences which he possesses and can set to work (ZARIFIAN 2002). The main goals of the companies which promote a competence approach are identified by WUSTERMANN (2001) as the improvement of the individual efficiency, the decrease of turn-over or the improvement of the technical competences. Some companies give as the main reason of their choice the necessity to develop new competences required by the enterprise (STREBLER, BEVAN 1996) while it can also be considered that the concept of competence may provide a common language and facilitate cultural exchanges (STREBLER et al. 1997). The strategic interest of competences has been emphasized in the 90's by the work of PRAHALAD and HAMEL (1992) on the *core competencies*, suggesting a new way to consider the competitiveness of a company.

In most of the recent approaches, a distinction is made between the competences of a person (called here *gained* competences) and the competences required by an activity (*required* competences) (FRANCHINI et al. 1999; HARZALLAH, VERNADATH 1999). In parallel, and under different labels, a difference is made between *technical* competences and *behavioural* competences (PILBEAM, CORBRIDGE 2002), similar to the *hard* and *soft* competencies introduced by McCLELLAND (1973).

The competences can be analyzed at the level of an individual, gathering all the techniques allowing to facilitate the emergence, maintenance and development of personal competences (AMHERDT et al. 2000), but also at a collective level (LE BOTERF 1998), or even at an organizational level (SANCHEZ et al. 1996). A good summary on the different views with which the competences can be considered can be found in the *competence cube* suggested in the Manufacturing System Integration (MSI) research institute (WESTON et al. 2003).

Being able to explicit how the competences of the human resources may be deployed in an industrial process requires an intermediary which can be found in the concept of *role* of an actor. Organizations can be considered as systems of interacting roles (KATZ, KAHN 1966), where a role is defined as a set of activities, or as an expected behaviour. A role can be linked to a workstation, or to an organizational position (SARBIN, ALLEN 1968). The interpretation of the notion of role in the enterprise leads to define the organization as a network of roles defined independently from the persons who operate (SINGH 1992).

The modelling framework suggested in (HERMOSILLO et al. 2003) and summarized in next section aims at correlating the concepts of role, competence and knowledge in a way which can be implemented within the enterprise, especially through the notion of business process.

3.2 Modelling Framework

The modelling framework that we suggest is centred on the following concepts:

- *Competence*, which results from a combined implementation of knowledge, know-how, abilities, attitude and behaviour. More precisely, it encapsulates the ability of an individual to perform an activity in a job-relevant area as well as what is required from an individual to realise effective performance (HERMOSILLO et al. 2003). The control of interactions in a process under all their forms - negotiation, production, regulation, execution, etc. - and whatever the activity to be assigned, requires individual and collective competences. We can identify six general competences categories: *technical* competences, *organisational* competences, *analyse* and *decisional* competences, competences of *interpretation* and *formalisation*, *adaptation* competences and *relational* or *motivation* competences, which can be directly related during industrial applications to the role classification described below (ibid.).
- *Role,* which encompass a group of functions to achieve a purpose, based on the application of role competences. Using the right person at the right moment for the right activities defining a business process significantly increases the probability that efficient, timely and high quality product and service will be realised. Based on MINTZBERG's work (1979), we identify four generic classes of roles which could be found in any kind of organisation: interpersonal roles (*symbol*, *connection*, *leader*); informational roles related to information flow (*monitor*, *diffuser*, *spokesman*); decisional roles referring to the decision-making (*contractor*, *regulator*, *resources distributor*, *negotiator*) and the operational roles related to the implementation of knowledge (*expert*, *operator*, *technical bond*) (HERMOSILLO et al. 2003).

- *Knowledge*, which is a fluid mix of framed experience, values, contextual information, and expert insight that provides a framework for evaluating and incorporating new experiences and information. In organisations, it often becomes embedded not only in documents or repositories but also in organisational routines, processes, practices and norms (DAVENPORT, PRUSAK 1998). Knowledge can be created by persons having a given competence (e.g. the experts of the experience feedback of Alstom), but experts can also require external knowledge in order to be able to apply their competences. It is for instance the reason why multidisciplinary teams are built in the Alstom application. Therefore, identifying the required and available knowledge is also a key point for allowing to efficiently use competences in an industrial process.

These concepts are related with the process modelling principles as shown in the general model of Figure 3.

In this model we propose to distinguish between competences *required* by an activity and/or *gained* by the actor. Each of these categories has various types of basic competences which are described in the next section. The actor uses several "informational resources" which allow him to perform his role. These resources are divided into three categories, namely data, information and knowledge. *Information* is a structured set of *data*, on which has been added a meaning or an interpretation. Associating *information* to a context in order to define application rules allows to build *knowledge*. It can be verified that these definitions are fully consistent with the Experience Feedback process implemented in Alstom.

Finally, a role is based on the application of competences, which are always related to a specific activity including one or more tasks which belong to a given process, with a mission to achieve.

This general model has been instantiated on a software database tool, named COCOROL (*co*nnaissances, *co*mpétences, *rol*es, in French), initially allowing the user to describe a process as a network of activities (execution or decision activities), where these processes can be described as either "as-is" (description of existing processes) or "to-be" processes (description of optimised processes). Secondly, it is possible for the user to describe decision activities in some detail according to the defined concepts (actors, roles, competences and knowledge). The matching of human resources to "as-is" and "to-be" processes can then be done. This allows the user and process experts to know about:

- required and available roles that can be assigned in the process,
- the kind of competences that are required and available (in the company) with reference to each activity, and when these resources are used in a given process,

- which are the data, information and knowledge sources and location, etc.

In order to be able to implement this conceptual modelling framework in industrial applications, we show in the next section how these concepts have been applied on the Alstom problematic.

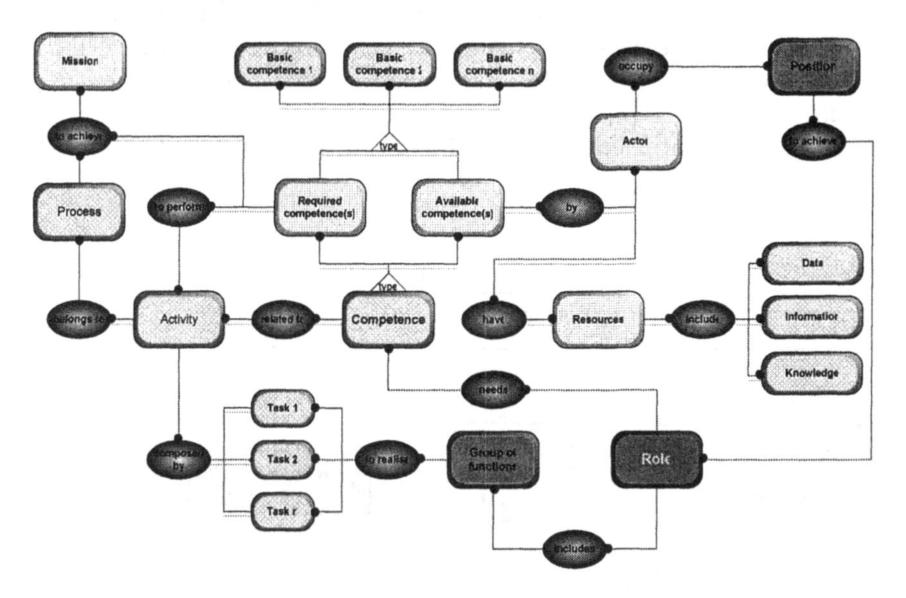

Figure 3. General model of the suggested framework

4. APPLICATION TO THE EXPERIENCE FEEDBACK PROBLEM

This framework has been tested on two representative activities of the EF process (see Figure 4): *"Analyse"* and *"Generate rules"* by a working group including researchers, project managers, experts and people from the Quality and Human Resources Departments.

For both activities, the same roles have been identified: *Leader* (project manager), *Monitor* (the expert in reliability, who merges external sources of information for the group) and *Expert*. These roles have then been more precisely described using the functional jobs of human actors concerning the analysed activities, for instance as summarised for the EF *"Analyse"* activity in Table 1.

The required competences for each role in each activity have then been listed. For that purpose, a list of typical competences has been built by merging a framework suggested by the researchers, based on comparable studies, and a framework defined by the Human Resource Department of Alstom, based on more operational issues. This framework lists:

- *behavioural* competences: organisational competences (compliance with rules, autonomy, responsibility, etc., interpretation and formalisation competences (ability to simplify problems, to structure information, etc.), adaptation competences (open mindedness, adaptability, stress tolerance, etc.)
- *technical* competences, regarding the various technical areas to be considered.

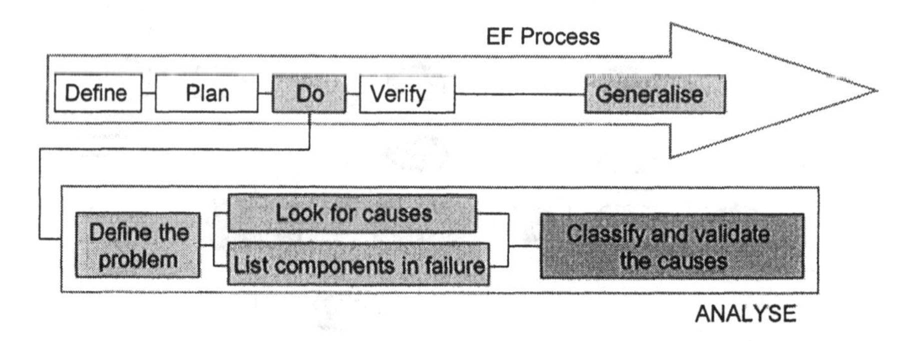

Figure 4. EF process activities

All competences (19 behavioural and 20 technical competences) have been assessed according to a five-level scale which degrees have all been precisely defined.

The result of this first step is a required competence profile for each role in each considered activity, represented on radar graphics. At a second step, the actors fill up the same form together with their supervisor in order to define their available competence profiles.

These first steps of course allow to make explicit some aspects of the considered activities which are important for the expert allocation, e.g.:

- The "*Generate rules*" activity requires the highest level experts from a technical point of view, but also people who are able to extrapolate the possible consequences of rules on many different aspects of a product life-cycle. As a consequence, they also need to have a very wide view on the company and its strategy regarding its products.
- On the other hand, and even if their technical competencies are important, "*Analyse*", like many collaborative activities, requires the actors to be open and tolerant, able to clearly explain their point of view and

understand the points of the others, which apparently are not always the dominant qualities of senior experts.

Table 1. Functional role description in the *"Analyse"* activity

FUNCTION	ROLE	ROLE DEFINITION
Product / Project Manager	Coordinator	Management / Communication with client
Reliability engineer	Monitor	Qualitative and quantitative data analysis + statistical data processing
Power engineer	Expert	Power technical analysis
Component validation engineer	Expert	Technical component analysis

The results of this study mainly consist in:

- A better understanding of the role of each actor in the experience feedback process.
- The identification of some divergences between required and gained competences for some individuals, mainly concerning the use of internal tools or standards of the company. This identification has allowed to define a plan for improving the knowledge of the actors on these points.
- A better identification of the technical and behavioural differences between the defined levels of expertise.
- Better positioning of the existing experts regarding the levels of expertise (who is close to the next level, what points have to be improved for allowing one to pass through the threshold, etc.)

As a consequence of the two previous points, it is now possible to complete the existing plan for managing the experts in the company, but also their evolution.

Unexpected points have also be found of interest after this study. For instance, the comparison between expected and real competence profiles may show that a person has not all the required competences, even if it can be stated that he perfectly holds his role. This may show that other competences than those identified can be applied in order to perform the role. It is in this case very interesting to explicit these competences, which may lead to new degrees of freedom in the allocation of people to tasks (i.e.: "I need a person who has competence A with level 2 *or* competence B with level 3.").

Finally, the managers of the company, involved in the evaluation of their subordinates, have found the competence referential much more comprehensive than the job description which was previously in use. Therefore, it has been suggested to describe all the jobs of the site using this framework.

Another lesson learned through this study was that, unlike what was initially feared, there has been no problem around the definition and use of the behavioural competences. On the contrary, many people have found interesting to at least see explicitly what was expected from them on that aspect. Defining explicitly required behavioural competences is not anymore a taboo in large companies, as soon as some support for improvement can be given.

Concerning the tools which have been defined, even if the "matching" between the actors (available competences) and the roles (required competences) is simplified by this first analysis, it remains a complex activity since, of course, there can not be a "perfect" matching between available and required competences for activities, especially if several projects are in progress at the same time. Different types of "matching indicators" are now tested in order to address this problem, with a view close to the "similarity functions" used in Case-Based Reasoning.

5. CONCLUSION

Within the Alstom company, a framework has been suggested in order

i) to better identify the needs of the actors involved in various activities of an Experience Feedback process
ii) to better know which are the characteristics of the available resources.

Even if this experiment has given promising results, an important amount of work is still required in order to efficiently support the allocation of experts to tasks, but also the management of their career which is of prime interest for Alstom in order to retain them within the company. Nevertheless, in a very sensible area (identification of human competences) it is encouraging to see that a project which is clearly explained and supported at a high level in a company can lead to operational results in a rather limited period of time.

REFERENCES

AMHERDT, C. H.; DUPUICH-RABASSE, F.; EMERY, Y.; GIAUQUE, D.:
Compétences collectives dans les organisations.
Laval: Presses universitaires de Laval, 2000.

AHA, D. W.; WEBER, R.:
Lessons learned links.
Wyoming (MD): AHA, D. W.; WEBER, R., 1999.
http://www.aic.nrl.navy.mil/~aha/lessons, 01.08.2003.
BICKFORD, C. J.:
Sharing lessons learned in the Department of Energy. Intelligent Lessons Learned Systems Workshop.
Austin, TX, 2000.
DAVENPORT, T.H.; PRUSAK, L.:
Working knowledge.
Boston, MA: Harvard Business School Press, Boston, 1998.
DELAHAYE, P.:
REX-FIAB: Un système de retour d'expérience sur la fiabilité d'équipements. 10ème Colloque National de Fiabilité et Maintenabilité.
Saint Malo, 1996, pp. 1015-1021.
FRANCHINI, L.; CAILLAUD, E.; NGUYEN, Ph.; LACOSTE, G.:
Planning and scheduling competences: towards a human resource management in manufacturing systems.
In: International Journal of Agile Manufacturing,
Bradford, 2(1999)2, pp. 247-260.
HARZALLAH, M.; VERNADAT, F.:
Human resource competency management in enterprise engineering.
In: 14th IFAC World Congress of Information Control in Manufacturing.
Beijing: International Federation of Automatic Control (IFAC), 1999, pp. 181-186.
HERMOSILLO WORLEY, J.; GRABOT, B.; GENESTE, L.; AGUIRRE, O.:
Role, skill and Knowledge: introducing human resources in BPR.
In: Production System Design, Supply Chain Management and Logistics: 9th International Multi-Conference on Advanced Computer Systems - ACS'2002.
Eds.: DOLGUI, A.; SOLDEL, J.; ZAIKIN, O.
Miedzyzdroje: Technical University of Szczecin, 2002.
KATZ, R.; KAHN, R. L.:
The Social Psychology of Organizations.
New York, NY: Wiley & Sons, 1966.
LE BOTERF, G.:
L'ingénierie des competences.
Paris, Les Editions d'Organisation, 1997.
McCLELLAND, D.:
Testing for Competence Rather than Intelligence.
In: American Psychologist,
Washington, DC, 28(1973)1, pp. 1-14.
MINTZBERG, H.:
The structuring of organisations,
Upper Saddle River, NJ: Prentice Hall, 1979.
PARADEISE, C.; LICHTENBERGER, Y.:
Compétence, competences.
In: Sociologie du travail, Editions Scientifiques et Médicales
Paris, 43(2001)1, pp. 33-48.
PILBEAM, S.; CORBRIDGE, M.:
People Resourcing – HRM in Practice.
New York, NY: Prentice-Hall, 2nd ed., 2002.

PRAHALAD, C. K.; HAMEL, G.:
The core competence of the corporation.
In: IEEE Engineering Management Review,
New York, NY, 20(1992)3, pp. 5-14.
SANCHEZ, R.; HEENE, A.; THOMAS, H.:
Dynamics of Competence-based Competition.
Amsterdam: Elsevier Science, 1996.
SARBIN, T. R.; ALLEN, V. L.:
Role theory.
In: Academy of social psychology.
Eds.: LINDZEY, G.; ARONSO, E.
New York: Random House, 2nd ed., Volume 1, 1968, pp. 488-567.
SCHEER:
ARIS Toolset documentation.
Saarbrücken: Scheer AG, 1999.
SINGH, B.:
Interconnected Roles (IR): A coordination model.
Austin, TX: Microelectronics and Computer Technology Corporation, 1992.
(MCC Technical Report CT-084-92)
STREBLER, M.; BEVAN, S.:
Competence Based Management Training.
Brighton: Institute of Employment Studies, 1996.
(Report 302)
STREBLER, M.; THOMPSON, M.; HERON, P.:
Skills, Competencies and Gender: Issues for Pay and Training.
Brighton: Institute of Employment Studies, 1997.
(Report 333)
WESTON, Richard H.; BYER, Nikita; AJAEFOBI, Joseph O.:
EM in support of team system engineering.
In: Proceedings of 10th ISPE International Conference on Concurrent Engineering: The Vision for the Future Generation in Research and Applications.
Eds.: JARDIM-GONCALVES, R., BALKEMA, J., CHA J., STEIGER-GARAO A.
Lisse: Swets & Zeitlinger, 2003, pp. 865-872.
WUSTERMANN, L.:
Recruitment, retention and return in the NHS.
In: Health Service Report, 32(2001)autumn, pp. 2-14.
ZARIFIAN, P.:
La politique de la compétence et l'appel aux connaissances dans la stratégie d'entreprise.
In: Vers l'articulation entre compétences et connaissances.
Nantes: Groupe de Travail Gestion des Compétences et des Connaissances en Génie Industriel (GCCGI), 2002, pp. 20-24.

PART FIVE

Management of Distributed Work

Learning for an Agile Manufacturing

Heinz-Hermann Erbe
Technische Universität Berlin, Institut für Berufliche Bildung, Franklinstrasse 28/29, D-10587 Berlin, Germany
Email: heinz.erbe@tu-berlin.de

Abstract: Agile Manufacturing is built around the synthesis of a number of independent enterprises forming a network to join their core skills, competencies and capacities to be capable of operating profitably in a competitive environment characterised by unpredictable and continually changing customer demands. Central to the ability to form networks is a cooperative learning of all members of an enterprise. This is an understanding of a learning enterprise. The objective is to produce a solid framework or structure for organisational learning. The introduction of organisational learning in individual as well as networked enterprises in order to prepare for virtual enterprises is discussed. The establishment of continuing improvement of work and learning processes stabilise enterprises and are a precondition for a successful network.

Key words: Informal learning, Co-operative work, Networking enterprises

1. INTRODUCTION

The environment for enterprises has changed essentially in the nineties. There are various reasons:

- Globalisation, which entails more competition and therefore forces enterprises to become more productive and market-oriented.
- Intensification of cost pressure caused by outsourcing programs of the large-scale enterprises.
- Ever faster changing market-conditions - more flexibility is needed to cope with it.
- Decreasing "time-to-market" time.

- Turning away from Taylorism and ever faster development of new technologies, which means that enterprises, their owners and employees, have to learn how to work with it and how to maintain flexibility within their organisation to integrate these new technologies.
- In industrialised countries, additional constraints from legislation like taxation and environmental care have increased.

Many enterprises, and particularly small and medium ones, in the United States as well in Europe are affected by these changes. Consequently, many firms not capable of coping with these changes have gone bankruptcy within the last couple of years. The question is therefore how to prepare these enterprises to be strong enough to survive this era of new competitiveness. Why are well-established enterprises failing in spite of innovative ideas for innovative products as well as organisational structures? What can be done to enhance their competitiveness long term?

- It turns out that the only long-term competitive advantage is the capability of enterprises to learn continuously for continuous improvement - including management and shop floor in order to become flexible for the changing environment.
- It has shown that flexibility, the ability to cope with and adapt to rapid changing quickly, is a core success factor, which needs to be improved.
- Forming networks of enterprises (extended enterprises) for an agile manufacturing, although they are still independent and competitors.

Agile Manufacturing is understood here as the synthesis of a number of independent enterprises forming a network to join their core skills, competencies and capacities to be capable of operating profitably in a competitive environment of continually, and unpredictable, changing customer demands. They do not own significant capital resources and that will help them to be agile.

Central to the ability of enterprises to form networks is besides trust building among each other and cultural attributes (FREITAG, WINKLER, 2000) the deployment of suitable information and communication technology (ICT) and the deployment of nimble organisational structures to support highly skilled, knowledgeable and empowered People. Networked task processing needs collective competencies. NULLMEIER (1999) discusses how ICT can promote or hamper networking and developing a collective competence. The implementation of information and communication technology top-down do not guarantee for necessary organisational change (BOEKHOFF, ERBE, 1999). However, a prerequisite for networking, supported or not through ITC, is the ability for cooperative work (WEBER, 1999). A collective objectivizing how to solve tasks by team-working, to build a collective mental model, is not natural for individuals, and has therefore to be trained, firstly in an individual enterprise. The step to networking cross-border of enterprises is a qualitative leap regarding team-working. The objectivizing of tasks is then mostly not possible face-to-face but instead guided by ITC. Here a suitable ITC platform comes into play together with an organisation preserving the autonomy of the cross-border team. Both can be interpreted as a learning-process, the sculpting of learning networked enterprises as an extension of the well described learning organisation (WATKINS, MARSICK, 1993).

2. TRANSFORMING TRADITIONAL ORGANISED ENTERPRISES TO AGILE MANUFACTURING IN NETWORKS

2.1 Cooperative Work and Learning

Economy interprets cooperation as all kinds of collaboration in industry and commerce, particularly within the national economy, based on the division of labour. Cooperation aims to increase the competitiveness of different and independent enterprises. Therefore strategic alliances, networks or extended enterprises develop, and in the end, virtual enterprises (SCHUH et al. 1998).

Psychology interprets cooperation as the individual benefit of collaboration and the tasks and objectives enhancing the likelihood of common work.

Sociology stresses the common values and objectives as a presupposition of collaboration and how to organise it. Cooperation can be considered as a benefit maximising, egocentric perspective, or with an aspect of mutual optimising the benefits. GLANCE, HUBERMAN (1993) discuss in this context the dynamics of social dilemmas. Consider a group of friends arranged to meet in a fine restaurant with an unspoken agreement to divide the bill evenly. What do each individual order? Selfishly the best to maximise her benefit or cooperating for the long term common good? It depends of course how often the same group will come together in the future. GLANCE, HUBERMAN (1993) found, with borrowing methods from statistical thermodynamics, a meta-stable equilibrium point when the number of cooperating members in the group is small. When the number increases, a highly unstable situation occurs until it settles in a true equilibrium point. These and further results of their considerations suggest practical ways to restructure organisations, and in particular, networks of small enterprises to secure cooperation.

WEBER (1999) considers cooperative acting in the context of work and calls it an exchange on equal terms between free and independent individuals. Pro-social acting means the support and encouragement of the other. Mutual representations are considered as indicators of a cooperative willingness to act. The members of a manufacturing team put the existing individual knowledge and the new obtained experiences cooperatively in an objectivized shape when coping with their tasks. Cooperative acting yields in a qualitative leap respecting the work organisation, for the individual as well as for the overall economic result. A collective leeway for planning and decision making, the complexity of a common task and the dependence of the different individually executed jobs, foster a common orientation of the tasks.

Innovation is not only the ability to combine existing experiences, knowledge and technologies effectively in order to develop products and services to put them on the market, but also the ability to change the organisation appropriately to the demands of the market.

Formal as well as informal learning within task-solving is not only to be understood solely as training for each individual staff member but also as a mean to develop the capacity for innovations within the organisation.

Continuous innovation requires a continuing education, continuing learning - this is where the leaning organisation comes into play. Not only at the management level one has to learn continuously, at the shop floor level this continuous learning is particularly important, particularly as the knowledge about the manufacturing technology is present here.

The following aspects have to be considered: actually there is individual learning (formal and informal, Figure 1), but how to shift to organisational learning and further to get from organisational learning to a "network-learning"? The subjects of learning are the individuals in an enterprise; organisational learning is a qualitative leap from individual learning: it is the ability of an organisation or enterprise to gain insight and understanding from experience through experimentation, observation, analysis, and a willingness to examine both successes and failures.

Learning in Organisations (SENGE 1990; WATKINS, MARSICK 1993) suggests an improvement of the learning environment, a delegation of competencies nowadays owned by the management. Furthermore, responsibilities have to be delegated in order to achieve more flexibility and to improve the quality of processes, products and services.

Attempts to implement such environments have been already tried, but (almost) exclusively in big international companies (YORKS et al. 1999), whereas the small enterprises often are financially unable to afford consulting services.

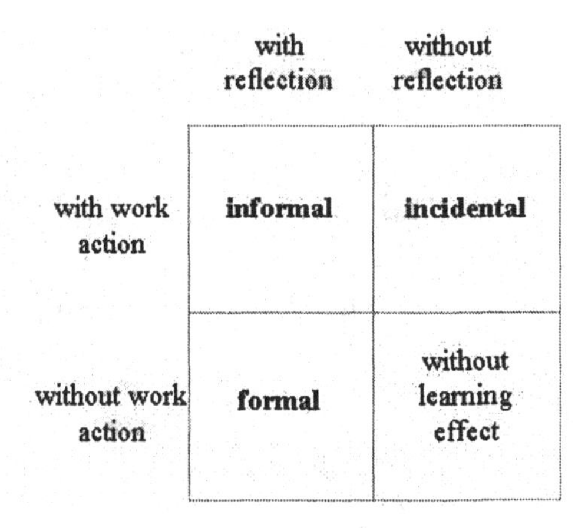

Figure 1. General aspects of learning

Figure 2 defines learning steps inside an individual enterprise. These steps were carried out in a small enterprise of model-making for sand casting. The challenge was to cope with new customer demands on sophisticated model-making using CAD-data the customer would transmit on-line. New workers trained for working with CAD-data and numerically controlled machines had to be integrated into the existing workforce of highly experienced handicraft model makers. Therefore one decided to learn for a cooperative work. A facilitator from outside fostered the learning process inside the enterprise (ERBE, KIM 1998).

Not only here but mostly in enterprises after some time the question arises: How to keep the knowledge and experience at the shop floor level? The current problem is that incentives within enterprises are forcing good and skilled workers to move vertically in the organisational structure, which means they are "encouraged" to leave the shop floor level. But since the view is taken that innovations need to take place at the shop floor level, these employees need incentives to stay.

2.2 Cooperative Work in Customer-Supplier-Networks

Networks between suppliers and between suppliers and customers should be designed as platforms, not as chains. A part of a chain can break and destroy or severely damage the whole network. The metaphor "platform" means that all enterprises in the network have access to all relevant information, regarding the order of a customer, the work-capacities of the partners in the network and their specific core competencies in order to discuss the distribution of tasks accordingly.

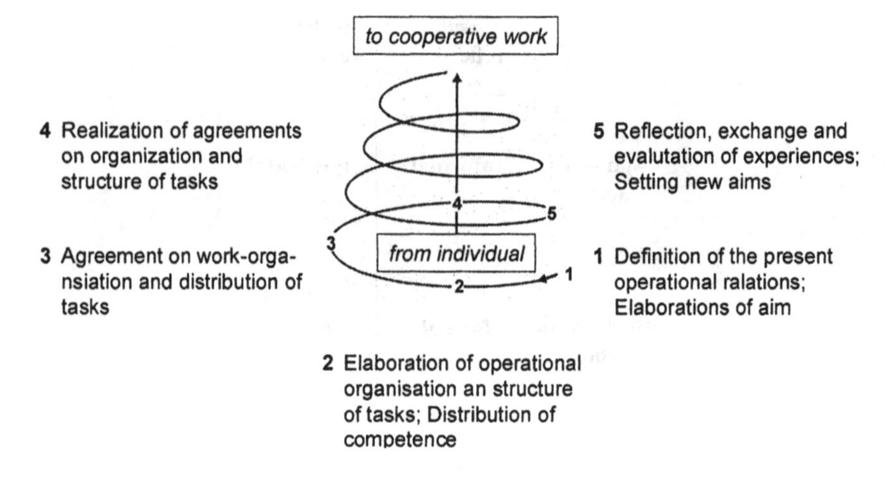

Figure 2. Learning steps from individual to cooperative work of management and shop floor members of an individual enterprise (Source: ERBE 2003, p. 270)

At first the management of the individual enterprises have to understand or to analyse the advantages and/or disadvantages of networks. Secondly they have to understand that all personnel of the enterprise have to be involved. Then, a lot of mental barriers must be overcome in order to make networks work effectively. Learning processes have to be developed wherein all employees are involved. It needs to be found how this will effectively work in view of the particular conditions confronting small enterprises.

Respecting customer orders not all partners are necessarily involved in processing them. That depends on the particular order regarding capacities and special equipment available. Therefore sometimes only few enterprises of the network are generating a "virtual" enterprise for processing the customer order. Figure 3 illustrates the generation and dissolution of a virtual enterprise based on a stable platform of networked enterprises (SCHUH et al. 1998).

The virtual enterprise executes cooperatively the tasks belonging to the order of a customer, respecting the core competencies of the networked enterprises and using computer networks of information and communication technology. These technologies should be customised to the specific needs in networks. CAMARINHA-MATOS, AFSARMANESH (2000) discuss cooperative systems as a set of autonomous agents (computational and human) interacting which each other "through sharing their information, decision making capabilities and other resources, and distributing the corresponding workload among themselves, in order to achieve common or complementary goals". They emphasise the social and organisational issues and their consequences at technical levels, the importance to identify the consequences in terms of the working structure, the processes and the roles played by all social intervening actors. As small and medium sized enterprises are characterised "by a strong human-centred decision making philosophy, strong feelings of autonomy and information privacy" it is the challenge to overcome these cultural attributes to a new culture of willingness to build confidence and "a new idea of ownership that goes beyond the borders" of each participating enterprise within the network. BOEKHOFF, ERBE (1999) stress, that besides the implementation of technology, the success of an enterprise and networks as well, is determined mainly through the capability of management and workforce for cooperative work. It is a misunderstanding to assume that technology is an external force that would have deterministic impacts on organisational properties and structure.

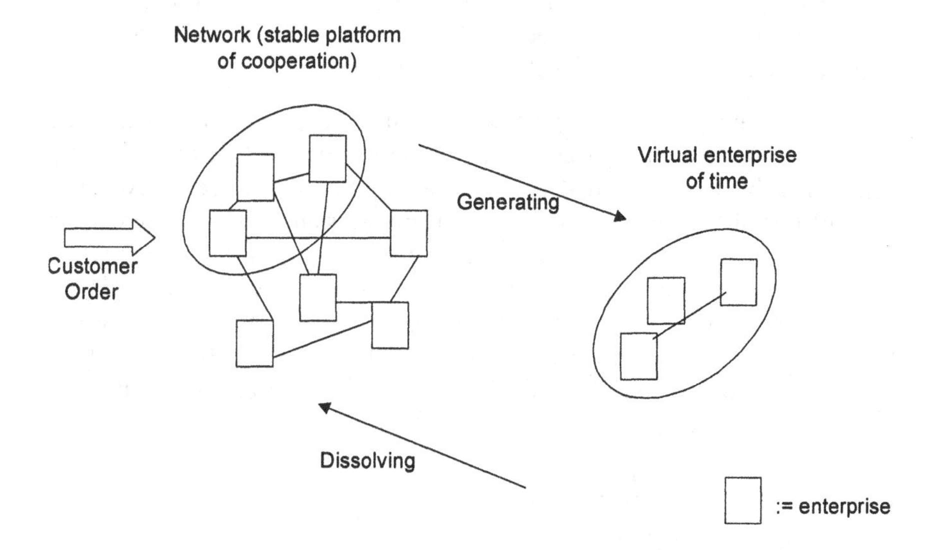

Figure 3. Generating a virtual enterprise from a stable network

2.3 Learning to Work within Networked Enterprises

Networked enterprises can be achieved through a cooperative learning using business processes crossing the border of individual enterprises, and leveraging knowledge and experiences across boundaries by learning how to empower People. Tasks of the workforce have to be enhanced and not restricted to well defined set of core competencies. That hampers the success of cross boundary teams. Perfectionism needs to be avoided as it acts as creativity brake. Three levels of learning have to be considered:

- *Learning level (1):* within an enterprise: preparing cooperative work crossing the border of the individual enterprise through training of cooperative work of management and workforce within the individual enterprise.
- *Learning level (2):* within the stable networked enterprises: educating basic qualifications for tasks in networked enterprises; exchange of knowledge and experience crossing the border between professions and enterprises (dynamic modelling of business processes); scenarios of cooperation for learning.
- *Learning level (3):* concerning the virtual enterprise (real, border crossing business processes): considering actual business processes; dynamic assignment of competence respecting the actual demands of the market and the customer for quality, cost and delivery time; quick solving of occurring problems through direct cooperation of the involved personnel; the management level will be involved in problem solving if and only if solutions are not possible within the given frame.

After finishing the level (1), teams of members of management and workforce of networked enterprises will be set up for the levels (2) and (3). Personnel involved in the actual border crossing business process have to be informed on all details, have to understand all tasks and must be able to carry out all due tasks. The restrictions to tasks belonging to the core competence of the respective networked enterprise, where a team member is employed, hampers the success of a virtual enterprise. At least every involved person must have an overview of the sequences of the actual business process to make a cooperative work effective. The following learning goals are connected to the levels:

- independent thinking and flexible acting;
- knowledge and skill which extends across the profession, the department and the specific enterprise;
- cooperating within the specific enterprise and cross-border of enterprises in task related teams;
- thinking entrepreneurial and acting responsible on all levels of the enterprises;
- indicators for growth – enterprises within a global environment;
- flexible cooperation – working in virtual structures;
- quality management of networked and virtual enterprises, QM systems cross-border of enterprises;
- implementation and effective use of communication and information technology;
- logistics among cross-border of enterprises.

This concept was used to establish and support a network of 10 metal-processing enterprises in the surrounding of Plauen, Saxonia, Germany. The network was called "Maschinenbau Vogtland (MAVO)". The spiral of Figure 2 served as the learning model, first to promote cooperative work in individual enterprises and thereafter to foster the step from network to virtual enterprises.

Staff members of the Innovation Centre Plauen as external facilitators kept the process going (ERBE, KIM 1998). Other experiences with establishing networks are reported. The Berkshire Plastic Network, Pittsfield, MA, was established in 1986. It is a consortium of more than 30 independent companies, representing virtually every discipline in the design and production of moulds, components and plastic products. The mould-makers of the Berkshire region were short on qualified workers. Therefore they decided to launch an apprenticeship program. This fostered a trust building between the then involved small shops. Now the network is organised as a wheel with the members as spokes. If one spoke breaks the network will remain stable. The network runs a small office where customer requests come in. All network members get the information regarding the request, and they can bid to get the order, mostly together with part of other bidders, forming a virtual enterprise. One of the enterprises will serve as the contractor, the others are the subcontractors. The office with a president (an owner of one of the member enterprises) serves as a facilitator to keep the network stable.

3. CONCLUSION

The learning process proceeds in loops and passing one loop completely describes one complete action. Based on the experiences of the former loops, a learning process starts, which leads to a permanent development of not only a personal but also a collective mastery. Metaphorically speaking, the learning loops will move up like a helix. This process is only carried out by the individuals involved. But after implementing such a process, individuals may leave, the process will not collapse. This is the aim of the learning process within an enterprise, which combines the individual with the collective.

As was expected, apart from the Innovation Centre, also the participating university got a better understanding of requirements of small enterprises, because in its teaching contents and teaching structure it is often far away from economic needs in these enterprises. So far it was a goal to develop new forms of cooperation in continuing education.

The technical support of cooperative work in teams is insufficient until now. Criteria for structuring planning- and decision making- support usable by teams are missing. For teams of networked enterprises there is no technical support at all. It should be analyzed if standardised communication- and information technology (internet, intranet, etc.) could be adapted for the specific needs of networked enterprises or if a customised technology has to be developed.

Missing is as well a structured help for the documentation of evaluated results of token decisions to support the learning process and to disseminate experiences. This can help to develop competence in teams.

REFERENCES

BOEKHOFF, H.; ERBE, H.-H.:
Organisationales Lernen: kritischer Erfolgsfaktor für virtuelle Unternehmen?
In: Industrie Management,
Berlin, 15(1999)6, pp. 13-16.

CAMARINHA-MATOS, L. M.; AFSARMANESH, H.:
Cooperative Systems. Challenges in Virtual Enterprises.
In: Proceedings Esprit project 22647, Prodnet II.
1999.

ERBE, H.-H.; KIM, J.:
Das Lernende Unternehmen, berufsbegleitende und berufsübergreifende Weiterbildung in Fa. Modellbau Roth GmbH.
Internal Report.
TU Berlin, 1998.

ERBE, Heinz-Hermann:
Learning for an Agile Manufacturing.
In: Human Aspects in Production Management.
Eds.: ZÜLCH, Gert; STOWASSER, Sascha; JAGDEV, Harinder S.
Aachen: Shaker Verlag, 2003, pp. 267-273.
(esim – European Series in Industrial Management, Volume 5)

FREITAG, M.; WINKLER, I.:
Mechanisms of Coordination in Regional Networks.
In: Preprint 7th International Conference on Multi-Organizational Partnerships and Cooperative Strategy.
Leuven, 2000.

GLANCE, N. J.; HUBERMAN, B. A.:
The outbreak of cooperation.
In: Journal of Mathematical Sociology,
London, 17(1993)4, pp. 281-302.

NULLMEIER, E.:
Personalentwicklung für informationstechnisch vernetzte Arbeit.
In: FH-Technik & Wirtschaft-Magazin,
Berlin, 2000, pp. 141-144.

SCHUH, G.; MILLARG, K.; GÖRANSSON, A.:
Virtuelle Fabrik – Neue Marktchancen durch dynamische Netzwerke.
München: Carl Hanser Verlag, 1998.

SENGE, P.:
The Fifth Discipline: the art and practice of the learning organization.
NewYork: Doubleday, 1990.
WATKINS, Karen E.; MARSICK, Victoria J.:
Sculpting the learning organization.
San Francisco, CA: Jossey-Bass Publishers, 1993.
WEBER, W. G.:
Kooperation in Organisationen unter arbeits- und sozialpsychologischern Gesichtspunkten – vom individual-utilitaristischen zum prosozialen Handeln?
In: Kooperation in Unternehmen. Sonderband 1998 der Zeitschrift für Personalforschung.
München: R. Hampp Verlag, 1998.
YORKS, L.; O'NEIL, J.; MARSICK, V. J.:
Action Learning.
In: Advances in Developing Human Resources.
Ed.: SWANSON, R.
Baton Rouge: Academy of Human Resource Development, 1999.

Competency Development in Distributed Work Environments

Pamela Meil and Eckhard Heidling
Institute for Social Science Research (ISF), Jakob-Klar-Strasse 9, D-80796 Munich, Germany.
Email: pamela.meil@isf-muenchen.de

Abstract: New company strategies are characterised by fluid company boundaries in which the integration of production and service processes take place outside of its borders. These open structures lead to increasing forms of distributed work, often organised into projects. This paper examines the shift in competency requirements associated with distributed work, and the individual and organisational measures that are necessary to assure their development and retention.

Key words: Geographically distributed work, Competency requirements, Critical situations, Experienced-based learning

1. INTRODUCTION

Today stringent customer demands, accelerated time-to-market schedules, and high levels of quality have put companies under pressure to adapt and modernise their product development processes. The ability to succeed over the long term on global markets depends in large part on keeping up with dynamic and constantly changing conditions. One way to meet these challenges is the organisation of product development and production processes in cooperative, often international, company networks which can react flexibly to new developments on the market. A central characteristic of this new type of work organisation can be seen in the increasing appearance of forms of "distributed work". Distributed work can be understood as the organisation of work across tasks, process chains, or production/service networks. In distributed work, employees from different departments, sites, and often countries, cooperate on a single task, a chain of tasks, or a network of tasks. One of the most important forms that distributed work takes is cooperation in temporary project teams, often supported by the use of modern information technologies. (DiMAGGIO 2001)

A number of new dimensions enter the work process under conditions of distributed work which require companies to organise the creation and transfer of know-how and competencies differently. The question arises as to how to develop competencies necessary for international project work, how to bind them together with existing skills, and how to use them effectively in processes of product development. These issues were examined in 3 companies from the automobile, aerospace and automobile supplier industries, which all organise their development processes in distributed project work. Based on research undertaken at these companies during recent years, this paper addresses following three issues:

1. how competency requirements have shifted as a result of distributed work, and which new competencies are necessary to carry out international project work;
2. what new demands are placed on the project participants and especially the project leaders in the framework of distributed work; and the difficulties involved in developing the appropriate competencies in existing organisational contexts;
3. recommendations for a new content and organisation for the development of distributed work competencies.

2. NEW COMPETENCY REQUIREMENTS IN DISTRIBUTED WORK

Distributed work is characterised by its complexity. New work situations, broadly defined tasks, changing actors and social frameworks – all in a limited time frame – make up the world of distributed work. Considering the contrast to traditional company-bound forms of work, it becomes clear that in such open work processes, a wide range of new demands arise for the work content and the competence development of engineers and other skilled workers.

In these forms of distributed work one of the major challenges is to utilise all of the competencies that are available in a process chain or a production or service network. There are two difficult aspects to this challenge: One is the shift in both the *intensity* and *diversity* of competencies that are required for distributed work environments. The other is activating very different types of competencies at different phases of the process.

2.1 Shifts in Intensity

Interviews in the 3 companies in our study revealed that a shift in competency requirements had occurred in a number of dimensions for project work in engineering and development. Table 1 lists five different dimensions of competency categories. The left hand column gives examples of the expectations for engineers that were characteristic of the organisational structures for product development prevalent up to now. The right hand column shows the shifts in *intensity* in competency dimensions that have taken place as a result of distributed work organised in international projects.

In the *technical dimension*, there continues to be a demand for high levels of technical expertise. However, what is changing is the additional requirement for system integration and monitoring the contributions from partners. This means that there is a need for a much broader range of technical knowhow outside of individual specialties, or even company boundaries. In the area of communication, the competency shift involves the ability to prioritise and filter large amounts of information to focus on the most significant parts. Also required is a more far-reaching coordination and understanding across fields, as well as the ability to negotiate, sometimes in conflicting situations.

Organisational competence in the era of distributed work reveals a growing complexity for work plans and work packages as well as increased difficulty in putting teams together to carry out the work plans. Project leaders are responsible for their project result and progress, but they have no authority in the traditional line organisations and hierarchies that still exist in companies. Therefore to organise the capacities they need, they often have to engage in quite a bit of convincing and diplomacy. Interview partners referred to the process of acquiring a team to organise the work as „horse trading." This term describes the informality as well as the difficulty involved in getting the capacity necessary to carry out work plans. This type of team formation presents a real challenge for the everyday work of project leaders in distributed work.

Particularly in the area of *management competence,* the project participants, and especially the project leaders, have experienced a large shift and intensification of their competence profile. Generally speaking, distributed work makes it necessary for project leaders to make decisions, take responsibility for fulfilling goals, and for disciplining project members much more autonomously than before. At the same time, this leadership role and risk taking is expected without the corresponding authority of hierarchical position. Thus formalised or ritualised procedures are thus replaced by informal interactions and more instinctual behaviours.

Table 1. Shifts in Competence Intensity

Traditional	New
Technical Dimension	
Coordination and execution of inner company processes	Inter-company system integration; monitoring partner contributions
Communication Dimension	
Formal Documentation; Consultation inside company	Prioritising material for overview; Consultation between companies
Customer presentations	Media competence
	Conflict negotiations
Organisation Dimension	
Design and formulation of work packages	Design and formulation of complex work packages
Team construction	Team construction under tight personnel availability
	Inter-departmental coordination of teams
Management Dimension	
Meet schedules	Heightened time and cost pressures
Monitor work content; Competence to evaluate work as basis for decision-making	Monitor work content and costs
	Ability to assert oneself in conflict situations; Conflict resolution; Risk taking
	Leadership without hierarchy
Intercultural Dimension	
Technically oriented international exchange	Intensive exchange beyond the technical dimension
Limited number of actors	Diverse actors
	Long and more frequent foreign assignments

Finally, more *intercultural competence* is demanded in distributed work processes. This does not mean merely learning the habits and qualities of other nationalities as broad generalities. For distributed work, a shift and increase in intercultural competence means having openness and flexibility in interactions taking place at the level of the work process and dealing with the varying bodies of knowledge and experiential backgrounds as well as the occupational and cultural differences that make up the cooperative base of distributed work. What emerges is an understanding that finding solutions to problems is not only an objective process, but also a negotiated one.

In effect, the intensification of international cooperative work leads to an interaction of different company and occupational structures which creates the need for the development of new working and social identities, at least in the context and duration of the work project. In order to carry out the diverse tasks embedded in project work, the participants, and in particular the project leader, has to find a way to manage and use the distributed competencies found among the employees.

2.2 Shifts in Diversity

Related to this point, one of the most challenging aspects of distributed work, particularly in a leadership position, is being able to activate different types of competencies in the various phases of a project at the appropriate time. (see Figure 1) To demonstrate what is involved in activating the competencies demanded in distributed work environments, it is necessary to look at the product development process more closely, in particular the various phases that a project passes through from the conceptual phase to the start of production.

Figure 1 show, from the example of a typical car development process that different stages in a project phase correspond to diverse demands for competency needs. At the beginning of a project, the participants shape and form the work process; they are in an *exploratory phase.* In this phase, the project group is engaged in a process of discovery and thus openness, flexibility, and creativity are called for. As the boundaries for the division of labour and delivery get fixed, a process of *negotiation* begins. At this second stage, the project leader is called upon to build a consensus, evaluate the various actors in the group, and attain a fair and effective distribution of resources. For all of the project members, there is a need to know when and how to compromise, while simultaneously recognising and representing company interests and constraints. In this framework, project participants have to find a basis for negotiating which brings them both closer to their goal while at the same time representing their particular interests. When particularly difficult problems arise or when negotiations break down, *mediation* is necessary. For the project leader, this requires the competency to resolve conflicts and keep the members goal-oriented. In the course of the project, a number of assumptions or plans will need revision. Given the very strong orientation in the *adjustment* stage often leads to critical situations. The project leader needs to have the ability to deal with changes and find new solutions. The final stages of the project involve a certain amount of *consolidation* of the project's many parts. This requires the competency to be goal-oriented, see the complete picture, think in process terms and obtain high performance levels from project participants under extreme time pressure. Finally, *problems solving* is required during most of the project's running time. In project organisation, it is the participants who have to be in a position to make intelligent and timely decisions even in cases where information or data is incomplete, contradictory, or communication is conflicting (HINDS, KIESLER 2002).

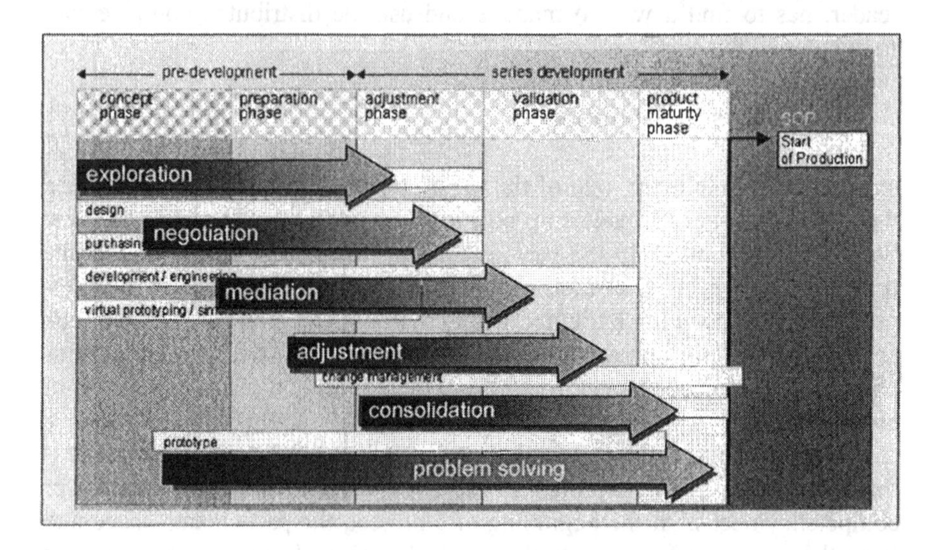

Figure 1. Product phases and competency needs

3. COMPETENCIES BEYOND BOUNDARIES

An important factor related to new forms of distributed work is how knowledge gets generated and transferred. Knowledge is created in concrete, spatially bounded learning contexts and work processes. The implementation of knowledge, however, takes place in different spaces, both within and across company contexts. This leads to a complex interaction between knowledge production tied to a particular place and the exchange of results which takes place across large distances. Out of this situation, a fundamental tension arises in which competencies are generated over a long period in territorially and socially embedded systems, but are used and exchanged in temporally limited and spatially disparate dimensions that go way beyond company boundaries. Projects have a key role in this exchange process, because they represent the links between the inside and outside. Thus, projects become the pool where newly reintegrated, quasi-connected pieces of knowledge come together. In this way, project participants take on an important linking function in regulating knowledge and information flows between the different company sites (HEIDLING et al. 2004).

3.1 Opening the Container Model

One of the difficulties for competency development in processes of distributed work is that the traditional way in which skills are acquired and experience is developed occurs in a "closed" space (MEIL 2000). Highly skilled employees are products of nationally-based institutions of training and further training where they acquire their formal skills and specific areas of expertise. Then, the continued development of knowledge takes place within the company, usually within hierarchically organised functional departments. We call this form of competency development the "container model" (see Figure 2).

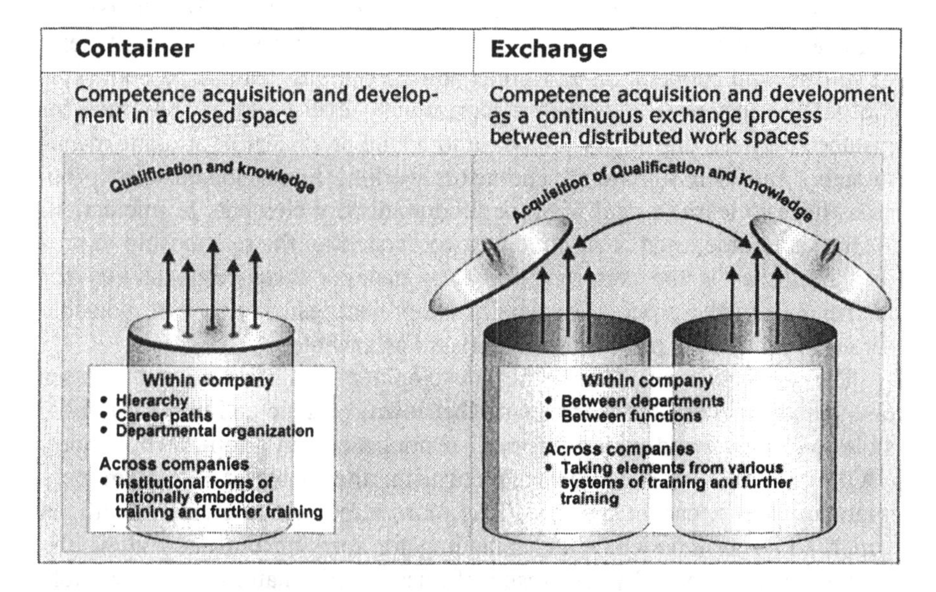

Figure 2. Container model of competency acquisition
(Source: MEIL, HEIDLING 2003, p. 182)

The way knowledge is generated in this closed system can be counterproductive for the requirements that face skilled workers and engineers in processes of distributed work. In distributed work processes, competence acquisition and development is a continuous process of exchange between departments and functional areas within companies. Across companies, it entails the exchange between different training traditions, cultural backgrounds and working habits. Therefore, the type of expertise and experience acquired in container type organisational structures does not provide the appropriate preparation for the dynamic, conflict-laden, broad-based knowledge and experience necessary in distributed work contexts.

3.2 Operating in Unclear Situations

Another critical aspect of operating in distributed work is the integration of different working styles, creating a new basis for interaction (BRANNEN, SALK 2000). In a project, new actors from different companies or departments, often with different cultural and training backgrounds, have to come together, and within a fixed period of time, have to complete something. "Normal" forms of hierarchy or authority do not exist in these systems, and the process can get very complicated due to the complexity of the products, and also due to conflicting interests. However, it should also be recognised that conflict is not the only medium of exchange, since the overall goal of the participants is basically the same and the technical basis for understanding is also the same. Thus, the interaction is characterised by a simultaneity of conflict and cooperation as well as power relations that can be both symmetric (for instance, in terms of hierarchical level), and asymmetric (for instance between customers such as automobile producers or airplane manufacturers and their suppliers). The actors working in distributed work processes have to learn to deal with these contradictory elements of interaction. At the same time, trust is an important component of the relationship in project groups as is the certainty that each member takes responsibility for his/her part of the process, especially the project leader who is responsible for ensuring that the project moves forward as a whole.

The complex network in which development and innovation processes take place in distributed work create their own dynamic and make it impossible to calculate and plan a priori all contingencies in a top-down manner. (BÖHLE, MEIL 2003) Moreover, recognising the source of a problem is no guarantee that it can be avoided. It is more important to create conditions which allow an interactive understanding for carrying out tasks along the process chain. Especially as a result of missing information and incomplete data, the employees themselves have to be in a position to make timely and informed decisions on their own. In these open and dynamic work processes, an entire set of new demands arise for the competence profile of the employees. One increasingly important demand lies in being able to act effectively in unclear situations.

When an incidence arises that deviates from pre-determined plans or problems occur that fall outside the boundary of eventuality, the situation in the development or production process moves from being typical or foreseeable to critical. Critical situations have a variety of causes ranging from external risks (last minute changes from the customer) to poor internal communication (contradictory priorities) or incomplete information (vague contractual details), problems of communication or leadership, or the unwillingness to try new solutions. For distributed work processes, critical situations occur so often that they take on a degree of normality. The company representatives expressed it such that: "critical situations are almost the norm – it is basically typical to have a critical situation."

Naturally, companies use every means at their disposal to learn from previous critical situations in order to plan better and take steps to prevent them (ANTONACOPOULOU 2002). Given that critical situations are often viewed as the outcome of a lack of clarity in the specifications that are provided, the response is frequently to employ technical tools and models to plan the control process to an even greater level of detail. Yet, despite a relatively sophisticated range of planning tools and models, and large financial investments in this process, the results are less than satisfactory. Especially in projects where innovative or new developments are involved, unplanned and critical situations arise frequently, and it can be precisely the wrong approach to try to plan in greater detail and leave less autonomy to the work process. Rather than preventing critical situations, this approach can have the undesired result of hindering inventive solutions. What is generally missed in purely technical orientations is the need to improve the experience and learning base of the project participants and particularly the project leader to deal with critical situations and to be oriented to constructive problem solving. This might sound trivial. However, the approaches that would apply to this type of learning contradict with strongly embedded views concerning planning against unforeseen contingencies as a solution, and a purely scientific-logical approach to problems and problem-solving (BÖHLE 2002).

4. RECOMMENDATIONS FOR DEVELOPING DISTRIBUTED WORK COMPETENCIES

4.1 Contents for Distributed Work Qualifications

Besides the usual technical expertise and "soft skills" (such as communication and cooperation) that are included in the training packages in companies today, there are three additional aspects that need to be considered to work effectively in a distributed work environment.

a) *Mediation and negotiation*: A fundamental problem at the centre of distributed work projects is that actors represent different departments or companies and therefore in certain situations have conflicting interests. When a technical product or problem the object of interaction is, it can easily seem that the grounds for discussion are neutral and the solution technically mediated. This can obscure the conflict that is surrounding the problem. Project leaders have to learn how to manage conflict and contradictory demands for interaction. The goal is to balance different interests and roles and to keep sight of a common end result while recognising the potential and source for conflict. Normal training trajectories and job experience do not prepare project leaders for this task.
b) *Process competence*: Process competence is a central element in distributed work because it is necessary for the acquisition and utilisation of knowledge and experience in interdisciplinary, inter-company and international work processes. Process competence guarantees that the various working steps and tasks of the various actors lead to the coherent development or construction of an end product. Naturally, given the extreme complexity of development processes, which are increasingly organised in inter-company process chains, the expectations for the project participants to have an overview and understanding of the entire process grow accordingly. An extremely important aspect for development work, which is highly conceptual, is the ability to make mental images of projects and processes. Project participants from various companies are responsible for one piece of a much larger product, whether it be a car or an airplane. They not only have to be in a position to envision the steps of development of their piece, but they have to foresee the integration of their piece in the total product. Thus, an important aspect of distributed work revolves around anticipation and openness, in contrast to more categorical and formalised ways of thinking. Openness is especially significant for distributed work because of the variety of different perspectives and ways of thinking that exist in a project group, and because the development process is not linear, but rather process-oriented. Anticipation is important because the end results can be years away from steps taken at a given time, steps that are nonetheless critical to the final result. Having process competence encompasses the ability to create analogies to past practice and experience, envisioning, in an abstract sense, steps to development, foreseeing the integration of the pieces into a total product or process, and being able to anticipate end results.

c) *Negotiated culture*: When working in distributed work environments, the working situation is characterised by its heterogeneity. Project participants or those feeding results into the project come, at the minimum, from different departments, and very often from different companies and different countries and therefore cultures. Through the interaction in mixed groups, new work identities emerge. To facilitate this process, the project leader has to achieve a common ground of compromise and understanding, open the project to create new experience spaces and develop a framework in which all group members are involved in a process of innovation. The openness and flexibility that this requires and the amount of negotiation, as opposed to the setting of objective criteria that is involved, is highly underestimated.

4.2 Instruments

One of the main messages of this paper is that the competencies necessary for distributed work processes are not effectively transmitted in existing training programs, embedded as they are in national and internal company contexts. Furthermore, the recognition that distributed work is highly complex and displays a large number of critical situations, has led to the unfruitful conclusion that only technical solutions and more planning can master the work process. The position here, however, is that it is precisely the subjective orientation and experience-based approach to work that has to be strengthened to carry out distributed project work, especially in the areas of development and engineering. Two instruments are needed to promote the type of competency development called for in distributed work: experience-based learning methods and an appropriate organisational framework to accumulate project work know-how.

Experience based learning methods possess the following characteristics:

1. They allow experience gathering.
2. They promote problem-solving.
3. They induce process competence and associative thinking.
4. They make it possible to learn from mistakes (without hurting company interests).

All of these components have to derive from learning methods or tools which stay close to the actual product development process and that are found in project structured distributed work environments. The tools can be scenarios, simulations, games, or self-organised learning methods. Important is that they are developed out of and linked to real work experiences. The more abstractly conceptualised the tools are, the less effective they will be.

The second pillar for accumulating work project know-how is an organisational framework that:

1. Creates career paths that are geared to accumulate experience and process competence.
2. Gives project leaders back-up support from superiors when negotiation processes require it.
3. Establishes forums for experience exchange and the opportunity (and time) to participate in them.

5. CONCLUDING REMARKS

For forms of distributed work to be effective, it is necessary to utilise the entire range of competencies that are available in a value chain or network of production or service processes. This involves designing the right kind of cooperation processes between individual employees or groups, who come from differing occupational and cultural traditions. It also involves generating qualifications which allow the development of a comprehensive "process" competence. The new competence profiles of project participants, and in particular project leaders, encompass the ability to negotiate, to react effectively in critical situations, and to create the basis for the emergence of new work identities. Thus, distributed work entails decisive challenges for company organisation and qualification, and requires new methods and instruments for the generation and maintenance of competencies.

REFERENCES

ANTONACOPOULOU, T.:
Time and reflexivity in Organisation Studies.
In: Organisation Studies,
Berlin, 23(2002)6, pp. 857-862.

BÖHLE, Fritz:
Vom Objekt zum gespaltenen Subjekt.
In: Subjektivierung von Arbeit.
Edts.: MOLDASCHL, Manfred; VOSS, Günter.
München, Mering: Hampp Verlag, 2002, pp. 101-133.

BÖHLE, F., MEIL, P.:
Das Unplanbare bewältigen.
In: Tagungsband der Fachtagung "Projektmanagement in Seiten des Wandels".
Eds.: BUTS, C.; PAPESCH, G.; WILHELMS, G.
Augsburg: Universität Augsburg, 2003, pp. 36-47.

BRANNEN, Mary; SALK, Jane:
Partnering across borders: Negotiating organizational culture in a German-Japanese joint venture.
In: Human Relations,
Thousand Oaks, CA, 53(2000)4, pp. 451-487.
DiMAGGIO, P.:
Introduction: Making Sense of the Contemporary Firm and Prefiguring Its Future.
In: The Twenty-First-Century Firm.
Ed.: DiMAGGIO, P.
Princeton, Oxford: Princeton University Press, 2001, pp. 3-30.
HEIDLING, E.; MEIL, P.; ROSE, H.:
Neue Anforderungen an Kompetensen erfahrungsgeleiteten Arbeitens und selbstgesteuerten Lernens bei industriellen Fachkräften unter Bedingungen verteilter Arbeit.
In: Bewältigung des Unplanbaren.
Eds.: BÖHLE, F.; SEVSAY-TEGELHOFF, N.
Wiesbaden: Westdeutscher Verlag, 2004. In print.
HINDS, P.; KIESLER, S. (eds.):
Distributed Work.
Cambridge, MA: The MIT Press, 2002.
MEIL, Pamela:
Blick über die Grenze – View across Borders: Approaches for Meeting New Demands for Skill in Different National Contexts.
In: Industrielle Fachkräfte für das 21. Jahrhundert.
Edts.: LUTZ, Burkart; MEIL, Pamela; WIENER, Bettina.
Frankfurt/M., New York, NY: Campus Verlag, 2000, pp. 129-157.
MEIL, Pamela; HEIDLING, Eckhard:
Competency Development in Distributed Work Environments.
In: Human Aspects in Production Management.
Eds.: ZÜLCH, Gert; STOWASSER, Sascha; JAGDEV, Harinder S.
Aachen: Shaker Verlag, 2003, pp. 180-186.
(esim – European Series in Industrial Management, Volume 5)

New Approach for Global Education - Simulating Supply Chains by Applying World Wide Web

Karl-Robert Graf[1], Siegfried Augustin[2] and Konstantinos Terzidis[3]

1) *Fachhochschule Karlsruhe, Fachbereich Wirtschaftsinformatik, Moltkestrasse 30, D-76133 Karlsruhe, Germany.*
   Email: Robert.Graf@fh-karlsruhe.de
2) *Schulstrasse 36/2, D-80634 München, Germany.*
   ProfAugustin@aol.com
3) *Technological Educational Institute of Kavala, Department of Information Management, P.O. Box 1194, GR 65404 Kavala, Greece.*
   Email: kter@teikav.edu.gr

Abstract: For a long time simulation tools have been used for implementing logistic functions and connections in teaching and vocational training (AUGUSTIN, GRAF 1995; Riis 1995; WIENDAHL 1996; LANG, JUNG 2001). In recent years, considerable progress has been made with regard to the following different ways and means used in logistics and computer science: Transformation of the logistics management into a fundamentally enlarged supply chain management, establishment of communication, set-up of business tools like B2B (Business to Business), B2C (Business to Customer), the Internet, e-commerce (Electronic Commerce), etc. for the efficient functioning of a supply chain, and development of new programming software for the open and distributed implementation of training modules in local and global networks. As part of a project, traditional aspects are going to be enhanced by current methods and concepts so that new interdisciplinary work methods can be applied in business networks. For this purpose the training medium "business simulation" is going to be brought to the state-of-the-art in computer, communication and network technology. Within the scope of a joint project of private industry partners (Siemens AG) and different universities (Montanuniversität Leoben, Austria; Technological Educational Institute of Kavala, Greece) a training concept is being developed, completing traditional logistics aspects (planning and control of intra-company material, purchase orders and information flow) by the previously mentioned new requirements.

Keywords: Logistics, Supply chain management, Simulation, Blended learning, E-Learning, Personal development, Web based training

1. INTRODUCTION

For a long time simulation tools have been used for implementing logistic functions and connections in teaching and vocational training (AUGUSTIN, GRAF 1995; RIIS 1995; WIENDAHL 1996; LANG, JUNG 2001). In recent years, considerable progress has been made with regard to the following different ways and means used in logistics and computer science (see Figure 1):

- Transformation of the logistics management into a fundamentally enlarged supply chain management,
- Establishment of communication, set-up of business tools like B2B (Business to Business), B2C (Business to Customer), the Internet, e-commerce (Electronic Commerce), etc. for the efficient functioning of a supply chain,
- Development of new programming software for the open and distributed implementation of training modules in local and global networks.

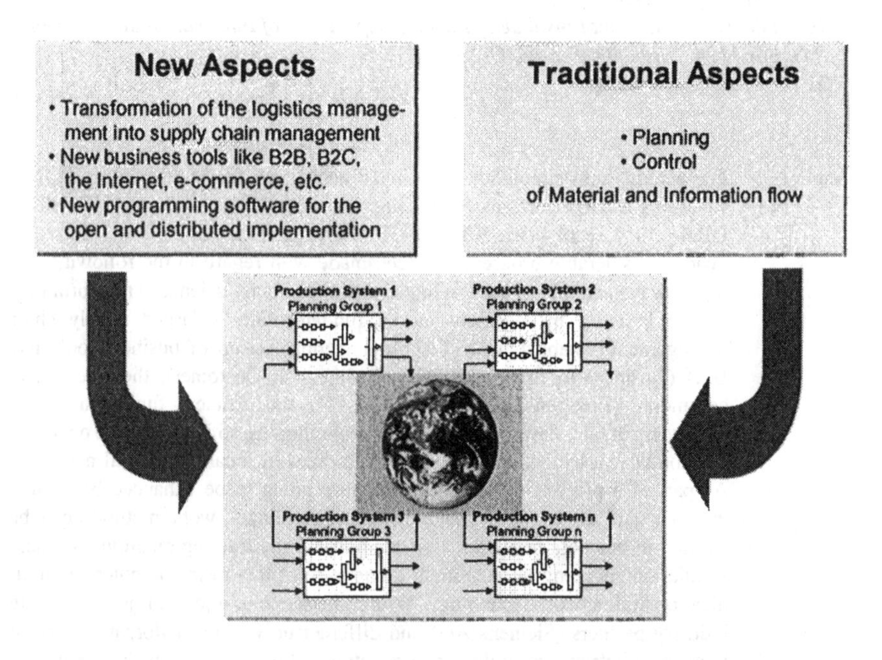

Figure 1. New aspects vs. traditional aspects

As part of a project, traditional aspects are going to be enhanced by current methods and concepts so that new interdisciplinary work methods can be applied in business networks. For this purpose the training medium "business simulation" is going to be brought to the state-of-the-art in computer, communication and network technology.

Within the scope of a joint project of private industry partners (Siemens AG) and different universities (Montanuniversität Leoben, Austria; Technological Educational Institute of Kavala, Greece) a training concept is being developed, completing traditional logistics aspects (planning and control of intra-company material, purchase orders and information flow) by the previously mentioned new requirements.

The goal of this project is the establishment of a simulated model system in which several value-added partners (located in various geographical areas) of a supply chain are involved (THALER 1999). The exchange of merchandise between those partners is going to be implemented via B2B, B2C, e-commerce, or similar solutions on a virtual market-place. The virtual market place and its conditions need to be set up in a simulation model that allows the value-added partners to plan and reorganize their businesses leading and interactions to the creation of an operating supply chain.

2. DESCRIPTION OF THE SIMULATION MODEL

The value-added partners shown in the simulated model system are planned by one training group each, so they can act and cooperate on a common virtual market, independent of their geographical location. This enables various training groups from different fields and universities to participate in the business simulation at the same time.

All groups are starting with an identical model company in which three products are manufactured. The in-house manufactured products consist of parts and components, and the purchased parts. The participating groups plan the production system simulated in the model (see Figure 2). The planning comprises:

- the release of purchase orders,
- production orders,
- production capacities for the working units and
- control methods.

At the starting point the participating groups work in an isolated model world. Parts to be purchased are obtained from an anonymous source of supply under various conditions. At the other side there is an anonymous sales-market with its primary and supplementary requirements. So the production units and planning groups design their system for markets that are almost nearly independent of each other (see Figure 3).

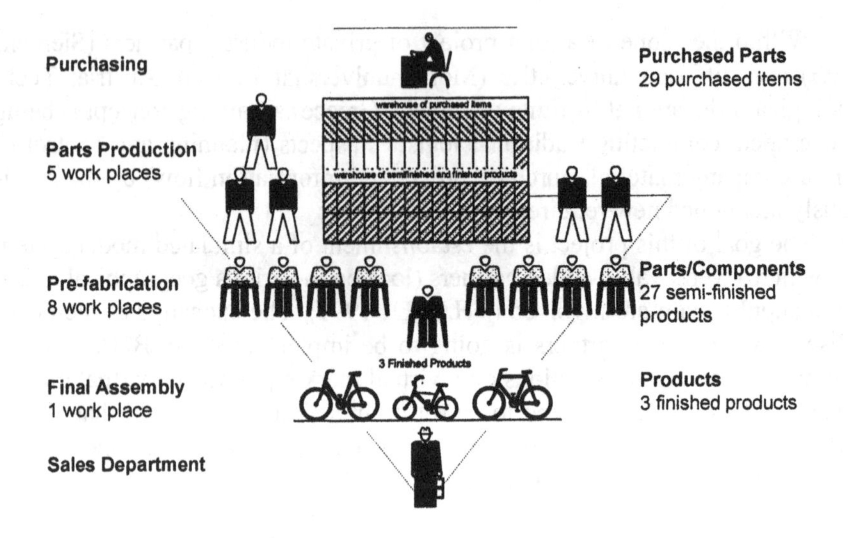

Figure 2. Elements of the production system

In the following, the groups have the chance to carry out successively their exchange of goods on a newly installed central market. The production units and planning groups design their systems on a common open market (see Figure 5).

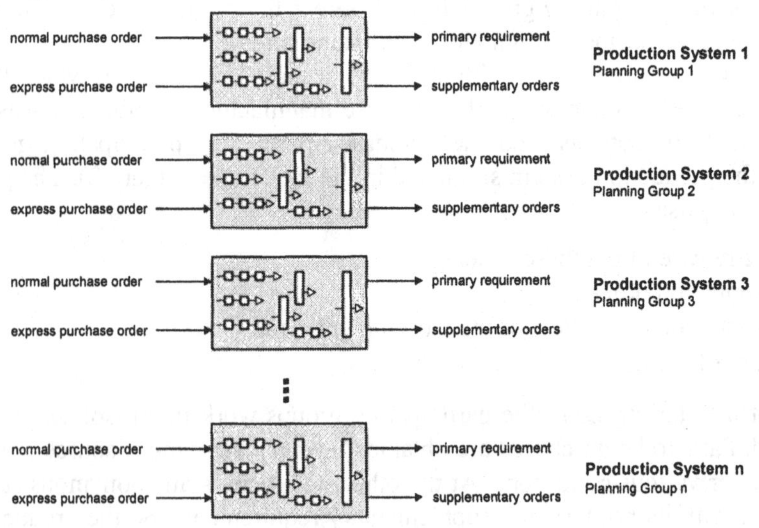

Figure 3. Isolated model structure: Traditional View of Logistics and Production Systems and the TCP Training Concept

In accordance with principles of Supply Chain Management, the procurement and distribution market is transparent and allows for flexibility new customer and supplier relations. Each group can act on the common market as a customer, or supplier for parts, semi-finished products, and final products (see Figure 4).

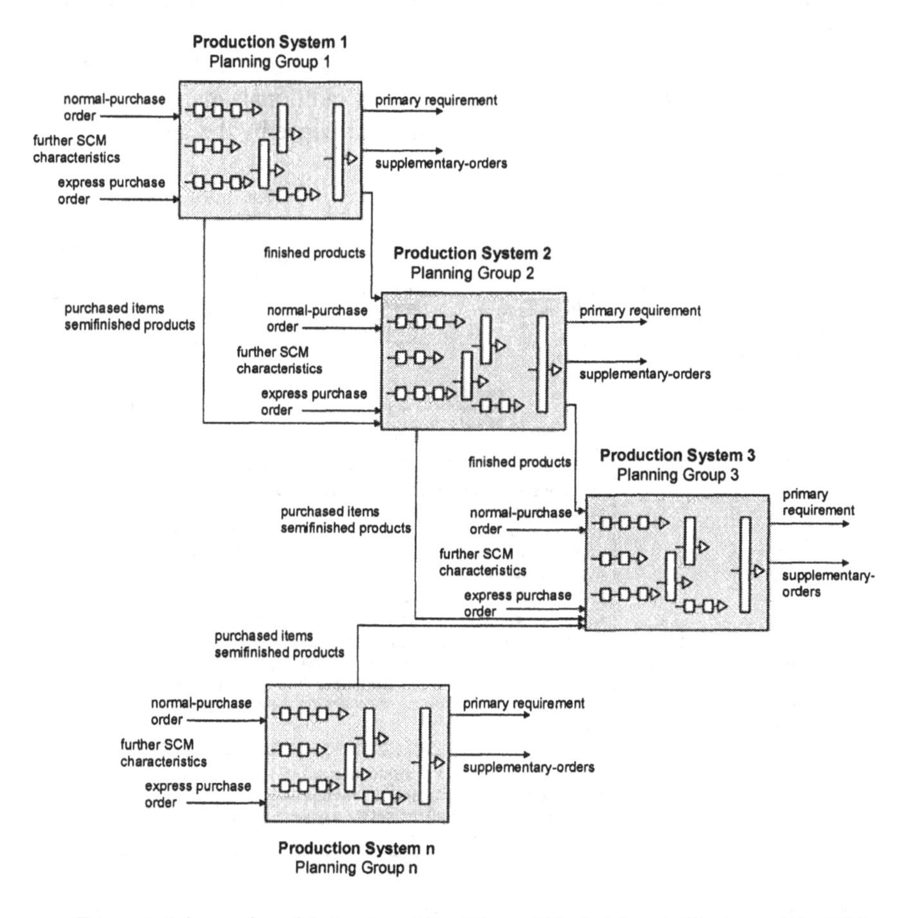

Figure 4. Advanced model structure: New View of Market Supply Chains and the SCS Training Concept
(Source: GRAF, TERZIDIS, AUGUSTIN 2003, p. 28)

Various characteristics within the newly established, Internet-based business relations are possible for the creation of customer-supplier relations (business relations). A concept for three characteristic stages has been set up (see Figure 5).

- Stage 1: *Business Relations 1 (Market Place) for Parts, Semi-finished and Finished Products:* Open to all suppliers and customers. Suppliers and customers alike determine the conditions according to their company situation and market observations.

- Stage 2: Business *Relations 2 (B2C, B2B, A2A (Application to Application) for Parts, Semi-finished and Finished Products:* The production units coordinate exclusive supply strategies. The conditions result from their company situation and the holistic formation of the supply chain to the advantage of all participants.
- Stage 3: *Business Relations 3 (Traditional Market) for Parts, Semi-finished and Finished Products:* Anonymous (traditional) market for suppliers and customers with somewhat limited influence by the groups.

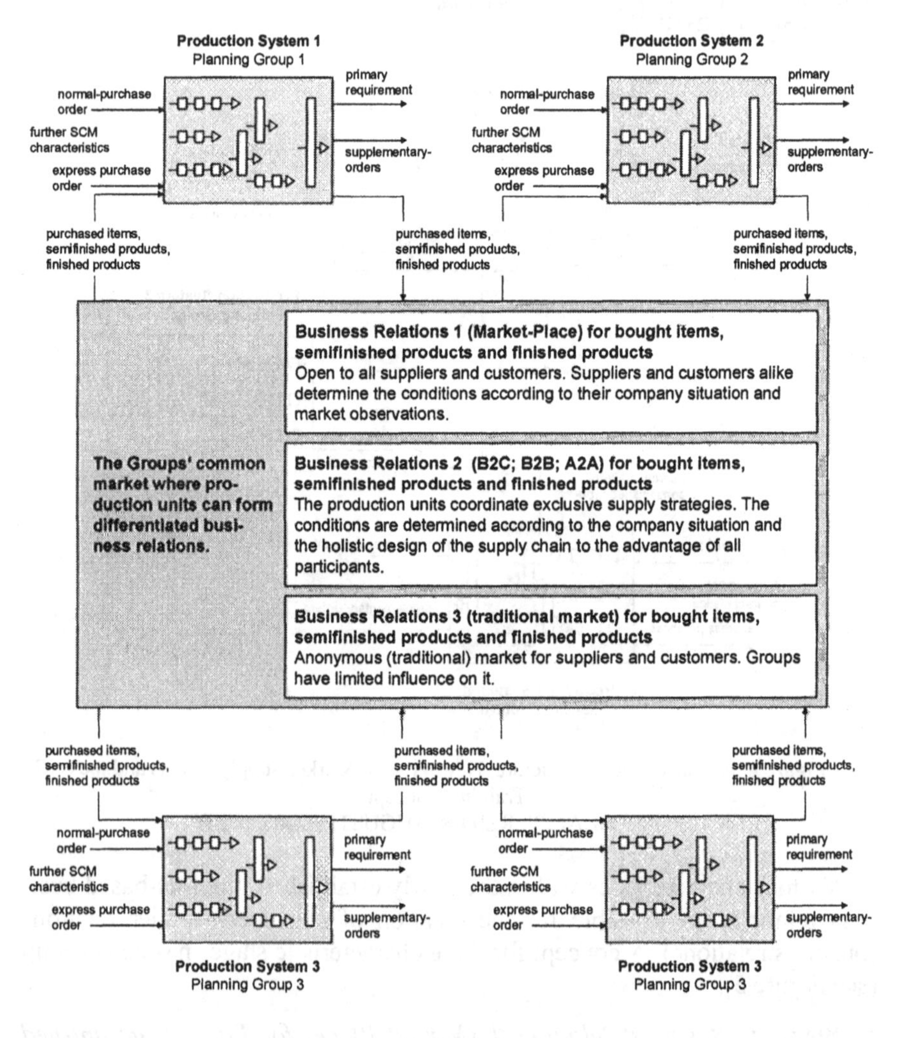

Figure 5. E-Business scenarios
(Source: GRAF, TERZIDIS, AUGUSTIN 2003, p. 30)

Business relations 1 and 3 have been technically implemented in the current system. Business Relations 2 are going to be realized in this time.

The traditional implementation of the business simulation was as follows: Regionally separate business simulations at various universities and locations showed a model design that was completely independent of the others. In the same way the different groups within a business simulation acted as suppliers and customers on an anonymous market and had very little exchange and contact with each other (see Figure 6).

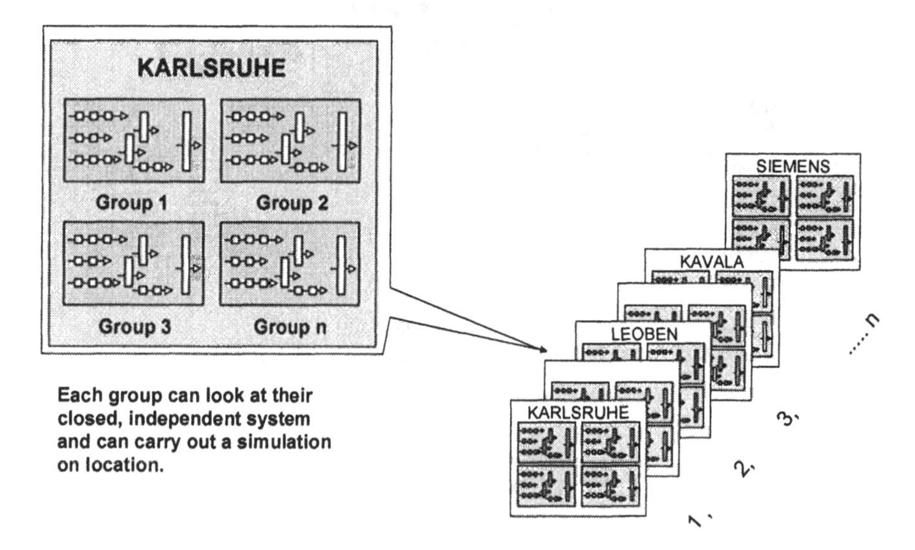

Figure 6. Proven structure

In contrast to the traditional model, Web integration between different business simulations and their participating groups open up completely new methods and possibilities. It allows the groups to act independently of their location - since the simulation of the model companies, as well as the exchange of products, take place via the Internet on a central server in a virtual environment (see Figure 7).

Each group, regardless of their integration into a particular business simulation, has the chance to participate on the common market. The time frame and location of a simulation game thus become irrelevant, as each time the groups that are currently present interact with each other group present on the market.

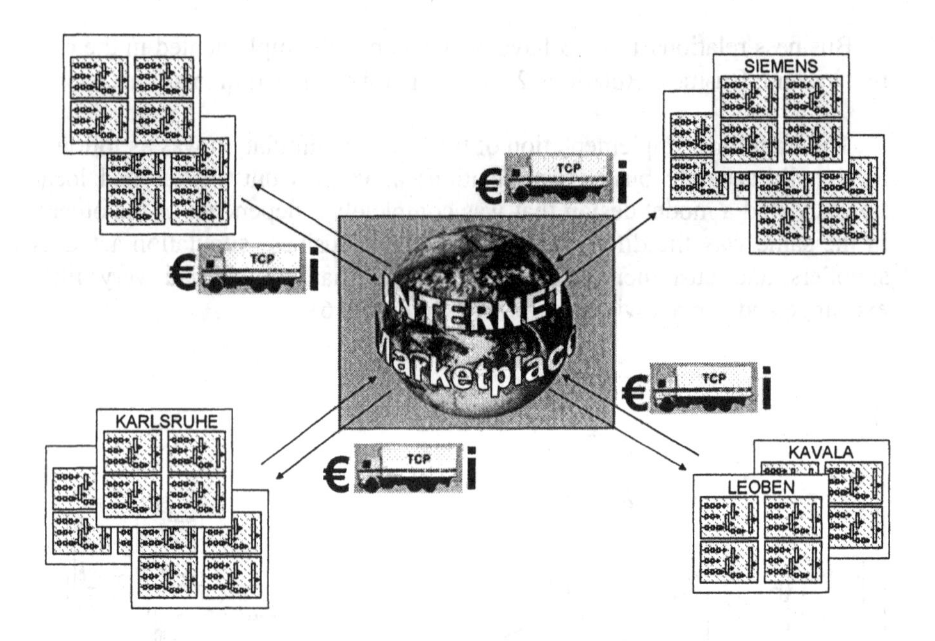

Figure 7. Web integration
(Source: GRAF, TERZIDIS, AUGUSTIN 2003, p. 31)

3. IMPLEMENTATION AND PROJECT STATUS

The core of this logistics training concept is a computer simulation of various industrial companies (value partners) for which a new simulation model, adjusted to the previously mentioned conditions, has been developed based on a well-established simulation concept (TCP - Training Centre Production (JUNG 2000; DGP 2004). The simulation of the production system and marketplace resides on a central application-server that participants can access via the Internet, or an Intranet and an interposed web server. According to the requirements in information technology, the implementation is carried out via the programming languages and scripting languages Java, HTML, XML, and SVG (see Figure 8).

The simulation is run via web portal. The currently available portal functions are (see Figure 9):

- data transfer via XML file,
- data input via Applet,
- simulation,
- user administration,
- market-place activities,

- result evaluation,
- simulation manual and simulation documentation,
- links.

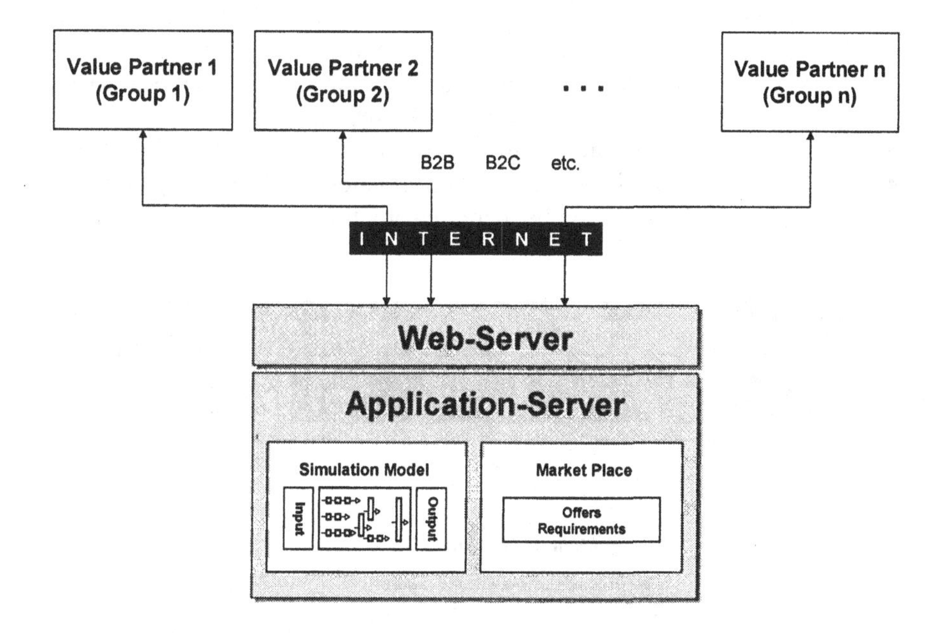

Figure 8. Simulated model system

Different language versions currently exist in English, German and Greek. All the information material necessary for the business simulation has been integrated into the web portal. The participants can access this material any time. The material includes: an electronic manual, all the forms necessary for the business simulation and field specific presentations.

4. CONCLUSION

In summary one can say that the idea to implement a global teaching and training concept with this new simulation system has been realized. The concept allows simulation and training in the field of extensive logistics methods and measures using new information technologies combined with new logistics concepts for supply chain management. Modern media and especially the internet, open and blend new opportunities for e-learning. E.g. as special training aspects by means of web integration can be mentioned:

- Creation of a dynamic cooperation among groups
- Groups are formed beyond the borders of universities and departments

- Training and work of the groups is independent of regional restrictions (internet connection necessary)
- Multilingual implementation is enabled via translated versions of the simulation tools
- Product and cash flow are simulated via the internet
- Information flow takes place according to the current methods on the free market and is trained in the business simulation
- Access to manuals, training materials, and assistance via the internet
- Web integration provides additional information for the participants
- Number of participants can be increased any time

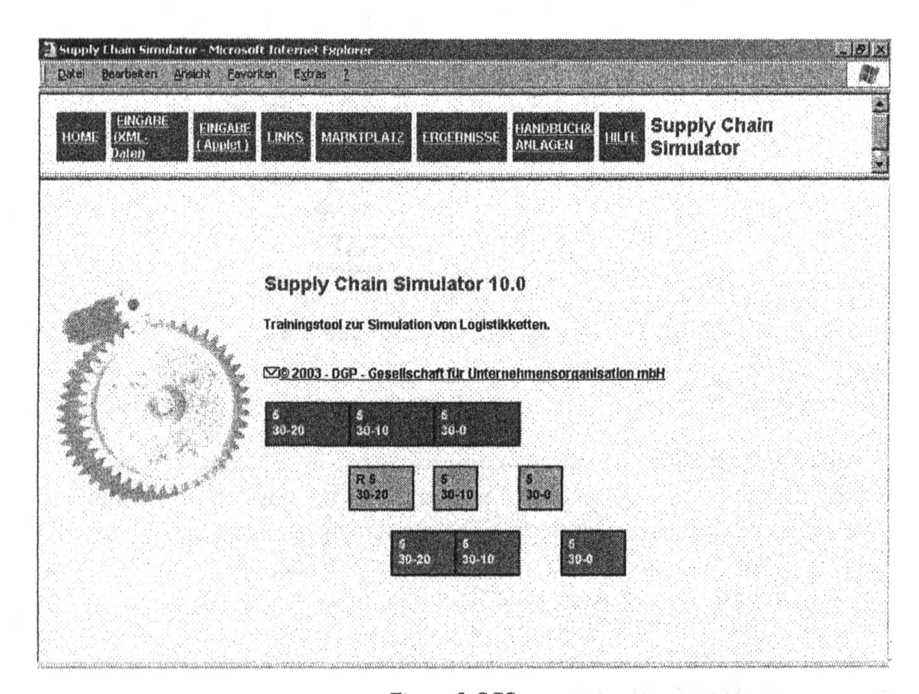

Figure 9. SCS access

The applied server technology with its possibility of a central simulation of production plants, combined with standard methods of Internet communication, allow, for the first time, a geographically independent exchange of purchased parts and components, semi-finished products, final products and funds. This enables an international and extremely realistic business simulation. The design and implementation of cash flow, the information logistics, as well as the contract conditions are based on the same methods and technologies used in "real life", whereas the manufacturing of semi-finished and finished products is simulated by the simulation system.

A continuous and global participation of educational institutions allows the creation of an international, simulated market. The multi-lingual versions of the concept and the cooperation of different universities in Germany, Austria and Greece brought this open concept to an international level for the first time. The first courses outside of the German-speaking countries were held at the Technological Educational Institute in Kavala (TEI), Greece.

However, not only universities all over the world might profit of such a concept. A LEARNTEC forum has shown particular interest in the problem of "Digital Divide", a program the UNESCO refers to as "Reach the Unreachables" with the goal to train people in less developed countries using modern media, such as, the new information and communication technologies, especially the Internet (KAPPEL 2002). The existing training concept can contribute greatly to that program. Furthermore it can be combined with a new approach for recruiting project teams for supply chain projects (WIENDAHL 1996; AUGUSTIN, KERN 2002).

REFERENCES

AUGUSTIN, Siegfried; GRAF, Karl-Robert:
Planspiele bei der Einführung kontinuierlicher Verbesserungsprozesse.
In: etz Elektrotechnik und Automation,
Düsseldorf, 124(1995)13-14, pp.12-17.

AUGUSTIN, Siegfried; KERN, Eva-Maria; HORNSTEIN, Elisabeth von:
Management von Supply Chain Projekten – Einsatz von Planspielen zur Optimierung der Projektbesetzung.
In: Logistikplanung und –management, 8. Magdeburger Logistik-Tagung.
Hrsg.: SCHENK, Michael; ZIEMS, Dietrich; INDERFURTH, Karl.
Magdeburg, 2002, pp. 270-282.

DGP – Gesellschaft für Unternehmensorganisation:
Logistikplanspiele.
Karlsruhe: DGP, 2004.
http://www.logistikplanspiele.de, 05.04.2004.

GRAF, Karl-Robert; TERZIDIS, Konstantinos; AUGUSTIN, Siegfried:
Global Education Integrating Simulation and the World Wide Web for Creating Supply Chains.
In: Current Trends in Production Management.
Eds.: ZÜLCH, Gert; STOWASSER, Sascha; JAGDEV, Harinder S.
Aachen: Shaker Verlag, 2003, pp. 25-33.
(esim – European Series in Industrial Management, Volume 6)

KAPPEL, Hans Henning:
LEARNTEC 2002: Internationalität wächst.
In: Frankfurter Allgemeine Sonntagszeitung,
Frankfurt, 27.01.2002, p. 70.

LANG, Sabine; JUNG, Klaus-Peter:
Planspiele für die Praxis.
In: Logistik heute,
München, 23(2001)3, pp. 50-52.
RIIS, Jens O.:
Simulation Games in Production Environment – An Introduction.
In: Simulation Games and Learning in Production Management.
London: Chapman & Hall, 1995, pp. 3-12.
THALER, Klaus:
Supply Chain Management.
Köln: Forbis, 1999.
WIENDAHL, Hans-Peter:
Fähigkeit zum Wandel und kurze Reaktionszeiten bestimmen den Erfolg.
In: Industrie-Anzeiger,
Stuttgart, 118(1996)34/35, pp. 28-31.

Overcoming Cultural Barriers in Distributed Work Environments

A comprehensive, internet-based concept

Ralf Lossack and Matthias Sander
University of Karlsruhe, Institute for Applied Computer Science in Mechanical Engineering (RPK), Kaiserstr. 12, 76131 Karlsruhe, Germany.
Email:{lossack, sander}@rpk.uni-karlsruhe.de

Abstract: This paper presents an approach on how to overcome intercultural barriers in European-Chinese engineering projects via an Internet portal. It is the result of one of the work packages of the EU project DRAGON. In consideration of the main users' requirements and the special preconditions that Chinese-European collaboration implicates, portal functionalities are conceived that support the project participants throughout their intercultural collaboration process.

Key words: Distributed product development, Cultural differences, Internet-based support

1. INTRODUCTION

One of the main challenges companies face today is the globalisation of business processes. Many companies respond to this challenge by establishing cross-national co-operations. Thus, internal and external partners worldwide are integrated in value chains that aim at benefiting from the local advantages of the specific locations. Recently, this policy has been applied in the field of product development. This study illustrates that many problems in product development between distant located collaboration partners are based on cultural differences. Consequently, adjusting cultural differences is of pivotal importance for the success of these projects.

Within the scope of the European Commission project DRAGON (Development of an inteRActive EnGineering Portal for Open Networks;

DRAGON 2004) cultural related problems during a collaborative product development process are addressed using collaborations between Chinese and European companies. One main objective of the project is to develop an internet-based service called Cultural Repository (CR). It aims to support enterprises to overcome barriers originating in different cultural backgrounds before and during the duration of a joint project. The business process as shown in Figure 1 is focus of this paper.

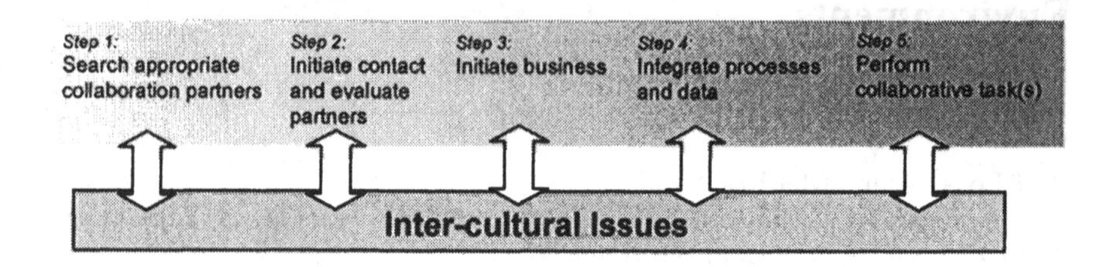

Figure 1. The cultural influenced collaboration business process steps covered within the project
(Source: GRABOWSKI et al. 2003, p. 194)

2. CULTURAL REPOSITORY (CR)

The procedure of approaching the CR objectives consists of three phases; the first phase comprised the analysis of user's requested support, phase two and three are a conception and a realisation/evaluation phase.

2.1 Results of Analysis Phase

During an analysis phase structured interviews were conducted in order to analyse success factors for intercultural collaboration and requested support. The analysis and evaluation of interview results showed that the following requirements have to be considered when creating the CR:

1. *Sensitisation* is the basic need to make people aware of the fact that culture is a crucial factor for success or failure of international collaborations.
2. *Provision of general information* about the cultural background of a collaboration partner is another important functionality.
3. The service has the task to form a common platform facilitating *cultural knowledge transfer* and *knowledge sharing* among users.

4. The analysis brought up the importance of *personal relationships* for the success of intercultural projects. Therefore the portal should facilitate the development and maintenance of personal networks.
5. The portal should also improve *joint understanding of terms and processes* among project partners in order to prevent misunderstandings leading to wrong expectations.

Moreover, the CR concept faces two main challenges that result from the basic characteristics of culture. The dynamic character of the culture leads to the first challenge. Though some characteristics of a culture may seem to remain the same over centuries, it should not be forgotten that culture is constantly changing. The other challenge is due to limitations to what can be achieved by means of an Internet portal when dealing with cultural issues. These limitations are partly due to the medium and partly due to the fact that cultural interactions take place between the human beings.

At this point, it can be ascertained that the CR cannot fully replace personal relationships between partners. Instead, it can help to determine what should be regarded as important in relationships and in which areas personal contact is indispensable.

2.2 Results of Conception Phase

Based on the results of the analysis phase, four functionalities of the CR were defined to meet the terms specified in the requirements:

1. Cultural Guidelines,
2. Cultural Data Base (Search & Find Functionality),
3. Agent Functionality,
4. Knowledge Management (experience sharing).

Figure 2 briefly outlines the contents of the CR functionalities. Furthermore, this the figure shows that the functionalities have access to a Culture Database. This database is split into two parts: an Information Base and a Knowledge Base. All information which is provided by the CR is stored in these databases. The contents of each function is described in detail in the following sections.

Table1 displays the contribution of the functionalities to the fulfilment of the requirements, as described above.

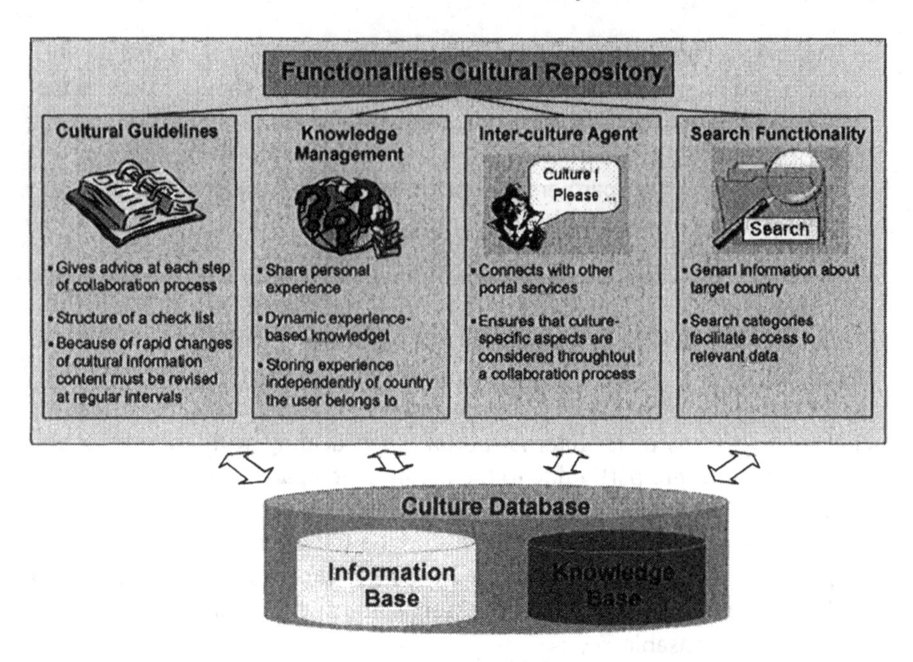

Figure 2. Functionalities of the cultural repository

Table 1. Contribution of CR functionalities to fulfil the predefined requirements

	Sensitation	Provision of general information	Knowledge transfer	Personal relationship	Joint understanding of terms and processes
Cultural Guidelines	[x]	[x]			[x]
Search & Find Functionality	[x]	[x]	[x]		
Agent Functionality	[x]	[x]			
Knowledge Management	[x]	[x]	[x]	[x]	

2.2.1 Cultural Guidelines

In order to sensitise companies which are involved in or intend to set-up an intercultural co-operation, the function *Cultural Guidelines* (CG) is being developed. The CG gives advice about what are the important cultural aspects and potential barriers to be considered for each step of a collaboration process. For example, before initiating business (see Figure 1) with a new co-operation partner, both sides of the partnership will go through a negotiation phase. When negotiating with a person from a different cultural background, certain rules of conduct should be followed in order to avoid misunderstandings or failure. Beyond sensitisation, the CG provides information how the encountered intercultural barrier can be overcome.

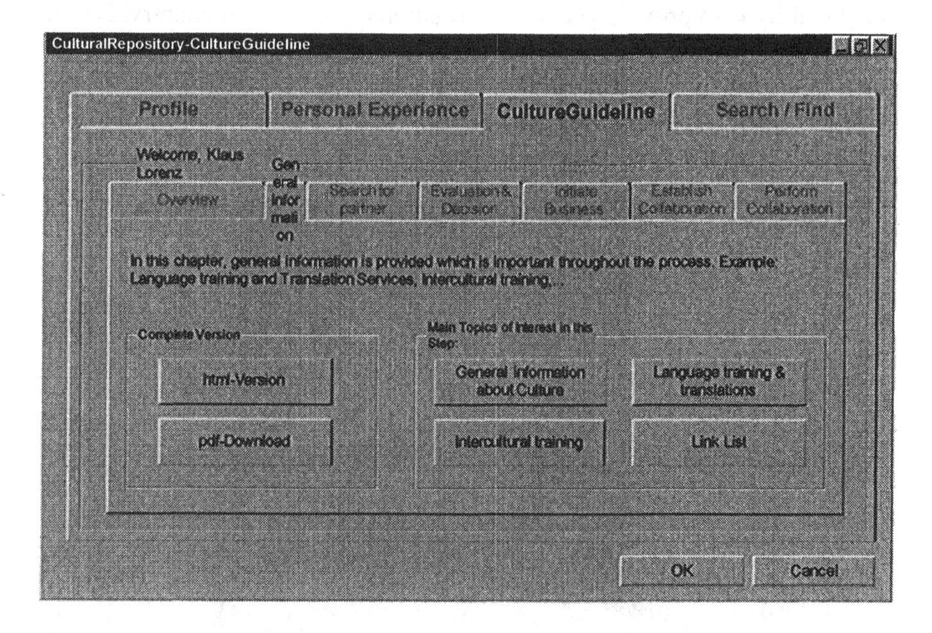

Figure 3. User interface of cultural guideline

The structure of the DRAGON collaboration business process is used to organise the information provided by the CG (see Figure 3). This set-up allows a targeted browsing as the user is able to navigate by following the chronological cycle of a co-operation. In addition, this structure enables the inexperienced user to learn about important cultural obstacles in a collaboration by simply following the co-operation business process. Whereas a user more experienced in collaborations can directly chose a business step relevant for him.

As cultures differ from nation to nation, the arising cultural barriers are different for each culture. Thus, for each culture guidelines will be provided. The information aggregated in the guidelines was collected during the analysis phase, and completed with current literature. Due to the rapid changes of cultural information, the content of this service must be revised at regular intervals. This function offers a comprehensive, process-accompanying collection of information which covers the complete collaboration process suited for a specific culture. It supports the co-operation partners to develop a common understanding for processes and terms.

2.2.2 Search & Find Functionality

Mirrored on the requirements the main focus of the *Search&Find* (SF) functionality is to provide general information about a target country. Beside cultural information like religion or cultural aspects, it also holds information about the geography, legal and political system, relevant business information for investments, etc. The information is stored in an underlying database.

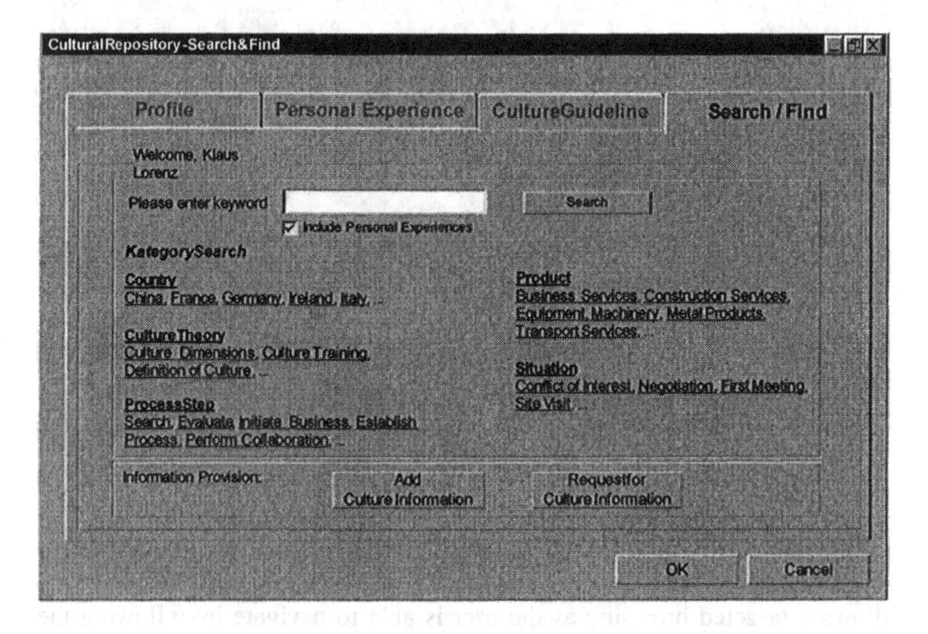

Figure 4. User interface of search & find functionality

The user can directly access the database to request or enter information. In order to find desired information, the user can choose from given categories and/or enter key words (see Figure 4). As cultural information is subject to constant change, the data accessed via the SF functionality have to be regularly reviewed and updated. Compared to the CG, this functionality provides more general information about a culture detached from the co-operation process.

2.2.3 Agent Functionality

The *Inter-culture Agent* (IA) is intended to assist the user of the DRAGON-portal regarding cultural issues at crucial steps during the complete lifecycle of the DRAGON collaboration process.

Beside the CR, the DRAGON portal comprises other components like a *Specification Modeller Component* (SMC) or *Request & Navigation Component* (RNC). These components support the user of the DRAGON-portal during the various steps of the collaboration process. Hence, the IA is triggered from components other than the CR. The kind of assistance provided can vary depending on the current position in the lifecycle. In general, it functions as a reminder for the user to consider cultural issues and cultural influences for his collaborative work, thus sensitising the user for cultural aspects. It suggests references for cultural categories and topics to the user, which are relevant for the current process step. The IA retrieves the information sending a request to the database.

A more advanced alternative of the inter-culture agent could provide the user with more detailed and tailored information. Instead of the user searching the cultural database the agent retrieves the information sending a request to the database. For example, at certain steps the cultural peculiarity of a specific country or group can have a crucial impact on the proceedings of the DRAGON component SMC and therefore should be considered for the further planning, monitoring and control of the course of the process. The requirements for a product, which are defined in the SMC, can be influenced through culture. A washing machine designed for the Indian market does not need a 90° washing program. Indian people mainly wear clothes made out of fabric, which should not be washed at that temperature. In this example the Agent functionality is activated when the user begins to define the requirements in the SMC. Then the Agent provides suited support for the specific step. The inter-culture agent will inform the user to consider cultural peculiarities when specifying the product requirements. If the user has already defined the target country and the product or product group the agent can provide the more specific information from the data base, i.e. that a washing machine in India does not need a 90°-washing program.

The user always decides whether to accept or not the support of the IA. The type of support given by the Agent depends on the triggering event of the current DRAGON-component, the available data in the cultural database and the degree of specification by the user.

The IA will only be triggered at specific occasions in order to avoid an overuse.

2.2.4 Knowledge Management

The *Knowledge Management* (KM) functionality facilitates the management, exchange, and sharing of personal experiences gained during an intercultural collaboration process. Experience is valuable, stored, specific knowledge that was acquired in a problem solving situation (BERGMANN 2002). Thereby, experience is in contrast to general knowledge as provided in the CG and Search & Find Functionality. Through the KM, enterprises are enabled to share and transfer the culture related experience of their employees in order to avoid mistakes caused by cultural differences in future product development processes. The KM supports the intention of the Cultural Repository to take into account the dynamic change of culture through updating experience and entering new experience into the database.

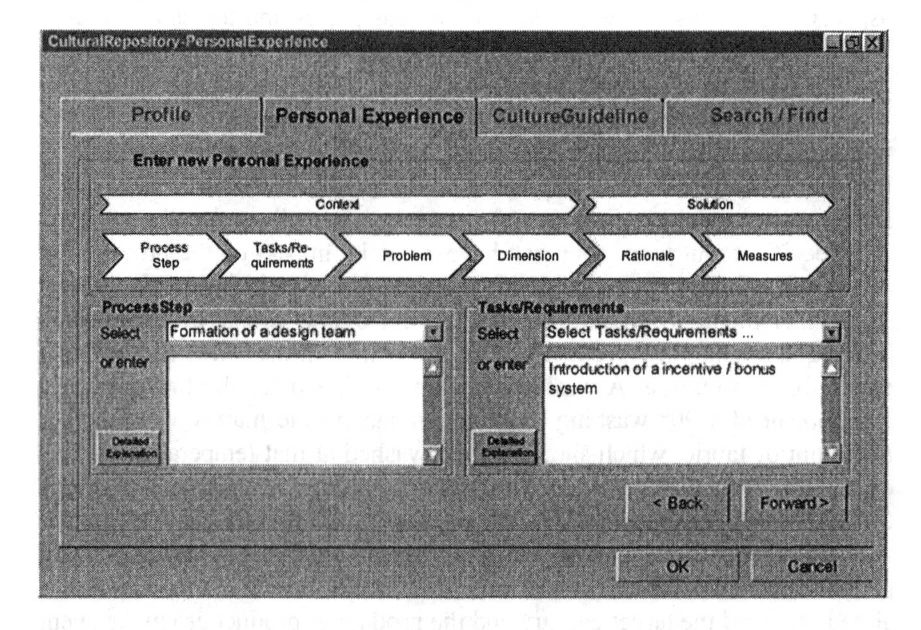

Figure 5. User interface of knowledge management

The method of *solution patterns* according to SUHM (1993) is used to store the experience of a user. A solution pattern consists of a context and a solution. The solution consists of a rationale and a measure. In the rationale it is described why the cultural problem occurred. The measure delivers methods how to overcome the problem. In order to be able to retrieve the solution from the knowledge base, the solution is allocated to a context. The context describes the situation in which the user collected his experience. It consists of the process step during which the cultural problem occurred, the description of the problem and the concerned part of culture where it occurred.

Using the structure of a collaboration process to classify the stored experience makes it possible to attribute experience to a relevant step of a collaboration process. This classification contributes to the purpose of the DRAGON portal to ensure a process-orientated support for enterprises during build up and process of a common collaboration and allows the user to find knowledge for a certain process step in the collaboration.

With the assistance of cultural dimensions (HOFSTEDE 2001) it becomes possible to assign the part of culture, which caused the problem and to store experiences independent of the country the user belongs to (see Figure 5). This approach enables the portal user to find a solution for a cultural caused problem for a certain process step in a collaboration. It informs about which part of a culture caused the problem.

2.3 Realisation / Evaluation Phase

2.3.1 Data Sharing and Provision in a Distributed Work Environment

The CR aims to support enterprises which are part of a cross-national cooperation, i.e. the users of the CR are part of a distributed work environment and can be spread over several countries. From there it is important to allow an access to the functionalities of the CR independent of time and place. Further the database of the CR can contain links and references to relevant information which is stored in variable distributed data sources, e.g. homepages on the Internet.

Internet Technology is employed to achieve both, an access to the CR independent of time and place and access of the CR to relevant information in distributed data sources. The CR is part of the DRAGON Engineering Portal and makes use of the data management and user administration capacities provided by the portal. The core of the portal is formed by an information platform called Data Information Broker (DIB). The DIB provides a transparent view on distributed data sources and contains fundamental functionalities to access the portal and to access and administer distributed data sources. Web services are used to link the CR to the DIB.

2.3.2 Scope of realisation

Within the scope of the DRAGON project, the focus for the realisation of the concept is laid upon the KM-, CG and CD-functionalities. At this point of time, the first prototypes are being developed. In parallel an evaluation phase will take place where the suitability of the Cultural Repository in dispersed intercultural engineering projects is assessed. Follow-up projects will be initiated in the near future in order to further advance the realisation of the concept.

3. CONCLUSION

Although there is an increasing awareness of the importance of intercultural differences, the practical implementation of these issues in international projects is often insufficient. The CR offers an approach on how intercultural aspects are integrated into a portal-based working environment for dispersed projects.

In order to fulfil the users' request for up-to-date information about the target country, the CR offers *cultural guidelines* and a *culture database. Agent* functionality interconnects the Cultural Repository with other DRAGON-components.

To meet the challenge that culture is constantly changing, CR creates a platform for knowledge transfer between project participants across cultural boundaries. Thus, it does not only refer to static information that becomes rapidly obsolete and worthless. The *Knowledge Management* component also takes into account that culture takes place between human beings by supporting the built up and maintenance of personal networks.

ACKNOWLEDGEMENTS

The results of the presented work were achieved with the valuable collaboration of Dr. Ralf Lossack from Institute for Applied Computer Science in Mechanical Engineering (RPK) of Karlsruhe University and Barbara Bumeder and Eva Dietz from Siemens AG, Corporate Technology, Munich.

REFERENCES

BERGMANN, Ralph:
Experience Management
Berlin: Springer Verlag, 2002.

DRAGON:
Development of an interactive Engineering Portal for Open Networks.
Karlsruhe: Institute for Applied Computer Science in Mechanical Engineering, 2003.
http://www.dragon.uni-karlsruhe.de, 05.04.2004.

GRABOWSKI, Hans; LOSSACK, Ralf-Stefan; SANDER, Matthias; BUMEDER, Barbara; DIETZ, Eva:
Overcoming Cultural Barriers in Distributed Work Environments.
In: Human Aspects in Production Management.
Eds.: ZÜLCH, Gert; STOWASSER, Sascha; JAGDEV, Harinder S.
Aachen: Shaker Verlag, 2003, pp. 194-200.
(esim – European Series in Industrial Management, Volume 5)

HOFSTEDE, G.:
Culture's Consequences.
Beverly Hills: Sage, 2001.

SUHM, A.:
Produktmodellierung in wissensbasierten Konstruktionssystemen auf der Basis von Lösungsmustern.
Aachen: Shaker Verlag, 1993.

Developing a Web Enabled Gaming Approach to Mediate Performance Skills in Interorganisational Learning and Collaboration to Engineers

Klaus D. Thoben and Max Schwesig
Bremen Institute for Industrial Technology and Applied Worksciences, Hochschulring 20, D-28359 Bremen, Germany.
Email: Thoben@biba.uni-bremen.de

Abstract: As a consequence of dynamic markets - supported by developments such as globalisation and the current 'explosion of knowledge' - organisational capacity for learning is being identified as one of the key abilities for organisations to survive. As products are getting more complex, it often requires various enterprises with certain key competencies to produce a product in collaboration. Thus, especially interorganisational learning gains importance. Because of the high knowledge intensity within the actual product development and the production, particularly engineers are obliged to constantly acquire, share and transform knowledge into new products. Having examined an existing simulation game for Concurrent Engineering – COSIGA, which serves as a basis for further developments, we have identified key elements that are used to simulate organisational / interorganisational learning. By integrating these elements in COSIGA, we have developed a two level web based group simulation game aiming to mediate performance skills in the domain of organisational and interorganisational learning to engineers.

Key words: Simulation gaming, Organisational learning, Interorganisational learning, Company collaboration

1. PROBLEM

As a consequence of dynamic markets - supported by developments such as globalisation and the current 'explosion of knowledge' - organisational capacity for learning is being identified as one of the key abilities for organisations to survive. As products are getting more complex, it often requires various enterprises with certain key competencies to produce a product in collaboration. Thus, especially interorganisational learning gains importance. Because of the high knowledge intensity within the actual product development and the production, engineers are obliged to constantly acquire, share and transform knowledge into new products. As a consequence of the stated developments, the way of working and the educational requirements that engineers have to face have changed as well. Performance skills about organisational and interorganisational learning and trust building competence are becoming vital. As shown, engineers need to know about organisational and interorganisational learning and how to apply this knowledge and the related skills in a working situation. Appropriate tools to mediate such skills are simulation games.

2. STATE OF THE ART AND RESEARCH APPROACH

According to RIEDEL, PAWAR, BARSON (2001), a simulation is based on a model representing a real life system to be learned. Simulations provide the opportunity to train and practice skills and knowledge without the risks involved with real life situations.

2.1 Existing Simulation Games

Existing games in the field of organisational learning and the closely related knowledge management like "KM QUEST" (SHOSTAK et al. 2002) or "ESCIO" (ADELSBERG et al. 2002), focus certain aspects of knowledge handling and knowledge sharing. Existing simulation games in the field of product development like "COSIGA" (PAWER et al. 1995), "City Car Simulation" (GOFFIN, MITCHELL 2002) and "GLOTRAIN" (WINDHOFF 2001) concentrate on the mediation of certain approaches or emphasize on important success factors in product development or even distributed production. As we were involved in the development of the COSIGA simulation game, which will serve as a basis for further developments, this gaming approach is examined in the following section in greater detail.

2.2 The COSIGA game

The COSIGA simulation game focuses on the mediation of concurrent engineering (CE) principles and practices and enabled us to gain valuable experience about the simulation of CE and product design and in a gaming environment. Being played by five individuals in the same room (co-located) or in a distributed group (virtual) and using the Internet and telecommunications, the game aims to realistically simulate the collaborative and co-operative process of new product development inherent in a concurrent engineering approach. The players interact in a product development scenario where they have to specify, design and produce a simple truck for a specific market. The five players represent the typical roles of a product development process: Project manager, Marketing manager, Designer, Production manager and the Purchasing manager. The learning goal of COSIGA is to show players 'how to' communicate, co-operate and work collaboratively to achieve a common goal (PAWAR et al. 1995). COSIGA primarily deals with concurrent engineering and communication, but it does not consider interorganisational collaboration and learning as one essential part of today's manufacturing. As shown, performance skills about organisational and interorganisational learning and trust building competence are becoming vital. Thus, the new game emphasizes the active experience and reflection of key processes and challenges of organizational/ interorganisational learning.

3. ENHANCING COSIGA

After having regarded the COSIGA simulation game, which serves as a basis for further developments, we now identify key elements that are used to simulate organisational / interorganisational learning. By integrating these elements in COSIGA, we have developed a web based group simulation game to mediate performance skills in the domain of organisational and interorganisational learning to engineers. This research approach is illustrated in Figure 1.

3.1 Identifying Processes and Challenges within Organizational Learning

We interpret organisational learning from a multi level perspective, comprising of the individual, group, organizational and inter-organizational level (NONAKA 1994), since the point of view enables us to regard the main levels of action within an enterprise. According to this perspective, our working definition of individual level learning focuses on individual knowledge acquisition and is linked to the approach of self directed learning.

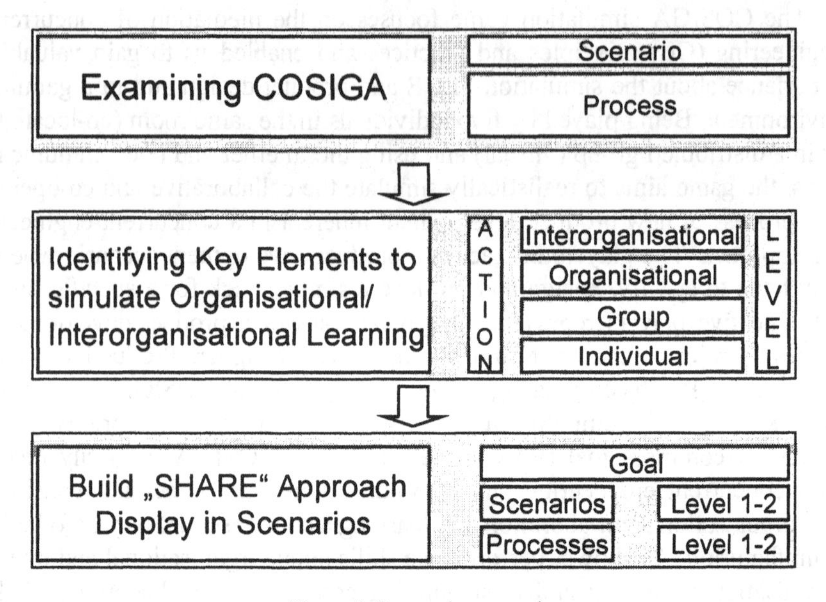

Figure 1. Research approach
(Source: THOBEN, SCHWESIG 2003, p. 256)

In the sense of group level learning, we follow MULHOLLAND et al. (2000) who defines group level learning as domain construction within communities of practice learning. By sharing vocabularies and practices, group communication and coordination during complex tasks is improved.

As MULHOLLAND et al. (2000) and SUMNER et al. (1999), we interpret organisation level learning as perspective taking. As organisations are typically composed of multiple interacting communities, each with highly specialized knowledge, skills and technologies, knowledge intensive firms require these diverse communities to bridge their difference to create a new shared perspective. By doing that communities recognize, use and evaluate the perspective of other groups. As a consequence of that, they view and evaluate themselves from another perspective in order to then create a shared cross community perspective. They are enabled to question work routines in order to reshape and thus improve their efficiency. Especially group level learning and organisational level learning are affected by "people barriers" like proprietary thinking, scepticism towards the sharing of knowledge and various fears (BARSON et al. 2000).

Interorganisational learning can happen in two ways: either through the transfer of existing knowledge from one organization to another, or through the creation of new knowledge (LARSSON et al. 1998). As the learning input comes from other organisations or joint interorganisational efforts, the intraorganisational learning activities continue to process knowledge as described above. In contrast to organisational learning, the participating organisations have to overcome certain organisational boundaries, like space, time, diversity, structure and distribution of knowledge and results (BOSCH-SIJTSEMA 2001). This makes interorganisational learning much more complex and causes much more effort. The interdependencies between the different learning levels are illustrated in Figure 2.

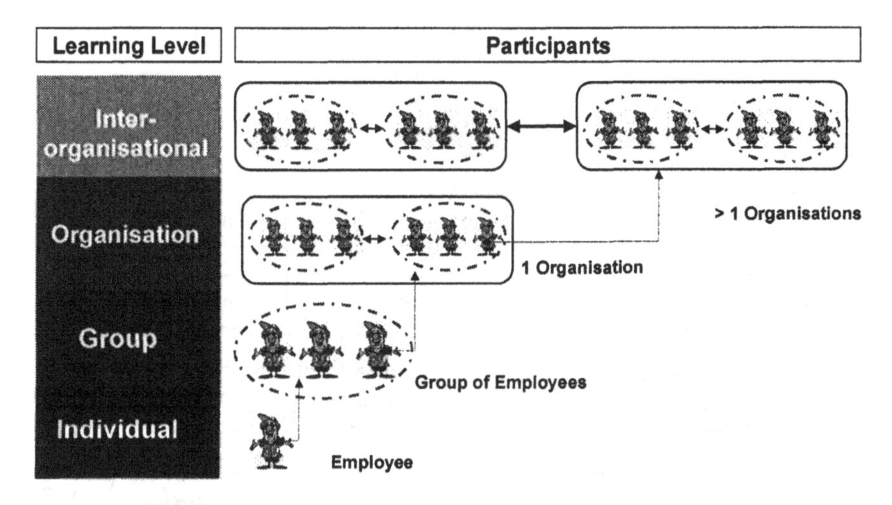

Figure 2. Interdependencies between the different levels of organisational learning

3.2 The Game Scenarios

As in COSIGA, the key process of the game is the joint experiencing of the development of a product in a virtual engineering working environment. In the first level, the players act as employees of an organisation that covers the basic economical functions: Design, procurement and sales/ services. Each department is made up by the particular department head and two employees. Since the game emphasizes the simulation of operational management processes within a company, the strategic position of the CEO cannot be played. As essential product of the modern watercraft industry and the "fun society", the Jetski was chosen as the central product of the first scenario in order to increase player's motivation, which indirectly improves the overall learning outcome. As initial evaluation and validation of the four step COSIGA product development process with target users have been very encouraging (RIEDEL, PAWAR, BARSON 2001), it will be adapted in the new game.

Thus, the players have to specify, design and produce a Jetski in one company. Each department is responsible for the successful completion of at least one sequential step within this product development process. Figure 3 illustrates the whole structure and the processes of level 1.

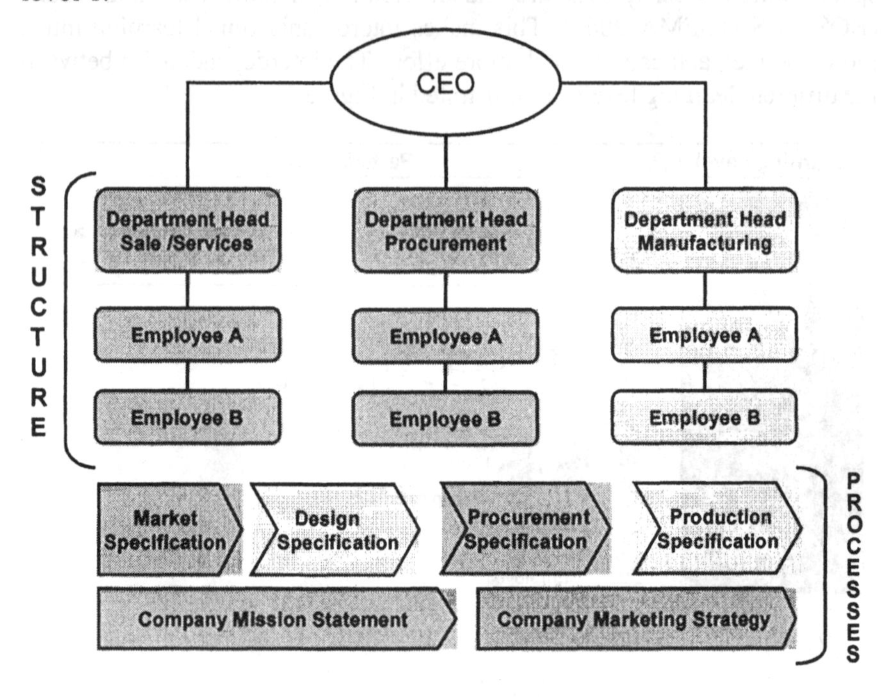

Figure 3. Game structure and processes of level 1

As the simulation of the realistic complexity within product development in a game would overextend the players, the following three-step process ha been developed. It consist out of three sub steps: First, basic information is provided to the players. By choosing between three different options, each having different durations and costs, the information is converted / enriched and then acts as basic information for the next step within the product development process. This process is applied throughout the game and can be illustrated by regarding the completion of the market specification: At the beginning of the game, basic market information is provided to the players. After having inserted it in a special template, the players can choose between either spying competitors, employing consultants, or buying a market study. After the players have chosen an option, new information in generated, which acts as input for the design specification.

As a sequential product development process does not involve the whole company, idle departments are given specific tasks, which they have to solve in teamwork to simulate group level learning. The players thereby have to structure their efforts to solve the tasks and present their outcome in front of the whole company. Particular specific knowledge has to be retrieved from the internet. Among those tasks is e.g. the development of a company mission statement to enforce the shared company culture. In order to simulate organisational level learning realistically, information about the particular costs and durations of these options is distributed unequally among the departments, the players have to cooperate and to communicate to get this essential information to be successful. Following their role descriptions, some players act non collaborative to illustrate "people barriers". Together, the players experience the destructive effect of such behaviour. The forced communication between the departments supports the process of organisational level learning, as the members of the different departments have to look into the perspective of other players/ groups in order negotiate successfully to get information. In order to improve the personal relations to others, the players can choose between various trust building measures (f.e."go dining" "schedule real world physical meeting" etc.). Table 1 summarizes the game characteristics within level 1.

Table 1. Game characteristics within level 1

Game Characteristics	Intended Effect	Simulated Process / Challenge
Unequally distributed information	• Increased interdepartmental communication to practice knowledge exchange, cooperation	Organisational level learning/Perspective Taking
Extra task for idle departments	• Increased intradepartmental communication to practice knowledge exchange, cooperation	Group level learning
Non collaborative roles	• Experience destructive effect of proprietary thinking, • Experiencing of the value of trust • Start loosing scepticism towards knowledge sharing	People barriers

Within the second level, the players deepen the acquired knowledge and skills in the interorganisational production of an extended product: a cell phone enriched by certain services. A cell phone was chosen as the product because of its common use in everyday life and its eligibility to present an extended product. Within the game, the players are acting as employees of three companies, that together form a consortium. This consortium consists of two manufacturing companies and one service providing company, which acts as the consortial leader. Each player is heading a department in one of the companies.

Again, the players have to complete particular specifications to then finally produce the cell phone and conceptualise services. To enable this interorganisational effort, the players first have to negotiate their collaboration contract in order to then specify, design and produce the cell phone. While the simulated service company takes consortial leadership and conceptualises services, the two simulated manufacturing companies develop, produce and assemble generic cell phone parts. Each company has the responsibility to complete a consortia wide part of the product development process. While the service company manages the consortial contract agreement and the market specification, one of the manufacturing companies arranges the procurement process; the other is responsible for the final assembly of produced generic cell phone parts. Figure 4 illustrates the structure and the processes within level 2.

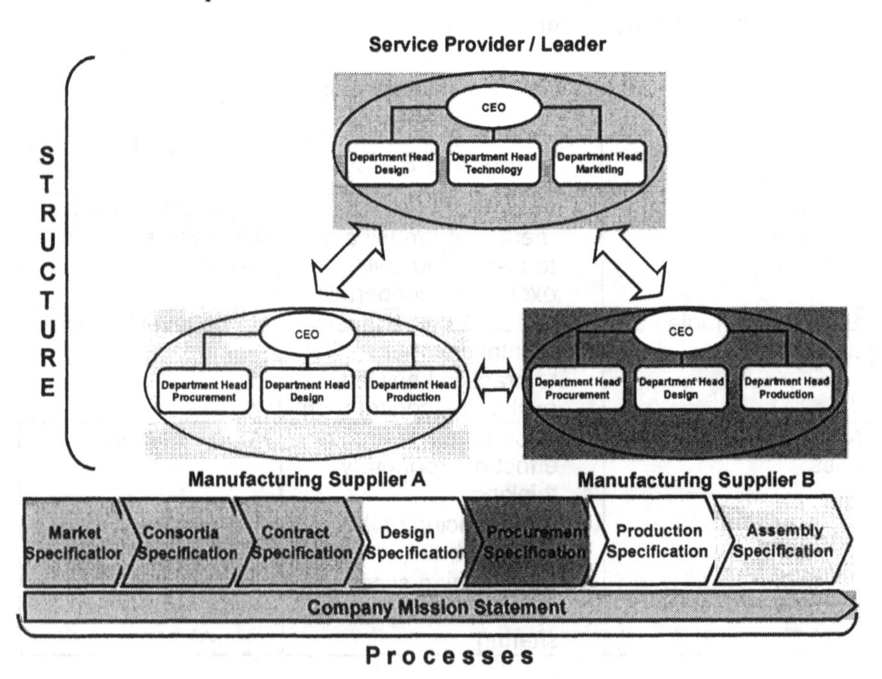

Figure 4. Game structure and processes within level 2

Again, necessary information will be distributed unequally, the partners have to cooperate to enable constant flow of information that will then lead to a constant flow of (virtual) material and parts. To simulate interorganisational collaboration realistically, organisational boundaries like space, time, diversity, structure and distribution of knowledge and results will be addressed as well. To simulate physical distance, the arrangement of physical meetings is much more expensive. As the simulated companies are located in different time zones, synchronous communication is limited to certain time phases. Due to their different natures, the companies of course maintain different cultures that will be simulated by providing different mission statements to the particular company members. In order to simulate the structural boundary, each company will have a different "IT standard". At the beginning, data flow between the companies will be disturbed /distorted. The players will have communicate / to collaborate in order to identify the lack of interoperability as the source of problems to then agree on a common standards. During the production process, the mentioned essential knowledge concerning information processing will be distributed unequally again. The game characteristics of level 2 are summarized in Table 2.

4. OUTLOOK

We have created a web based group simulation game focussing the learner's experience of company collaboration as well as organisational and interorganisational learning processes and challenges. Therefore, we have combined a gaming approach used in the simulation game COSIGA with identified processes and challenges in (inter)organisational learning. The game is currently being evaluated and validated. Furthermore, it is planned to develop an adjustable game that is able to simulate different kinds of vertical and horizontal collaborations. An accordant system architecture is in development. Additionally, modern wireless technologies will be integrated to enable an easy implementation in working and learning environments and to realistically present of future ubiquitous learning environments for engineers.

ACKNOWLEDGEMENTS

The authors wish to acknowledge, that this work has been partly funded by the European Commission though the IST Project GEM (Global Education in Manufacturing; No. IST-2001-32059).

Table 2. Game characteristics in level 2

Game Characteristics	Intended Effect	Simulated Process / Challenge
Unequally distributed information	Increased intercompany communication to practice knowledge exchange /perspective taking	Boundary of distributed information
		Interorganisational level learning
Distributed consortium responsibilities	Increased intercompany communication to practice knowledge exchange /perspective taking	Interorganisational level learning
		Boundary of distributed information & results
Intercompany IT interoperability	Practice Perspective taking to identify the problem, solve the problem by intense communication and collaboration	Structural diversity
Limited communication	Experience communication & cooperation under time pressure, find strategies how to cope with it	Boundary of time
Limited physical meetings	Experience geographical distribution, Identification of balance between physical meetings and usual communication	Boundary of space
Each company receives different mission statements	Experience interorganisational negotiation processes, find a common goal by compromising	Boundary of diversity
Extra task for idle departments	Identification and application of strategies how to overcome such a challenge	Group level learning

REFERENCES

ADELSBERGER, H.; BLICK, M.; HANKE, T.:
Einführung und Etablierung einer Kultur des Wissensteilens in Organisationen.
In: Virtuelle Organisationen und Neue Medien 2002.
Eds.: ENGELIEN, M.; HOHMANN, J.
Köln: Joseph Eul Verlag, 2002. pp. 529-552

BARSON, R.; FOSTER, G.; STRUCK, T.; RATCHEV, S., PAWAR, K.; WEBER, F.; WUNRAM, M.:
Inter- and Intra-Organisational Barriers to Sharing Knowledge in the Extended Supply-Chain.
In: E-business - Key Issues, Applications, Technologies.
Eds.: STANFORD-SMITH, B.; KIDD, P. T.
Amsterdam et al.: IOS Press, 2000, pp. 367-373.

BOSCH-SIJTSEMA, P.:
Knowledge development in a Virtual organisation: an Information Processing Perspective.
Lund: University, dissertation, 2001.

GOFFIN, K.; MITCHELL, R.:
Teaching Innovation and New Product Development using the "City Car" Simulation.
In: Proceedings of the 13th Annual Meeting of the Production and Operations Management Society.
Miami, FL: Production and Operations Management Society (POMS), 2002.

LARSSON, R.; BENGTSSON, L.; HENRIKSSON, K.; SPARKS, J.: The Interorganizational Learning Dilemma: Collective Knowledge Development in Strategic Alliances. In: Special issue: Managing Partnerships and Strategic Alliances, in Organization Science Vol. 9 1998, pp: 285-306

MULHOLLAND, P.; DOMINGUE, J.; ZDRAHAL, Z.; HATALA, M.:
Organisational Learning: An Overview of the Enrich Approach.
In: Journal of Information Services and Use,
Amsterdam, 20(2000)1, pp. 9-23.

NONAKA, I.:
A dynamic Theory of Organizational Knowledge Creation.
In: Organization Science,
Linthicum, MD, 5(1994)1, pp. 14-37.

PAWAR, K.S.; THOBEN, K-D.; OEHLMANN, R.:
Developing concurrent engineering conceptual model and knowledge platform.
In: Proceedings of the second conference on Concurrent Engineering, Research and Application (CERA).
Washington, DC: CERA, 1995, pp. 487–497.

RIEDEL, J. C. K. H.; PAWAR, K. S.; BARSON, R.:
Academic and Industrial User Needs for a Concurrent Engineering Simulation Game.
In: Concurrent Engineering - Research and Applications,
Thousand Oaks, CA, 9(2001)3, S. 223.

SHOSTAK, I.; ANJEWIERDEN, A.; DE HOOG, R.:
Modelling and Simulating Process-oriented Knowledge Management.
In: Proceedings of the 3rd European Conference on Knowledge Management (ECKM).
Dublin: Management Centre International Limited, 2002, pp. 634-648.

SUMNER, T.; DOMINGUE, J.; ZDRAHAL, Z.; MILLICAN, A.; MURRAY, J.:
Moving from On-the-Job Training towards Organisational Learning.
In: Proceedings of the 12th Banff Knowledge Acquisition Workshop.
Banff, Alberta: University of Calgary, 1999.

WINDHOFF, G.:
Planspiele für die verteilte Produktion. Entwicklung und Einsatz von Trainingsmodulen für das aktive Erleben charakteristischer Arbeitssituationen in arbeitsteiligen, verteilten Produktionssystemen auf Basis der Planspielmethodik.
Bremen: University, dissertation, 2001.

A Platform for Technical Consultation of Service Providers in Rapid Prototyping

Claus Aumund-Kopp, Frank Ellebrecht, Holger Fricke, Holm Gottschalch and Christian Panse
University of Bremen, Institute of Industrial Technology and Applied Work Science, Hochschulring 20, D-28359 Bremen, Germany.
Email: {ak, el, got , pan}@biba.uni-bremen.de

Abstract: An e-business platform for communication and interaction with customers and partners was designed and built. On this platform customers from anywhere can communicate anytime about rapid prototyping services. Partners and suppliers anywhere can be subcontracted or can be integrated in a network for specific services. Networks can be developed, or at least, chains can be unfolded. Technical advice for customers can be performed on the platform and mobile with a virtual show-case on a laptop at the customer's site. A "knowledge cube" allows to allocate and connect information in three dimensions. – The new style of working in networks and interacting online with customers and partners presupposes some training and qualification as well as acceptance and motivation for the evolving characteristics of "networking" and "collaborating", which are described and defined. A comparison of activity patterns and communicative behaviour "today and tomorrow" is performed. The evolution of a new style of interacting and communicating in networks ("networking") implies an adequate understanding of and a new concept for organisation and enterprise as well as concepts for new forms of cooperation / collaboration.

Key words: E-Business platform, Networking & collaborating; Training & qualification

1. THE *AGEPRO* ONLINE PLATFORM

Customer requirements in product development and especially in rapid prototyping and rapid tooling have changed; they demand longer and complete process chains in even shorter lead times. To meet these demands of accelerated product development in interaction with customers and cooperating partners small and medium sized enterprises (SMEs) should extend their activities on the internet and develop internal e-business structures.

1.1 Description of the Platform

The *agepro* project built an online communication platform for SMEs in model and tool production which is closely oriented towards special needs of very fast Rapid Prototyping services. End users in five SMEs participated in software development during a cyclic-iterative process. In several workshops user expectations, needs and requirements were explored; programme specifications were defined; test versions of the platform were exposed to future users and their critique and suggestions were integrated in the next versions.

The platform is embedded into the company's website and opened by a browser interface. The *agepro* online platform assists rapid prototyping service providers in communicating and interacting with customers on the one hand and partners on the other, it allows and supports :

- customer acquisition and technical advice in a "virtual showcase"
- online requests of customers and specification and calculation of orders
- order status information by basic tracking functionalities,
- documentation for the company's internal use, collecting information on both successful and problematic orders and on customers
- allocation of important information with a three dimensional core competencies-employee-customer matrix called "knowledge cube".

The platform accelerates and improves the overall flow and storage of information, increases transparency of order control and production planning, harmonises transactions, improves training, and creates knowledge bases. However, no replacement of personal (face to face) customer contact is intended. In this respect, the platform will be no more than an additional communication channel.

1.2 Platform Use Scenarios

The communication platform will be used in different ways (Figure 1):

- a visitor looks into the virtual showcase in order to get information on a specific geometry of a model building process,
- a (new) customer demands information on the status of his order, especially on time limits,
- a technical advisor of the service provider visits customers at their site and exhibits masterpieces and demonstrates new processes in the virtual showcase

- internal staff has access to all kinds of information on the platform and may contribute or complete new information to the "virtual showcase", to the "lessons learnt" and to the "knowledge cube".

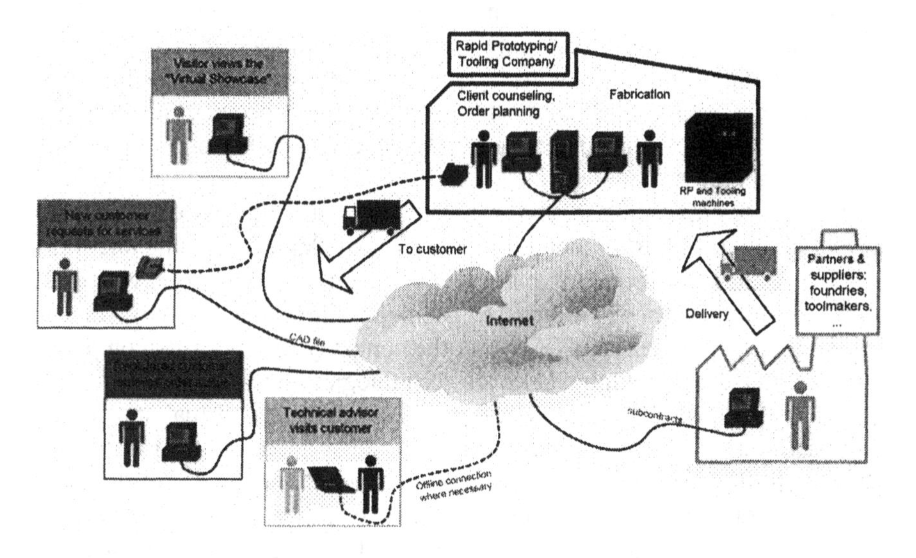

Figure 1. Different platform use situations

Figure 2 shows in a formal way which platform functionalities can be accessed by which type of user.

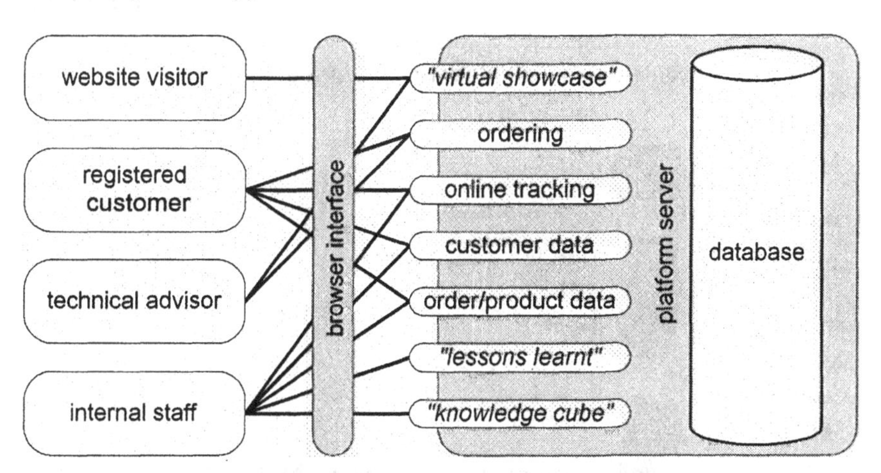

Figure 2. Dedicated functionalities for different types of users
(Source: AUMUND-KOPP et al. 2003, p. 189)

2. THE "VIRTUAL SHOWCASE"

When we visited model building companies we were always led to showcases containing masterpieces, sample parts and failures; new potentialities and challenges of rapid prototyping techniques are demonstrated proudly in this way and the limits of new techniques are discussed with these examples at hand. Technical consultation with a customer on special geometric features is easier and more precise with the object at hands or at least its 3 dimensional representation.

In order to support these demonstrations a "virtual showcase" was developed and integrated as a module into the *agepro* platform. On the one hand, it allows customers (from any place if there is www access) to have a close look at an exhibition of products or geometry similar to that they want to build and, on the other hand, it allows the company to exhibit its competencies – masterpieces and innovative RP-technologies – in the web.

Figure 3 shows the 2nd level of our example of the "virtual showcase". The user gets an overview of RP-methods and services and can choose an item for further information. The 3rd level – not shown here – offers more detailed information on the chosen process, e.g. stereolithography and its potentialities, or the user may select a specific sample part generated by stereolithography (Figure 4). The parts itself are on 4th level.

Figure3. Example of the "virtual showcase", menu and 2nd level, showing different rapid prototyping methods

Technical advisors and customers can discuss and analyse specific features of the sample part and get visual support in the virtual showcase. They may check on specific geometry or applicable coatings, looking at a corresponding example on the screen. Both may sit in different places locally but look at the same part in the "virtual showcase" and communicate via telephone or they may have a look at the advisor's laptop and sit together in the customer's office. From the customer's point of view it is easy to inform himself before getting into contact with a company to clarify details. From the advisor's point of view this avoids carrying some dozens of physical models to a customer.

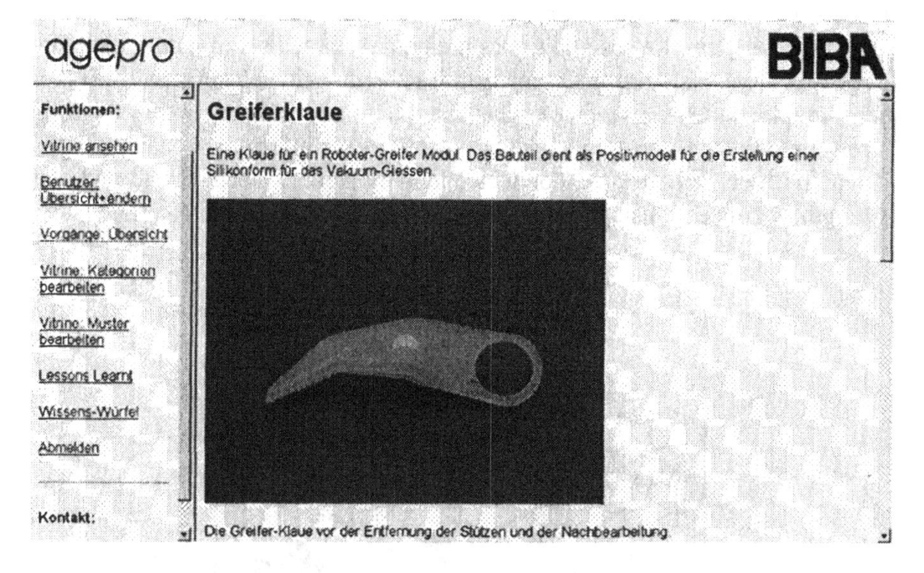

Figure -4. Example of the "virtual showcase", menu and 4th level, showing a sample part made by stereolithography

Another promising application of the "virtual showcase" is internal training. The "virtual showcase" can easily be filled with content. You can put masterpieces or failed attempts in it – the so called "lessons learned". Company staff can be trained by these examples; they can learn from mistakes and failures and may learn something about the limits (precision, tolerances) of an innovative RP-technology (however, failures should not be exposed to the public in the web). This relates a feature of the virtual showcase to the "knowledge cube", it supports a company's knowledge management : you can best learn from your mistakes.

Of course this tool must be kept up-to-date, everybody should contribute new masterpieces and lessons learned or observations on the limits of a new RP-technology.

3. THE "KNOWLEDGE CUBE"

3.1 Representation and Allocation of Knowledge by Defining Relations between List Elements

The "Knowledge Cube" is a component of the *agepro* communication platform, dedicated to store and display in a weighted manner relationships between customers, employees, and topics/core competencies of a company. The weights of relationships are intended to be helpful in analysing strengths and weaknesses of the company, in finding optimisation potentials, and planning organisational change.

Customers, employees and topics are held in three lists of a database. For each two elements from different lists it can be defined interactively (via a semi-graphical user interface) whether the relationship between them is strong, medium, weak or inexistent. In this way, information can be allocated in three dimensions, connected and related to each other, and the strength of connections can be expressed.

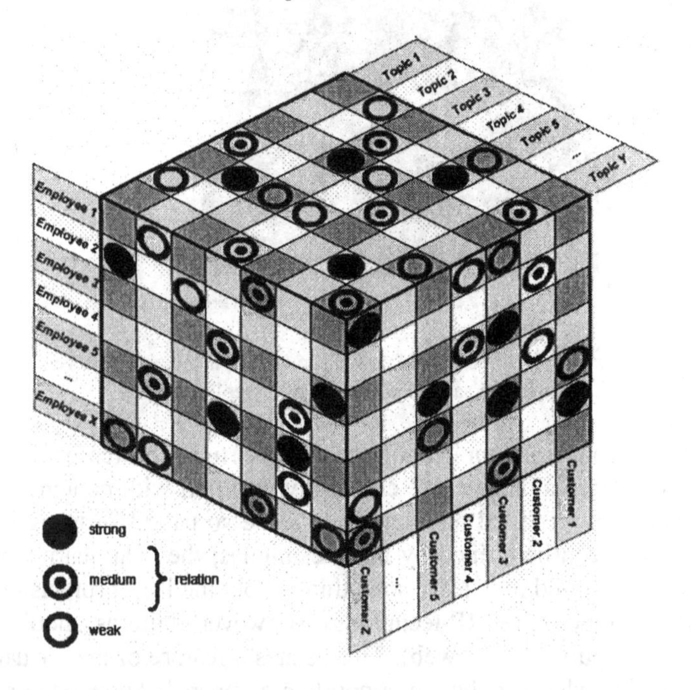

Figure 5. Representation of relation weights by the "Knowledge Cube" (schema)

We call this "Knowledge Cube" because we think that it is more than a mere information matrix: this component allows – in a specific incident or problem or event or interaction – to allocate and connect information in three dimensions to each other; it is the connection of diverse information in relation to an event or a specific question which generates *knowledge* from *information*; otherwise diverse lists of information could not contribute to knowledge, or the knowledge of employees would remain tacit. This means: the "Knowledge Cube" does not itself represent knowledge, but it can help in generating knowledge in the heads of employees.

Therefore, the cube must be very easy to use in work actions; it must be user-friendly, or it will not be used.

There are many applications of this platform component. Questions which could be posed to the "Knowledge Cube", are for example:

- If employee A retires in the near future. Who is the company to train in his special knowledge areas in order to avoid a gap? Where exactly and with respect to which customers and topics will be the knowledge gaps when the experienced employee leaves?
- Customer B is on the phone and demands to talk to employee C who is on vacation. Who else in the company knows this customer and his typical orders so that the call can be redirected to him/her?
- A new rapid prototyping technology is introduced in the company; this will be a new core competency or topic. Which customers might be interested in this for future orders? Which employees should be trained to work with that new technology? Which kind of orders and services could as well be performed by the new technology?

Knowledge is needed to give an answer to a question; it has to be generated in order to solve an actual problem. The lists of customers, employees and topics contain simple information; it is their connection and relation in work actions that generates knowledge in the heads of employees. The "Knowledge Cube" itself cannot represent, generate nor contain knowledge – it is a tool for managers and engineers to find connections and relationships. Therefore we think that the process of identifying, allocating, connecting and contributing information (maybe in a collective effort, discussion, workshops) is more important and productive for the knowledge of a company than the result: the cube in its specific developmental stage (to stay with the visualisation in Figure 5: with dozens of "dots" and "rings") in a continuous process.

3.2 The "Generic Knowledge Cube"

In a second step the above-described "Knowledge Cube" was enhanced by some concepts that made it more flexible and abstract.

- The semantics of the lists within the "Generic Knowledge Cube" is opened for other entities than just employees, customers and topics to be represented, for example projects or machines;
- Furthermore, the "Generic Knowledge Cube" can contain any number of lists, according to individual needs;
- Finally, it is possible to define weighted relations between elements of the same list; this allows for instance to detect close relationships between different employees, customers, etc.

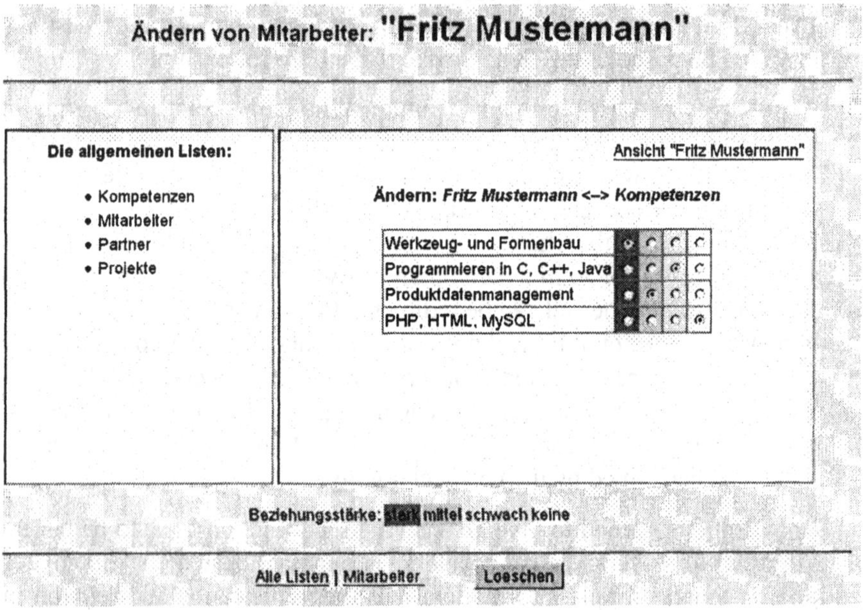

Figure 6. Screenshot of one item in the list of employees (a person) in the "Generic Knowledge Cube"

3.3 Future prospects and potentials

In discussions about the "Knowledge Cube", several new ideas were brought up that might be of use when thinking about the incorporation of the "Knowledge Cube" in new and extended contexts:

- to mark relationships between elements of lists with additional information, instead of just stating their weight: this could be comments, protocols, documents, pictures, or any other data;

- to introduce grouping structures for lists (like folders in file systems) together with appropriate inheritance mechanisms, which will ease the handling of long lists;
- to add timestamp and history functions that trace changes;
- to realise graphical representations of different kinds for the database content: tables, detail views, global views, dynamic graphs, ...

4. NETWORKING AND COLLABORATING

The communication platform and the virtual showcase open a wide space for communication and collaboration with customers on the one side and partners/subcontractors on the other. The platform was designed according to the needs and requirements of users in SMEs offering Rapid Prototyping & Rapid Tooling services. The functionalities of the platform support the diverse tasks and activities of technical advisors interacting with customers and partners.

The *agepro* project expects that the work processes and characteristics of work activities will develop as soon as the platform will be implemented. However, there will not be a sudden change to networking and collaborating. We expect a slow evolution of new characteristics of work. It takes time for the engineers to realise the potential of the new communication and collaboration space and to develop appropriate activity patterns and communicative behaviour (interacting with clients and partners anywhere and anytime). We hypothesise that within a year or two this change of activity patterns, attitudes and mindsets will have developed. The hypothesis is explicated (with respect to the work of technical advisors for Rapid Prototyping & Rapid Tooling services) in a comparison or even confrontation of "today" and "tomorrow". We formulated the features in an ideal type and polarised way (Table 1 and 2), thus we make new characteristics visible in the first stages of developments. Our assumptions are based on observations in the book "The rise of the network society" (CASTELLS 2000) as well as empirical work analysis, applying the VERA/RHIA method (LEITNER et al. 1993), interviews and extensive observations in the participating SMEs. The assumptions should be fruitful for analyses and explication.

The evolution of a new style of interaction and communication in networks (networking) implies a new concept for "cooperation/ collaboration". Processes and configurations in networks are different from established hierarchical structures; therefore evolving characteristics of networking necessitates different competencies and appropriate training and qualification. In *agepro* we found, that in terms of qualification and training no additional content, professional competence or knowledge is required for networking but rather a change of attitude, mindsets and communicative behaviour.

Table 1. Characteristics of networking and collaborating (according to AUMUND-KOPP 2003, p. 191)

	today : conventional processes	**tomorrow: e-business structures**
Acquisition	Customers come to service provider.	Customers are called on and visited; they are advised specifically to their requirements; acquisition and communication with customers on the long run. Attention is called to homepage and platform.
	Customers ask via mail, fax or telephone.	Customers interact on homepage, communication platform, virtual showcase; e. g. semi-automatic request → answer and generation of offer, client-specific advice.
Advice	Consultation of customer in own office.	Consultation with customer in his office.
	Receive and collect information at own desktop.	Take up and collect information at client's office, mobile data entry.
	Demonstrations with physical models.	Demonstrations with virtual exhibits.
	Information predominantly on paper.	Information, data predominantly digital.
Network	Company conceives itself as discrete and singular, behaves individually in relation to competitors, partners and subcontractors.	Company distinguishes itself and takes position with its core competencies in a network of cooperating partners.
	Few stable and long lasting relations.	Many weak and short range relations.
Knowledge Management	Experience and tacit knowledge is in the heads of employees – distributed all over the company's processes, cannot easily be found and combined.	Information about customers, orders and processes are documented; they are available anytime and can be networked and processed to knowledge.
	Info exchange and connection often happen by chance, are undirected.	Access to all kinds of information is direct and precise.
	Communication between a person and a specific other person.	Single person communicates with a group, e.g. online communities.

Special topics, arising from the transformation to e-business structures in Rapid Prototyping & Rapid Tooling services, should be incorporated into qualification activities and training:

- go out for mobile work, find new customers, identify their needs
- help customers to identify implicit problems and present solutions
- be pro-active and take initiative in communication with the environment of the company
- contribute to and share information with network – instead of protect and hedge
- expose your company's competencies and search for opportunities in the web
- demonstrate Rapid Prototyping/Rapid Tooling expertise with master-pieces in "virtual showcase"
- communicate with several partners at the same time in networks
- think beyond company walls, see partners instead of competitors
- empower technical advisors for self-dependent, mobile interaction
- identify and manage in-house knowledge about customers and previous orders
- make use of the company's core competencies, the employees' experience and know-how, and the lessons learned (mistakes) in Rapid Prototyping processes and on shop floor
- make use of methods to filtrate relevant information out of masses of data

All these are imperative formulations, directed to those, who want to start networking successfully. SME managers should lead their employees into networks and show the options of networking, thus they will accelerate the evolution of a new way of working.

ACKNOWLEDGEMENTS

The research and development project *agepro* was funded by the German Federal Ministry of Research and Technology (BMBF) in the framework programme "Innovative Arbeitsgestaltung – Zukunft der Arbeit" (project no: 01HT0131-135).

REFERENCES

AUMUND-KOPP, Claus; ELLEBRECHT, Frank; FRICKE, Holger; GOTTSCHALCH, Holm; PANSE, Christian:
Networking and Collaborating on a Platform in Design and Production of Models and Tools.
In: Human Aspects in Production Management.
Eds.: ZÜLCH, Gert; STOWASSER, Sascha; JAGDEV, Harinder S.
Aachen: Shaker Verlag, 2003, pp. 187-193.
(esim – European Series in Industrial Management, Volume 5)

CASTELLS, Manuel:
Der Aufstieg der Netzwerkgesellschaft.
Opladen : Leske + Budrich, 2001.

LEITNER, Konrad; LÜDERS, Elke; GREINER, Birgit: DUCKI, Antje; NIEDERMEIER, Renate; VOLPERT, Walter:
Analyse psychischer Anforderungen und Belastungen in der Büroarbeit – Das RHIA/VERA-Büro-Verfahren Handbuch.
Göttingen: Hogrefe Verlag für Psychologie, 1993.

The Evolution of Outsourced Operations - A Five-Phase Model

A Case Based Approach

Kimmo Pekkola[1], Riitta Smeds[1], Heli Syväoja[1] and Pekka Turunen[2]

1) *Helsinki University of Technology, Business Process Networks, SimLab, P.O. Box 9560, FIN-02015 HUT, Finland. Email: {kimmo.pekkola, riitta.smeds, heli.syväoja} @hut.fi*

2) *EC Project Solutions Oy, Särkiniementie 3A, FIN-00210 Helsinki, Finland. Email: pekka.turunen@ecprojectsolutions.fi*

Abstract: This article presents a framework to study the development of an inter-firm relation in the context of transformational outsourcing. Current literature is mostly focussed on studying the motivations and rationale for outsourcing. However, the value added of outsourcing can be realised on the operational level only after the outsourcing event. This occurs in a process between Outsourcer and Supplier. Considering the development of the operational inter-firm relationship at the strategic level before the outsourcing decision can provide opportunities to gain value, whereas neglecting this approach creates challenges.

Key words: Business, Process, Outsourcing, model

1. INTRODUCTION

Outsourcing is currently a central topic among IT companies as they strive for enhancing efficiency to survive in the current market environment (SAWHNEY 2003). It is also a topic of increasing academic interest (WILLCOCKS, KERN 1998). Most outsourcing research focuses on decision-making based on environmental analyses, or analyse the outsourcing event itself (KESSLER et al. 1999; KAKABADSE, KAKABADSE 2000; McIVOR 2000) rather than studying the development of the outsourced operations after outsourcing. Further analysis on the development of outsourced services is needed (KAKABADSE 2000). Longitudinal studies could reveal new knowledge on institutionalisation and adaptation in the context of outsourcing (KERN, WILLCOCKS 2002).

In this study, a longitudinal view on the development of an inter-firm relation and the implications on the business logic and organisation, leads us to analyse the process to achieve outsourcing goals. Two case examples are discussed.

2. RESEARCH METHODOLOGY

The research methodology of the study is a mixture of grounded theory building (GLASER, STRAUSS 1967; STARRIN 1997) and case study approach (YIN 1981). Theory-generating case study by EISENHARDT (1989) is very similar to the grounded theory approach, because it is based on material collected without hypotheses, and it formulates new concepts (HAHO 2002).

The data concerning the two examined cases was gathered using the SimLab business process simulation method (SMEDS 1994, 1997; SMEDS, ALVESALO 2003; FORSSÉN, HAHO 2001; HAHO 2002). In the context of this study, process simulation was utilised for business process development.

The following research procedure was followed in both cases presented later in this article a selected inter-firm business process was first modelled through real life examples. Secondly, the model was complemented with approximately 15 interviews and written material.

The researchers prepared a simulation day aimed for personnel from operative to management level. The simulation day's goals were to establish a common view to the process and develop the process through group work and discussion.

This methodology provided a lot of data concerning a specific inter-firm business process and its context. This study utilises mostly the collected contextual data without describing the specific processes or their details.

3. OUTSOURCING

Highlighting its operational appearance, KERN and WILLCOCKS (2002) define IT outsourcing as a process whereby an organisation decides to contract-out or sell the firm's IT assets, people and /or activities to a third party supplier, who in exchange provides and manages these assets and services for an agreed fee over an agreed period.

From an economic point of view the previous means transfer from a hierarchy (i.e. coordinating resources inside a company) to market oriented direction, but not necessarily to the markets (i.e. coordinating resources through price mechanism). Thus, another definition for outsourcing can be found through the organisation of economic activity. Two fundamental ways discussed by classics like Adam Smith (in The Wealth of Nations) and Ronald H. COASE (1937) are market and hierarchy. They differ by dimensions like coordination of production factors, essence of competition, contracting time frame and organisational structure.

In addition to market and hierarchy, a network can also be considered as a form of organising economic activity. It is long-term coordination between independent, specialised companies (WILLIAMSON 1975; JARILLO 1988; GRANDORI, SODA 1995) – a hybrid between market and hierarchy.

According to the previous classification FIEBIG (1996) has presented the following constellations for outsourcing: Simple outsourcing, Transfer outsourcing, and Joint Venture outsourcing. The Joint Venture constellation includes two forms of outsourcing: Group and Joint Venture outsourcing.

Table 1. Outsourcing type and market relation (adapted from FIEBIG 1996 and KIIHA 2002).

Relation	Characteristics	Outsourcing type
Hierarchy	Coordination of production factors is based on ownership	Joint Venture outsourcing, Group outsourcing
Network	Companies remain independent entities. Scope is several years. People and assets are transferred. Coordination of production factors is based on common interest.	Transfer outsourcing
Market	Scope is on transaction. Satisfactory market solution is available. Coordination of production factors is based on price mechanism.	Simple outsourcing

Group and Joint Venture outsourcing include establishment of a new business unit either internally (Group) or through the markets (Joint Venture). It results in hierarchical control through joint ownership.

Transfer outsourcing involves a complete transfer of employees and equipment from one organisation to another. The agreement for Service Provisioning is done for a pre-defined period of time, typically for several years. The result is a networked relation between the outsourcer and the supplier.

Simple outsourcing requires an "off the shelf" solution between the outsourcer and the supplier. This creates a pure market relation. The outsourcer simply stops certain internal activities and starts to buy them from the markets.

In addition, MAZZAWI (2002) differentiates two motivations for outsourcing: traditional and transformational. Traditional outsourcing focuses on transaction through leveraging economies of scale in non-core and non-complex areas. Transformational outsourcing stimulates and facilitates business change, and helps to create and sustain adaptive enterprise.

In this article, we focus on Transfer outsourcing.

4. THE MODEL

Based on two case studies, we present an evolutionary model to analyse the development of an inter-firm relation in Transfer outsourcing (see Figure 1). The model includes five phases: Internal Operations, Outsourcing Event, Resource Leasing, Service Provisioning and Public Service Provisioning. These phases are further presented in the following chapters.

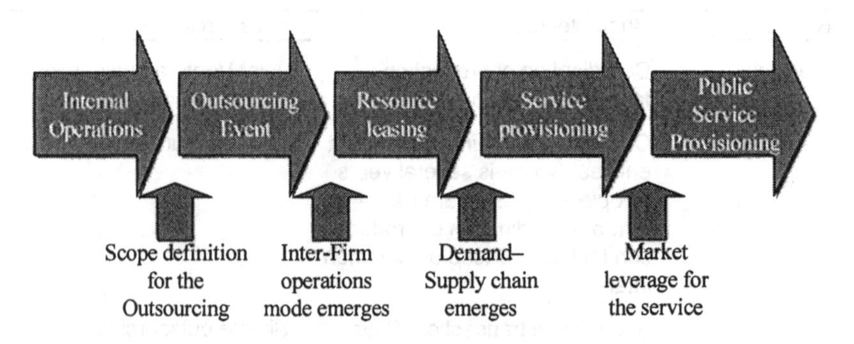

Figure 1. An evolutionary five-phase model to analyse the development of outsourced operations
(Source: PEKKOLA et al. 2003, p. 283)

4.1 Internal Operations

In the Internal Operations phase, the underlying motivation for outsourcing is reduction of cost (WILLIAMSON 1991) and cost control (WILLCOCKS, CHOI 1995). Cost considerations are often driven through strategic considerations (KAKABADSE, KAKABADSE 2000). The transformational view to outsourcing highlights organisational renewal and new business opportunity (MAZZAWI 2002; KAKABADSE, KAKABADSE 2003).

To maintain control on cost, organisations must know their own business and processes before handing them to suppliers (KAKABADSE, KAKABADSE 2000). The decision to outsource must be based on a clear need that guides the outsourcing process further on. Common definitions for outsourced operations and a shared view on the outsourcing process are essential for success (HIRVENSALO et al. 2003). Thus, operational maturity can have an effect on the outsourcing decision.

4.2 Outsourcing Event

The Outsourcing Event is critical regarding inter-firm operations in the future. KESSLER et al. (1999) highlight the human resource management point of view in aligning two organisations, in which well-coordinated communication is essential. Neglecting HR issues can cause fear with employee resistance (DORMBERGER 1998) challenging outsourcing success. This culminates in informing relevant stakeholders at the right time to prevent rumours and maintain motivation (HIRVENSALO et al. 2003).

4.3 Resource Leasing

In the Resource Leasing phase practices like open book accounting enable transparency and control in inter-organizational processes through cost and operational transparency (MOURITSEN et al. 2001). During this phase the demand-supply process between outsourcer and supplier develops, and has to be integrated to the processes of the outsourcer and the supplier through operational interfaces.

From the marketing point of view, transformational outsourcing means commitment for a long-term customer relation. According to STORBACKA and LEHTINEN (1997) the organisational interface is in such cases "a zipper" in which the outsourcer's and supplier's value producing processes are integrated in the most efficient way. Thus, organisational rearrangements are necessary to institutionalise the new operations mode and divest duplicate and low value activities.

4.4 Service Provisioning

In the Service Provisioning phase the supplier provides the service for the outsourcer only. Cost transparency is lost through service pricing. The operational demand-supply process between outsourcer and supplier has clear interfaces and organisational linkages.

Service consumption is essentially consumption of the service process (GRÖNROOS 2000), which can be defined as the process between order penetration point (OPP) in the supplier's supply process and value offering point (VOP) in the outsourcer's demand process (HOLMSTRÖM et al. 1999). These concepts become the tools for controlling the service process as outsourcer and supplier learn each other's processes and are able to increase efficiency and add value through measurement and re-evaluation.

4.5 Public Service Provisioning

In the Public Service Provisioning phase the supplier offers the service for several customers. This enables increased capacity utilisation, decreased competence development cost per client, stability through balanced customer base, and service process innovation in interaction with the markets.

The requirement for this phase is that the service has become well defined and can be provided through market transactions. The "zipper" orientation in previous phases is changed into more transactional orientation in which the firms' processes meet at fixed points. Exchange occurs through these points only.

5. CASES

In cases A and B, we studied outsourcing as an event and analysed the operational level processes between the Supplier and the Outsourcer (Client) one and a half year after the Outsourcing Event. Both cases were about business critical operations that were outsourced with a scope of three or more years. The cases are about industrial product design service (A) and data centre hosting (B). The former was a challenge while the latter one was a success.

5.1 Case A

In the Internal Operations mode, the design operations had not been a separate entity in the Outsourcer's line organisation. Instead, the working practices had varied between units. The motivation to outsource design was to decrease or avoid competence development costs and release internal development resources for core competence development.

The Outsourcing Event was considered successful, especially regarding human resources. About 60 designers and the equipment were transferred locally. However, operational definitions were later found to be loose, and contracting was not aligned with the operations and the strategic intent of the Supplier.

The Supplier thought it would receive a uniform and a relatively well-defined design service. Instead, it received a divergent service embedded with redundant support functions. The Supplier got rid of these functions through downsizing, which it found irritating.

The Resource Leasing phase was bypassed in Case A. The Outsourcer and the Supplier proceeded right away from outsourcing to the service-provisioning phase. It caused operational challenges through lost cost transparency and control, especially from the Outsourcer's point of view.

The Service Provisioning phase had most effect on the Outsourcer's product development projects that bought the outsourced service. The Outsourcer's product development managers had not been involved in the outsourcing, nor were they prepared for the new operations mode. The outsourced design service was considered as poor and costly among outsourcer's product development personnel, despite the fact that it was provided by their former colleagues. No organisational rearrangements had taken place in the Outsourcer's design related operations after the outsourcing. Service purchasing became a long process compared to the design time of an average product, which increased overheads in product development.

The realisation of the financial goals of outsourcing was questionable, because no clear goals or benchmarks existed. The lack of clear common targets challenged consistent managerial intervention and caused conflicts.

The outsourcing contract allowed the Supplier to provide the design service for the Outsourcer only. This means that market leverage could not be gained through Public Service Provisioning. This caused financial challenges for the Supplier and affected its competence development. Varying demand for design service increased Supplier's uncertainty.

5.2 Case B

During the Internal Operations phase, the Outsourcer's motivation for outsourcing its data centre was to decrease operational cost and to gain business agility and scalability. The scope for the outsourcing was defined as a set of IT services between the Supplier and the Outsourcer. This set can be changed through a predefined process. Financial goals were jointly agreed and measured during fixed intervals, and the goals and results were communicated to Outsourcer's and Supplier's personnel.

More than 300 people with equipment were outsourced globally. It was later found out that the operations did not develop in a homogenous manner in all locations.

During the Resource Leasing phase, all costs were transparent to both parties through open book accounting. The Supplier and the Outsourcer developed a shared web based tool to support and monitor the service delivery process.

After the Outsourcing event the legacy of the Internal Operations mode within the Outsourcer's personnel was strong. Therefore, the Outsourcer changed its organisational structure, and introduced new roles to better fit the inter-firm process and to gain control over the outsourced operations. Organizational restructuring was complemented with an internal marketing and training package to educate the personnel into the new operations mode. In the long run, this organisational change enabled the Outsourcer to achieve the goals set for outsourcing.

To further increase operational efficiency, both companies are currently introducing measures to control the demand-supply process of the outsourced IT services. They aim first to identify the process control points and start measuring lead-time for different process phases.

Especially the Outsourcer regarded the Resource Leasing phase as necessary for successful outsourcing. The transparent pricing had created openness in the inter-firm relationship and built trust. The outsourcer has been able to see the market price of formerly internal service. This has ensured a well functioning market oriented relationship.

In the upcoming Service Provisioning phase, price transparency will be lost as the current cost plus pricing will be replaced by service oriented single line pricing. Since the Supplier has received similar IT service operations earlier, it is already able to leverage its competence in the next, Public Service Provisioning phase.

However, the Supplier is not expected to provide new solutions for the market. Thus, the relation of the Outsourcer and Supplier is likely to remain "networked" also in the near future.

5.3 Summary and Discussion

Table 2 below summarises cases A and B and their evolution through the five-phase model.

The cases support the cost argument for outsourcing Internal Operations. In Case B, strategy was successfully linked with cost considerations. In Case A, the contract prevented the companies from achieving their financial goals. This highlights the importance for aligning contracting process with strategy and business process management in the future. In addition, the companies in Case B were able to set goals for the development of their relation before the outsourcing event, whereas in Case A the lack of goals caused challenges.

Often, the Outsourcer's motivation to outsource some operations is to control cost. According to KAKABADSE (2000), to maintain cost control, the Outsourcer must know before outsourcing its operations and processes, but not necessarily the internal costs of the operation. This sheds light on an interesting financial issue. Outsourcing can be a tool to establish a price that coordinates the use of resources. In cases where knowledge about the subject and the scope for outsourcing is asymmetric or does not exist, it might be difficult for the Supplier to set a price. As well it might be difficult for the Outsourcer to evaluate the price against the target, which often is set as a fixed percentage cost reduction, compared with the cost of internal operations.

The case companies of this study had the necessary management capability and tools to manage the Outsourcing Events. However, the adjacent phases caused challenges, especially in the learning of each others' processes and in introducing organisational change to realise the value potential of the new mode of operation beyond cutting cost. Thus, the inter-company processes and their development in outsourcing should be researcher further.

According to our two cases, transformational opportunity exists in the Resource Leasing phase. The processes in both companies must be developed to take control over the emerging inter-firm process and overcome the legacy of the Internal Operations mode. Otherwise overheads are likely to rise.

The new operations mode, a transformation, became institutionalised in the companies of Case B. This was achieved through the introduction of new roles that changed the previous production-oriented mindset increasingly towards service. Instead of production, the remaining personnel got more time to evaluate the real business need and plan for the most efficient solutions. In Case A, the mindset in the outsourcing company did not change, which lead to challenges on the operational level. Thus, in Case A, the lack of organisational restructuring reduced operational efficiency, whereas in Case B, organisational rearrangements enabled the companies to benefit from the new outsourced mode of operations.

Table 2. Comparison of cases A and B using the five-phase model.

Phase	Case A	Case B
Internal Operations	A design service embedded in the line organisation with strong linkages to adjacent product development phases. Aim to reduce/avoid personnel cost in competence development.	Global service. Aim to gain economies of scale with cost control through centralised service. Clear cost reduction goals agreed by both companies.
Outsourcing	Approx. 60 people locally	More than 300 people globally
Resource Leasing	Bypassed	"Current state" Open book accounting to maintain cost and operational transparency. Organisational restructuring resulted in a renewed demand-supply process.
Service Provisioning	"Current state" Challenges at Outsourcer due to lost cost and process transparency and control. No organisational restructuring at outsourcer.	"Next step – common goal" Cost and operational transparency will be decreased through service pricing.
Public Service Provisioning	Outsourcer prohibits through contract → high cost for the Outsourcer, low profitability for the Supplier. Divergent goals.	Exists already. Benefits through economies of scale.

Introducing process control points in the Service Provisioning phase can further decrease overheads in planning, speed up response time and clarify roles in decision-making. These hypotheses should be verified in future research.

In Case B clear goals, which were openly communicated between the companies, guided the development of outsourced operations in both organisations. In Case A, the lack of measurable goals challenged consistent managerial intervention.

The Public Service Provisioning phase can realise the financial goals of an outsourcing. Although the Outsourcer might want to secure its intangible knowledge through contractually restricting the Supplier, it should not ruin what was the goal for its outsourcing – cost benefits and innovation.

From an economic point of view, the model presented in this article (Figure 1) describes a possible transition path between two fundamental economic governance mechanisms, a firm and the market, through a networked relation. Whether the collaborating companies are aiming toward a market or a networked relationship is likely to have an effect on the operational development of outsourcing. Maybe the five phase model also differs depending on the target of outsourcing? This should be investigated in future studies.

6. CONCLUSIONS

This study has investigated transfer outsourcing and presented a five-phase model as an analysis tool. This tool can be used to plan and valuate the development of an outsourcing relationship. When applying the tool, the Outsourcer and the Supplier should have a shared goal about the nature of their relation.

Transactional motivations like cutting cost, cost awareness and cost control were clearly present in both studied cases. Beyond cutting cost is the ability to add value through business process transformation. However, transformation requires that the Outsourcer company is able to change the orientation of the remaining employees from production towards service.

ACKNOWLEDGEMENTS

We thank warmly the members of SimLab, who contributed to this article. The research was funded by Tekes - National Technology Agency of Finland, pilot companies of the R&Net and Co-Create projects, and Helsinki University of Technology, which we gratefully acknowledge.

NOTES

A previous version of this article is presented as "The evolution of a commercially viable service concept through a five step model" IFIP Working Group 5.7 International Working Conference 2003 Human Aspects in Production Management October 5-9, 2003 Karlsruhe, Germany.

REFERENCES

COASE, R. H.:
The Nature of the Firm.
In: Economica, Oxford, 4(1937)16, pp. 386-405.

FIEBIG, André:
Outsourcing under the EC Merger Control Regulation.
17 European Competition Law Review 123, 1996.
FORSSÉN, Minna; HAHO, Päivi:
Participative Development and Training for Business Processes in Industry: Review of 88 Simulation Games.
In: International Journal of Technology Management,
Geneva, 22(2001)1-3, pp 233-262.
HIRVENSALO, Antero; EVOKARI, Juha; FELLER, Jan; PEKKOLA Kimmo; TURUNEN, Pekka; SMEDS, Riitta:
R&DNet Final Report.
Espoo: Helsinki University of Technology, 2003.
(SimLab Publications, Report Series: 2)
GLASER, Barney; STRAUSS, Amselm:
The discovery of grounded theory: strategies for qualitative research.
London: Wiedenfeld and Nicholson, 1967.
GRANDORI, Anna; SODA, Giuseppe:
Inter-firm networks: Antecedents, mechanisms and forms.
In: Organization Studies, 16(1995)2, pp. 183-214
GRÖNROOS, Christian:
Palveluiden johtaminen ja markkinointi.
Helsinki: Werner Söderström Corporation (WSOY), 2000.
HAHO, P.:
Benefits of the Simulation Game Based Development Method in Business Process Development Project. Success Factors of a Good Development Method.
Helsinki: Helsinki University of Technology, Department of Industrial Management and Department of Computer Science and Engineering,
Espoo, Licentiate's Thesis, 2002.
HOLMSTRÖM, J.; HOOVER, W. E.; ELORANTA, E.; VASARA, A.:
Using value reengineering to implement breakthrough solutions for customers.
In: International Journal of Logistics Management,
Ponte Vedra Beach, FL, 10(1999)2, pp. 1-12.
JARILLO, J. Carlos:
On Strategic Networks.
In: Strategic Management Journal,
Chichester, 9(1988)1, pp. 31-41.
KAKABADSE, Andrew; KAKABADSE, Nada:
Outsourcing Best Practice: Transformational and Transactional Considerations.
In: Knowledge and Process Management,
Chichester, 10(2003)1, pp. 60-71.
KAKABADSE, Nada; KAKABADSE, Andrew:
Critical review - outsourcing: A paradigm shift.
In: Journal of Management Development,
Bradford, 19(2000)8, pp. 670-728.
KESSLER, I.; COYLE-SHAPRIO, J.; PURCELL, J.:
Outsourcing and the employee perspective.
In: Human Resource Management Journal,
London, 9(1999)2, pp. 5-19.

KERN, T.; WILLCOCKS, L.:
Exploring relationships in information technology outsourcing: the interaction approach.
In: European Journal of Information Systems,
Birmingham, 11(2002)1, pp. 3-19.
KIIHA, Jarkko:
Yritystoiminnan ulkoistaminen ja sopimusvastuu. Kauppakaari,
Helsinki, 2002.
MAZZAWI, Elias:
Transformational outsourcing.
In: Business Strategy Review,
Oxford, 13(2002)3, pp. 39-43.
McIVOR, Ronan:
A practical framework for understanding the outsourcing process.
In: Supply Chain Management,
Bradford, 5(2000)1, pp. 22-36.
MOURITSEN, J.; HANSEN, A.; HANSEN, C. Ø.:
Inter-organizational controls and organizational competencies: Episodes around target cost management/functional analysis and open book accounting.
In: Management Accounting Research,
London, 12(2001)2, pp. 221-244.
PEKKOLA, Kimmo; SMEDS, Riitta; SYVÄOJA, Heli; TURUNEN, Pekka:
The Evolution of a Commercially Viable Service Concept through a Five Step Model.
In: Current Trends in Production Management.
Eds.: ZÜLCH, Gert; STOWASSER, Sascha; JAGDEV, Harinder S.
Aachen: Shaker Verlag, 2003, pp. 282-288.
(esim – European Series in Industrial Management, Volume 6)
SAWHNEY, Anjum:
Outsourcing: A cure for the industry's bills?
In: Telecommunications International,
Dedham , MA, 37(2003)5, pp. 18-19.
STARRIN, B.; DAHLGREN, L.; LARSSON, G.; STYRBORN, S.:
Along the path of discovery: qualitative methods and grounded theory.
Studentlitteratur, Lund, 1997.
SMEDS, Riitta:
Managing Change towards Lean Enterprises.
In: International Journal of Operations & Production Management,
Bradford, 14(1994)3, pp. 66-82.
SMEDS, Riitta:
Organizational learning and innovation through tailored simulation games: Two process re-engineering case studies. Knowledge and Process Management.
In: The Journal of Corporate Transformation,
Chichester, 4(1997)1, pp. 22-33.
SMEDS, Riitta; ALVESALO, Jukka:
Global business process development in a virtual community of practice.
In: Production Planning and Control,
London, 14 (2003)4, pp. 361-371.
STORBACKA, Kaj; LEHTINEN, Jarmo R.:
Asiakkuuden ehdoilla vai asiakkaiden armoilla.
Helsinki: Werner Söderström Corporation (WSOY), 1997.

WILLCOCKS, Leslie; CHOI, Chong Ju:
Co-operative Partnership and 'Total' IT outsourcing: From Contractual Obligation to Strategic Alliance?
In: European Management Journal,
Oxford, 13(1995)1, pp. 67-78.

WILLIAMSON, Oliver E.:
Strategizing, Economizing, and Economic Organization.
In: Strategic Management Journal,
Chichester, 12(1991)1, pp. 75-94.

YIN, R.:
The case study crisis: Some answers.
In: Administrative Science Quarterly,
Ithaca, NY, 26(1981)1, pp. 58-65.

PART SIX

Service Engineering

Implementing the Service Concept through Value Engineering

Alastair Nicholson[1] and Katarzyna Zdunczyk[2]

1) *London Business School, Department of Operations and Technology Management, Sussex Place, London NW1 4SA, United Kingdom.*
   Email: anicholson@london.edu.
2) *Warsaw University of Technology Business School, ul. Koszykowa 79, 02-008 Warszawa, Poland.*
   Email: k.zdunczyk@business.edu.pl.KAP

Abstract: As products have become commoditised and technology knowledge made available to all, the possession of customers has become crucially dependent on the service content of what is delivered. Initially service was seen as an add-on aspect of business and to be highly personalised to customers and employees alike. But the significance of service to justify price, to support brand, to enrich the product, even to learn about the customer, means that the service concept and the service delivery must be managed even if such management is much less well defined than physical product and process management. This paper sets out to establish the possibility of designing both the content of service value and the service system which is needed to provide it. Following the analysis of service itself, the paper identifies the tools needed to engineer the service system to inspect it and position it within the commercial scheme. The McDonald's operation and others will be used to illustrate the concepts and their application. The principle of effective design and evaluation will be stated and defended as the fundamental means of achieving the value of service to both clients and the providing company. The conditions for successful adoption of the principles will be identified.

Key words: Service delivery system, Service design, Service value, Service process engineering

1. THE CHALLENGE OF SERVICE ENGINEERING

Most Western economies are now called "service" economies. As pointed out by FITZSIMMONS (1994), the percentage of employment in service roles increased to 78 % by 1992. The service economy, now becoming extended to the experience economy, is dominated by organisations which process customers rather than materials. This provokes a new set of challenges:

- of how to define what customers expect and want when that definition depends on the current circumstances which the customer finds themselves in;
- of how to provide the service economically (profitably) when only a few parts of the service can be charged for;
- of how to manage employees who are the fundamental service providers in a situation where we want both consistency overall and individuality as needed.

In fact, the challenges are not new. The great body of human work in engineering companies has always been service – supplied internally or externally in information, discussion, and interpretation. It is the formal accounting system which does not classify outputs into service content and making content. Now most of the banks and call centres are much more like factories in their behaviour than manufacturers who are keenly engaged in design discussion and after sales service.

What is new is the idea of managing service in a systematic way rather than in an ad-hoc way. The significance and pervasiveness of service makes it critical that it is managed and that requires a "model" of how it works, how the trade-offs involved are resolved, and how it leads to success. The starting point is to recognise service provision in terms of two loosely coupled systems:

- the customer satisfaction system;
- the service delivery system.

Unlike product manufacture (or product provision) the service system requires that satisfaction and provision take place concurrently and against undefined specification. In product provision, the terms of what is provided are defined in advance and any failure is debated around a specification which is broadly "on the producer's side". In the service system the customers move from expectations to satisfaction on their own terms. Although we may estimate both expectations and satisfaction, they are at best guesses.

The service delivery system is not a making process: it is a context within which the customer can acquire the elements of service wanted. Physically, it is buildings, atmosphere, and responsiveness of employees. It is an offer; it requires specification without knowing how precisely it could be specified. The art of service engineering is to steer the customer towards the service delivery system and steer the service delivery system towards the customer. It is an art rather than a science – but it is by implication a design and it can be monitored. The next two sections will explore these two interrelated transformations.

2. GIVING MEANING TO THE SERVICE VALUE

Although the customer may regard the service delivered in rather abstract terms such as responsiveness, courtesy, competence, etc., these attributes of service value must be acquired through the experience which customers have in passing through the service delivery system. They will often be acquired subconsciously but nevertheless valued for it. The service provided must therefore translate the specific encounters during that experience into factors which can be created and managed.

Taking the familiar example of McDonald's fast food chain, they stated their intention of delivering a service concept (SLACK et al. 1998). The concept was embodied in four criteria: *quality*, *service*, *cleanliness* and *value*. These are general terms but the ideas are vivid. By *service* the McDonald's organisation understood speed. Of course, representing a fast food chain, speed is of the essence. But by delivering more than speed they were adding to the total service package being delivered. They were adding *quality* – a necessity in food delivery, *cleanliness* – which is a surprise to the customer, and *value* – a more general term meaning, one suspects, getting a satisfaction at which the price is sufficiently low not to be questioned.

Some of these criteria could be discussed with customers, like the precise meaning of cleanliness: no litter, clean windows, clean uniforms. But McDonald's had no intention of pre-specifying what *cleanliness* meant to the customer. Cleanliness was something which satisfied and pleased as part of the experience. The required "outputs" from the service system are "qualities" which create the end customer value which, in turn, makes up the satisfaction.

The "black box" of the service system must embrace the service value content by the way it can operationalise the service experience. The scheme is illustrated in Figure 1, which shows customer expectations being connected into satisfaction by the way the mechanism of the service delivery system works.

Let it be stressed again that we have little direct control over either expectations or satisfaction. Both are unknown. Of course, we can influence the expectations through marketing – certainly for new customers – and we can apparently analyse satisfaction through questionnaires. But both these connections are suspect. However much we try, individual customers will misunderstand the marketing statement and customer feedback is notoriously biased towards complaints and not compliments. Therefore we have to take a view on how to identify and organise the service elements prior to designing the service delivery system.

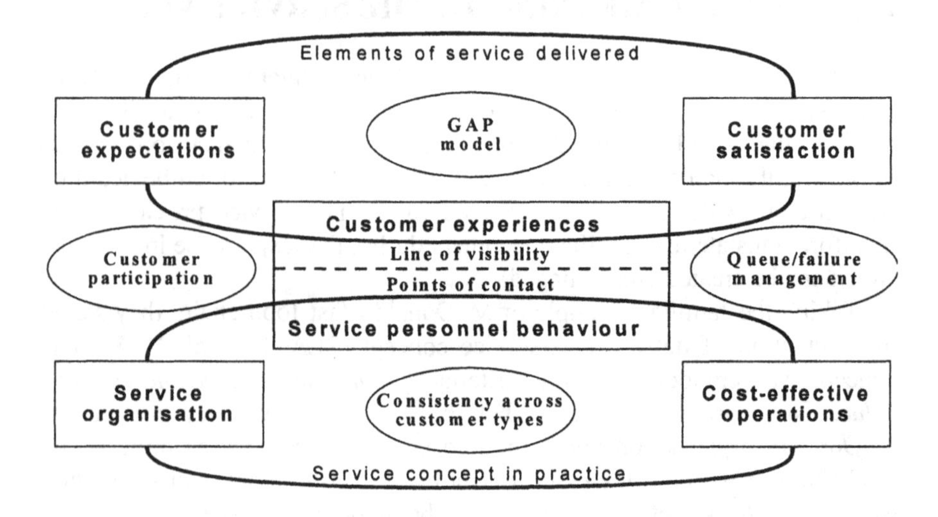

Figure 1. The service delivery system

3. THE CHALLENGE OF IDENTIFYING SERVICE REQUIREMENTS

The delivery of service satisfaction is much more obscure than the delivery of product satisfaction, which will be only approximately known at best. The requirement is to at least meet customers' expectations – expectations which are determined by previous experiences, and marketing statements. But here again expectations cannot realistically or sensibly be analysed in depth in advance. Thus the providers of service are caught in the trap of:

- engineering a service providing system
- without a definition of requirements
- which works on a concurrent creation and fulfilment of service needs

- capable of having standards but adapting to individual needs.

The first step towards defining requirements is to recognise all the elements which may form part of the satisfaction package. The package of value can be related to properties of the package such as: appearance, reliability, responsiveness, competence, and quality of the overall experience. The actual elements or encounters which create the impression and reality of that experience can then be identified in a list of factors which the service experience provides. The elements can now be classified into four categories: tangible, intangible (on whether the element is "acquired" directly by the customer); explicit, implicit (officially stated by the organisation or just assumed). See Figure 2 for an example set of elements from a hospital which specialises in hernia operations.

	Explicit	**Implicit**
Tangible	▪ Specialised in external hernia operations ▪ Unique technique with highest success rate ▪ Low total cost to patient	▪ Pleasant location and premises ▪ Common meals prepared from fresh ingredients ▪ Roommates chosen to have similar backgrounds/interests
Intangible	▪ Painless experience ▪ Quick recovery ▪ Patient physically and psychologically well throughout the experience.	▪ Socialising with other patients and staff ▪ "Country-club" atmosphere ▪ Outdoor and indoor activities available ▪ Patients treated like partners in the process ▪ Sense of belonging to a community

Figure 2. Elements of value – hernia hospital

The usefulness of making this classification is that it helps to identify all the elements which may contribute to satisfaction and it helps to link rather general abstract notions of satisfaction into specific characteristics of the service system. But the more fundamental advantage is that it provides a structure within which the organisation can manage the transformation of expectations into satisfaction. The elements which form the tangible and explicit are classically the factors which determine the official statement called "the service offer": the declared promise to the customer.

The tangible, implicit elements will typically be forced upon the organisation by circumstances and the intangible, implicit will be forced upon the organisation by competitors' practices. However, the transformation of expectations into satisfaction will be mostly achieved through the intangible, implicit elements which capture the mind of the customer as additional value wherein the whole "value for money" deal is fulfilled in a way which is private to both parties.

The challenge now is to create a service delivery system – a context and a set of working practices through which the satisfaction is generated.

4. THE CONSTRUCTION OF THE SERVICE DELIVERY SYSTEM

The differences between organisation of products and organisation of service is equally significant. For product manufacture we define technical processes for creating the object to tolerances which can be agreed and research matching options for reliability, volume and speed. The same broad results are being achieved, but in the service system we are not buying the machinery to achieve the transformation, we are creating a context within which the transformation can happen of its own accord. We are not processing the physical object to create a form or shape with properties which in use will serve a purpose. It is the perspective of the customer him/herself which is being transformed and which is the material, the machine and the end result!

Organising the conditions for successful service transformation is enabled through several major influencing variables:

- the facilities themselves, their décor, and their use of space,
- the employees, their standards, behaviours and responsiveness,
- the use of technologies and information, and
- the presence of other customers – both negatively and positively.

Organising the service system for effectiveness as a design for conditioning the transformation has several key technical properties which can be managed and indeed designed to create success:

- selection of appropriate "points of contact" where customers and the organisation interact,
- provision of the line of visibility, line of interaction and line of intervention where the customer can penetrate the organisation as appropriate,

- customer participation where the customer can become part of the contribution to the process,
- channelling of customers for differentiation of needs and effectiveness, and
- locating the points where queues of customers may develop and planning the means of avoiding queues (demand-supply management) or addressing the queue as a part of the experience.

Few organisations have pursued these distinguishing factors for designing effective service systems despite the broad recognition of the importance of service. But perhaps most significantly it is precisely through these factors that the current practices of an organisation can be investigated.

We will illustrate some of these themes as they apply to two organisations – McDonald's fast food and the hernia hospital referred to earlier. But first it is worth considering the steps which the customer (or customers) are likely to go through. This enables the recognition of the opportunities for value acquisition and service delivery. Such diagrams tracking a customer through service look no different from the process flow diagrams used for engineered products. See for example Figure 3 which shows the process flow of a patient through the hospital treatment for a hernia.

But such a process flow is not the design, it is the analysis of steps which the customer makes in using the design. The design is the physical layout, the atmosphere, the environment, the selection of points for employee encounter, the signage and the assisting technologies. This provides the architecture of a potential working design. And that design may have to accommodate very different customers, i.e. the same facilities may have to be viewed very differently by different groups. This is illustrated in Table 1 which shows how the facilities in a hotel may be recognised very differently by different customer groups. The process flow diagram thus is not a design but a means of knowing what possibilities of encounters exist. We do not derive the 'product', i.e. service value from this: we derive the possibilities for the customer to recognise the service.

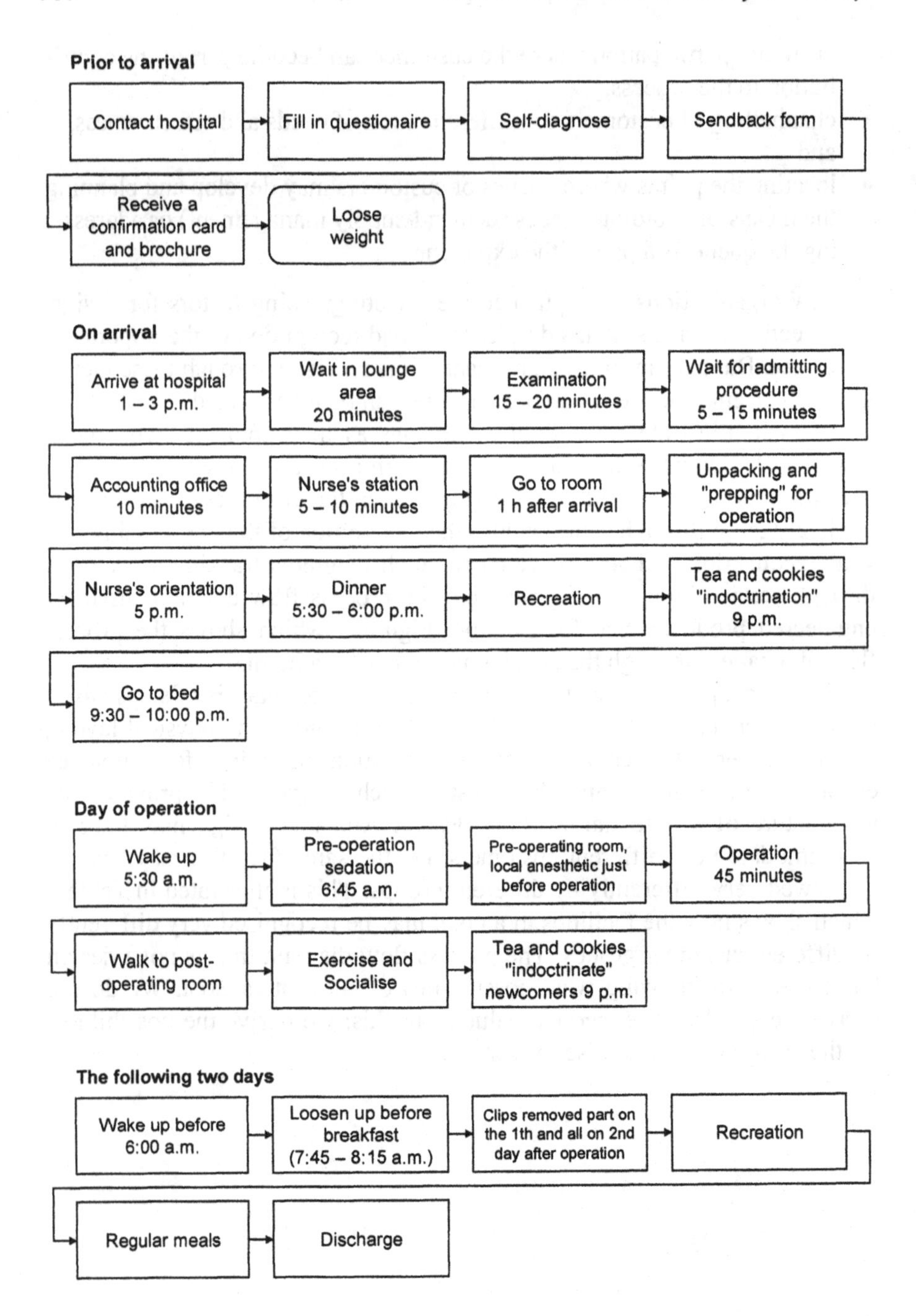

Figure 3. Customer progression through the system – hernia hospital

Table 1. Hotel service as viewed by different customer groups

Process	Business traveller	Wealthy tourist	Romantic break	Function room
Telephone to make reservation	Efficiency, Interested in corporate rates	Polite services	Special weekend rates, Complimentary champagne	Precise details of room specification
Arrive at hotel	Easy/fast access from road/rail/air	Easy/fast access from road/rail/air	First impressions important	Clear sign-posting
Check-in	Efficiency, Speed	Polite service, Helpful staff	Polite service, Helpful staff	Uninterested
Go to room	Well laid out, Executive desk. Bar room	Luxurious room, Bar in room	Nice setting, Good view, Bar in room	Clear sign-posting, Well set out room
Eat in restaurant	High quality food with speedy service	Quality food	Quality of experience	Food served hot, Quality experience
Go to business centre	Very interested, Fully serviced	Uninterested	Uninterested	Uninterested
Go to sleep	Comfortable bed, Quiet	Comfortable room, Quiet	Nice setting, Quiet	Uninterested
Check out	Efficiency, Instant attention	Polite service, Speed	Polite service, Speed	Uninterested

5. EXAMPLES OF SERVICE PROCESS ENGINEERING

Two examples will illustrate the way in which we handle the gap between the service delivery perspective and the customer satisfaction perspective. The presentation of that gap was vividly defined by the *gap model* for service delivery developed by PARASURAMAN et al. in 1985 (see Figure 4). The gap model illuminated how the distance between customer expectations and customer satisfaction is inevitable through the way in which unknown expectations have to be converted into specifications which in the course of delivery become distorted by events and communications.

It might appear that the gap model provides impossible knowledge barriers to be overcome. However, interpreted correctly, the gaps are inevitable: they are there to be recognised, developed and exploited by the service system. In an engineering sense (a six sigma sense) the gaps propose impossibilities. In a service sense they are the magic of the process if they can be recognised and coped with. The machines of the service system must work the passage of the customer through the system so that service and productivity are both being provided for without compromise.

McDonald's achieve this with remarkable skill. The layout and organisation of the space demonstrate cleanliness and flow. The counter provides the one point of contact (and hence economies) of the system, making the queue visible from both the customer's and the provider's perspective. The visibility of the kitchen provides conspicuous evidence of cleanliness to the customers as well as access for the managers. To support the speed of the process the employee manning levels on the shifts are chosen to cope with peak loading to maintain employee tempo to match the culture of speed assumed by the customer. In the detailed process organisation no excess movement in needed. All this in accordance with the McDonald's proclaimed standards of *"quality, service, cleanliness, value"*.

We thus reach a "requirement" of the service system – defined as the status or conditions of the delivery system – which can deliver satisfaction to customers consistently, possibly differentiated for different groups, in a reliable and profitable way!

It seems to be an insurmountable task to design a service delivery system that would deliver on the apparently conflicting, and each in itself difficult to achieve, requirements of transforming the customer expectations (which we do know precisely) into customer satisfaction (which is private to each individual customer) at a profit to the organisation. As we will demonstrate with examples, the task of designing an effective service delivery system, far from being a futile effort, does require a rare combination of engineering capability with a high tolerance for the uncertain.

What is needed is a careful consideration of the role of each of the service system elements at our disposal within the general context of a clearly specified service concept as translated into attributes of customer satisfaction. Furthermore, to strengthen the overall effect, it is necessary to identify and utilise any possibilities for interplay between the different elements of the system. By following this logic it is possible to arrive at a model of the service operation capable of delivering on all the requirements placed on it.

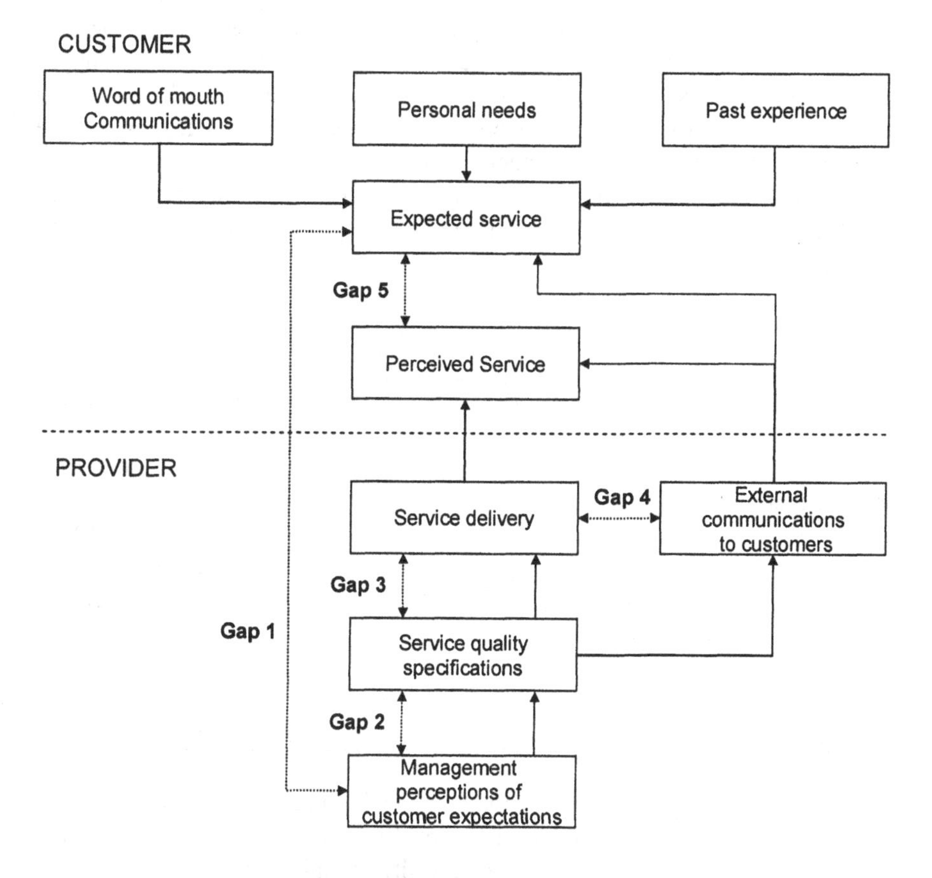

Figure 4. The Gap Model
(Source: PARASURAMAN et al. 1985, p. 43)

In the hospital system the design of adjacent facilities for operations and the involvement of all the staff in all processes unify the interfaces and motivations of both customers and employees. The support for the "country club" feeling in the hospital is underpinned by a rigid timing of procedures and encounters from entry to exit in which all uncertainties for the patient are removed by procedure itself thus removing an suppressing the principal customer worries. Queue management in both examples was avoided by "design". In McDonald's the queue is handled by rapidly adjustable staffing and flexibilities and always believing that it was preferable for the staff to wait (but not relax) than the customer (see Table 2). In the hospital queue control was managed at input to the facility by the management and pre-screening of arrivals.

Queue management, in its more general sense, takes the form of balancing the demand placed on the organisation by the customers with the supply capabilities (assets, people) of the organisation. Action can be taken on either side or both sides simultaneously to manage the trade-off between the quality of service provided as specified in the service offer the organisation communicates to the customers and the productivity requirement representing the pressure to serve the maximum number of customers and increase revenue.

Table 2. McDonald's Staffing

	Number of people	Grill	Windows	Drive-thru	Bin	Fry	Floaters	$ per hour volume guidelines
Minimum to open:	4	1	1	1	-	-	1	120
	5	1	1	1	-	-	2	150
	6	2	2	1	-	-	2	180
	7	2	2	1	-	-	2	210
	8	2	2	2	1	-	1	240
	9	2	2	2	1	-	2	275
	10	3	3	2	1	-	1	310
Fully staffed:	17	5	5	3	1	1	2	645

This is a delicate balance to manage, one which requires the separation of the concepts of revenue generation and profitability to the organisation, and a further distinction between the profit to the organisation and the 'profit' or value received as perceived by the customer. It is the concern with the maximisation of "profit" to both parties that ensures the success of capacity management efforts, as is well illustrated by the example of the hernia hospital's careful consideration of the implications of demand and supply policies on both the net profit to the organisation and the preservation of the service concept (see Table 3).

6. IDENTIFYING SUCCESS IN THE DESIGN OF A SERVICE

The successful linking of the service delivery system with the customer experience system has been proposed as a careful interweaving of customer processing and facilities architecture as worked by employees. We have proposed that despite the indirect nature of the relationship, it is still advantageous to approach the service provision through a "design" approach. The broad pattern of the links is

Service promise	→ Service experience	→ Value for money
Service system design	→ Service delivery by employees	→ Cost-effective outcome

These two streams of translation must work coherently to mutual advantage whilst being fundamentally different! The push for cost effectiveness will classically inhibit service delivery. Equally, the demands of a customer if fulfilled to the letter will destroy the economies of the operation. Success is therefore not at all measured by "profitability". It is not measured simply by sales and the trends on costs. Success means that the design of the system, as worked by the employees and as experienced by the customers, is in fact economic and efficient to both parties. Nor should satisfaction be measured by customer complaints or employee contribution be measured by hours. The real test of the service system requires a view of the right working of the service provision to be defined and this view to be testable and tested.

This reverses the classic role of service value as embodied by customers. It proposes that there is a view on the means through which the service system can deliver the elements of satisfaction and the appraisal by the organisation of success should be measured in terms of the ability to meet those targets. It means that we measure our ability to inspect the design on the one hand and are able to explain "profit" for the customer separately from profit for the service provider.

Such a position is well illustrated in McDonald's, where a store inspection sheet verifies that:

- evidence that the service value elements can be delivered is provided in a store visit report (see Table 4);
- the planning of the staffing system is proven by the ability to forecast;
- the provision of assets, facilities, and space is never the restriction on capacity requirements;
- the profitability of the delivered items is ignored in favour of an explanation of how relative value-added and cost is perceived by the customer as the basis for pricing.

Similarly, in the case of the hospital, "inspection" was of the ability to do the job, not on the outcome and the profitability for the patient was evaluated alongside profitability for the hospital.

Table 3. Managing demand and supply – the hernia hospital.

DEMAND	SUPPLY
▪ Seasonal variation through the year	▪ Reservation system based on scheduling according to variations in demand throughout the year
▪ No effort at promoting off-peak demand	▪ Slack maintained in the system - no more than 4 operations per day, - no weekend operations, - closed for two weeks.
▪ Constant backlog independent of supply capabilities	▪ Careful scheduling, customer participation, flexibility of employees.
▪ Word-of-mouth advertising only – any form of advertising might result in too much demand being generated	▪ Maintaining the "holiday resort" atmosphere sets limits to adding extra capacity.
▪ Reasonable rates: The Hernia Hospital: Average charges: $ 954 + av. round trip fares: $ 200-$ 600 + time to return to work: 1-4 weeks Other hospitals: Charges: $2000 - $4000 Time to return to work: 2–8 weeks	▪ Quick and efficient progression through the system: - Customer participation: ▫ self-diagnosis prior to arrival ▫ self-preparation for surgery ▫ patients indoctrinate patients - The unique hernia operation method - Strict schedule

Service can thus be provided through design principles and the success of the service design should be judged against the inputs of design intentions, not simply the outputs realised. This provides service engineers with a rich set of aims in all organisations in searching for the appropriate ways to deliver the vaguer elements of service as distinct from production.

When this is done well, the real value to the customer will be recognised by driving up demand and demand predictability through repeat buying and justifying higher prices. Equally, on the costing side we must only consider the total service system cost in valuing the provision on service and not dis-aggregate the elements of delivery and elements of cost into a fragmented analysis which will prevent understanding of the total experience engaged by the customer.

Table 4. McDonald's: Store visitation report (extract)

Service	Points
1. Management in backup position, following proper procedures, expediting service	3
2. Service atmosphere – smiles, hustle, teamwork, neat appearance	3
3. Greeting, order assembly, presentation, thank you, and asking for repeat business	3
4. Service accuracy – order accurately filled, change accurately made	3
Cleanliness, Merchandising, Outside	Points
1. Neighborhood – free of litter (1 block in each direction)	1
2. Landscaping – free of litter and well maintained	1
3. Painted surfaces – in good condition and clean	1
4. Parking lot – free of litter, clean, seal and stripe in good condition	1
5. Waste receptacles – clean, good repair and emptied as needed	1
6. Sidewalks – clean, sealed, and free of hazardous conditions	1
7. Windows and doors – glass and all associated areas clean	2

REFERENCES

FITZSIMMONS, J.A.; FITZSIMMONS, M.J.:
Service Management for Competitive Advantage.
New York, NY: McGraw-Hill, 1994.
JOHNSTON, R.; SLACK, N. D. C. (eds.):
Service Operations: The Design and Delivery of Effective Service Operations.
Warwick: Warwick Printing, 1993.
JOHNSTON, R.; CLARK, G.:
Service Operations Management.
London: Financial Times Prentice Hall, 2001.

LOVELOCK, C. H.:
Managing Services.
Upper Saddle River, NJ: Prentice Hall, 1992.
PARASURAMAN, A.; ZEITHAML, V. A.; BERRY, L. L.:
A conceptual model of service quality and implications for further research.
In: Journal of Marketing,
Chicago, IL, 49(1985)3, pp. 41-50.
SLACK, N.; CHAMBERS, S.; HARLAND, C.; HARRISON, A.; JOHNSTON, R.:
Operations Management.
London: Pitman, 2nd ed., 1998.

Service Management in Production Companies

Alexander Karapidis
Fraunhofer Institute Arbeitswirtschaft und Organisation, Nobelstrasse 12, D-70569 Stuttgart, Germany.
Email: Alexander.Karapidis@iao.fhg.de

Abstract: "Services are going to move in this decade to being the front edge of the industry" (Ante, Sager 2002, pp. 66-72).
In the past, with regard to product development production companies focused mainly on the intermediate product engineering process. In comparison, the service processes within the creation, modelling and marketing phase of the product development are less standardised. Moreover, the service shares in hybrid products are often hand-made styled and also less standardised. Service Management in production companies is focussing on customer expectations, service standards, service marketing and service communication. Especially the fit for service engineering approach helps to bring the service engineering process on a level of systematic standardisation comparable to the product engineering process. It gives answers how the partly "soft factors" driven service processes can be controlled by a strategic management approach. In this fit for service engineering approach the competence card is a key instrument. It helps to structure, measure and evaluate the service processes and gives organizational hints which skills should be improved for a better service engineering performance of production companies.

Key words: Service management, Service engineering

1. PRODUCT ENGINEERING vs. SERVICE ENGINEERING - ARE TWO WORLDS COLLIDING?

Product life-cycles are getting shorter, cost-value ratio is going down, products are getting increasingly complex – these are three among many reasons as an outcome of a highly competitive global market. Today, even complex products can be copied fast and from many companies world-wide. As an outcome of this development, unique selling points are much harder to find than in the past. The strategy to have intelligent and high effective and efficient processes in production engineering seems to be a key competence for companies to reduce the time-to-market, hidden costs and to offer high-quality products (see MILLER, VOLLMANN 1985, pp. 142). So even in traditional production companies significant services are required to support the physical products.

To recognize the differences between a product and a service, we first look at a short definition of service:

> "Services includes all economic activities of companies, whose output is not a physical product."

Product	Service
More tangible Like the package delivery of the TV	**More tangible** Like a TV
Storable TV's are storable	**NOT Storable** TV delivery is not storable
It is generally NOT bound to a specific date TV's can be bought every day	**Generally consumed at the time it is produced** The package delivery is only consumed when the TV is delivered
It is NOT bound to a person TV's can be made without a customer	**It's bound to a person** The package delivery can only be when there's a customer who bought the TV

Figure 1. The differences between service and product

But it is much easier to understand, in comparing attributes of products and services (see figure 1). Some services are critical factors for success and most of them strengthen the value of the product (BRÄNNSTRÖM, ELF-STRÖM, THOMPSON 2001, p. 1). As a result, it is difficult to draw a distinction between products and services, since nearly all products include services which are vital for their value (NORMANN, RAMIREZ 1994). Physical products and services usually meet at two levels: In product development (e.g. construction, marketing, data processing) and in the support of physical products (e.g. hybrid products = computer and support for 2 years).

Thus it is not a scientific debate about physical products vs. services but a fact that the integration of physical products and services into service packages are used to increase the competitiveness (see BULLINGER, SCHEER 2003, pp. 26). The VDMA (the federation of the engineering industries) stresses that many companies carry a high yield from services and not from their products. Unfortunately, as a result of several studies the importance of services in production companies is still underrated (e.g. EGGERS, WALLMEIER, LAY 2000).

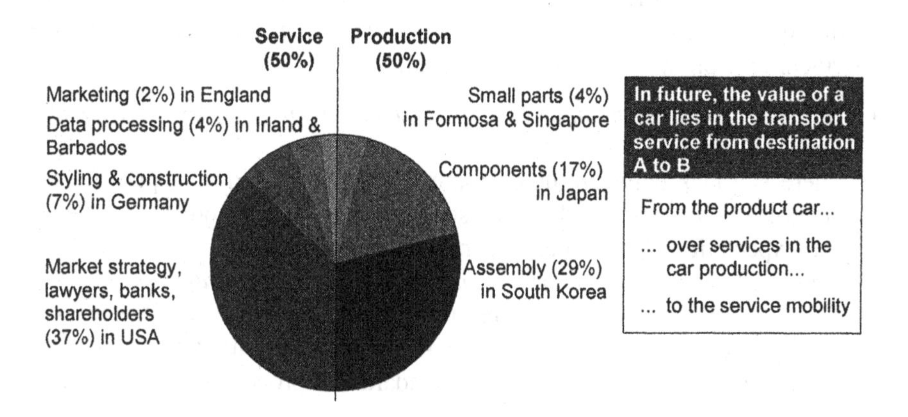

Figure 2. Service parts in car production (REICH 1996)

Nevertheless, this assumption raises the question, why production companies have on the one hand well-developed tools and methods to design products but often have hand-made processes to design services on the other hand.

Various methods and procedures have been worked out and developed properly for physical products. Standardisation aspects underlie a worldwide permanent and continuous process. The measurement of product engineering processes is well worked-out. On the other hand, almost every production company has a large fraction of service parts, not only in administration, distribution or marketing departments, but also in core production areas.

Despite many approaches, it is very difficult to bring the "two worlds" production and service together. Especially hybrid product development and engineering stresses the difficulties when a high standardised and systematic product development meets hand-made styled product services.

2. SERVICE PERFORMANCE MANAGEMENT IN PRODUCTION COMPANIES

In the last years, activities in service development in production companies have been dominated by marketing, quality management and customer relationship management. But these activities cover only some issues in improving the service performance of tourism companies.

ZEITHAML and BITNER (2003, pp. 31) stressed in their gap model of service quality four points which are the most important issues for service providers to avoid a low service performance:

- *Gap 1: The management of a company does not know the expectation of their customers*
  There is a lack between the customer expectations and perceptions. Production companies should offer services that satisfy consumers' needs and expectations but also ensure their economic survival.
- *Gap 2: The provision of services are configured inadequately*
  The expectations of clients are not satisfied, because of services not appropriate to customers' desires. This mismatch results from a shortage of standards in service design and standardized processes – especially in dealing with intangible assets.
- *Gap 3: The information about services does not correspond with the effective achievement*
  There is a gap in connecting the service design with the actual service performance by company employees and the organisation of work. This brings companies to contradictory situations between customers' needs and employees' service provisions. As a result, employees often have to act beyond organisations rules to satisfy customers' needs with the effect, that high service quality is difficult to reproduce. Consequently, customers are not satisfied and employees permanently have to go beyond their capacity (NELSON 1970, p. 311; GROVE et al. 2001, p. 84).
- *Gap 4: The assurance of the provisions of services does not correspond with the result*
  There has to be a fit between the external communication to customers and the service delivery of companies i.e. appropriate marketing communication especially with regard to cost-performance ratio concerns.

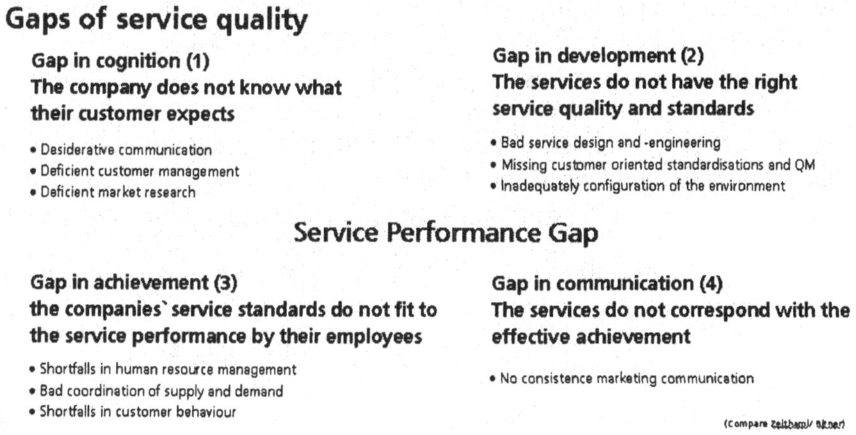

Gaps of service quality

Gap in cognition (1)
The company does not know what their customer expects

- Desiderative communication
- Deficient customer management
- Deficient market research

Gap in development (2)
The services do not have the right service quality and standards

- Bad service design and -engineering
- Missing customer oriented standardisations and QM
- Inadequately configuration of the environment

Service Performance Gap

Gap in achievement (3)
the companies` service standards do not fit to the service performance by their employees

- Shortfalls in human resource management
- Bad coordination of supply and demand
- Shortfalls in customer behaviour

Gap in communication (4)
The services do not correspond with the effective achievement

- No consistence marketing communication

(Compare Zeithaml/ Bitner)

Figure 3. Service Gap Model

For these reasons, most of the new service performance management methods and tools are focussing on the following points to bridge the gaps:

- The translation of visions and strategies in measurable factors of success seen by stakeholders and clients.
- The combination of key data for material and intangible assets.
- The provision of instruments to evaluate and improve the success of a company.
- The empowerment of managers and staff.
- The implementation of change in organizations.

The awareness that the success of services in production companies depends on the conceptual design and configuration is not very widespread. That implicates to see services as service products which have to be developed and managed like products as in e.g. the automotive industry. So, the question is to find methods, tools and instruments to close these gaps.

3. SERVICE ENGINEERING - CLOSING THE GAP BETWEEN STANDARDISED PRODUCTS AND HAND-MADE SERVICES

One argument for not focussing on standardised and systematic service design is that the production companies are traditionally strongly focused on their products. So, the attitude of the management towards service development and engineering departments is weak or even non-existing. Another reason for neglecting standardisation lies in the lack of knowledge of service design concepts. However, HOLLINS and HOLLINS (1991), RAMASWAMY (1996) and BULLINGER and SCHRAINER (2001) have developed substantial applied concepts, studies and tools for professional service development. FÄHNRICH (1999, pp. 18) stressed that the transformation from traditional product development methods is unsuitable to develop services. Moreover, interdisciplinary methods are necessary which combine human resources, organisational aspects and technologies. In addition to that, a formal order for steps in the process of service development is represented by the service engineering concept applied by the Fraunhofer Institute for Industrial Engineering (IAO).

The aim of service engineering is to develop high quality customer-oriented services.

Advantages for production companies lie in (see MEIREN, BARTH 2002, p. 11):

- *Competition advantages*: Especially in turbulent markets innovative services are critical factors for success
- *Building-up business segments*: Service development as an operative link between customer needs and physical product development
- *Success of the product in target markets*: By respecting customer and market needs systematically in service development, the value of products raises
- *Time-to-Market*: By standardising service development processes time and costs for products decrease
- *Knowledge Management*: Product know-how and attributes can be better transferred between different development projects
- *Customer Relationship Management*: Some studies show that reasons why production companies lose clients mainly lie in problems with the service quality
- *Transfer of service know-how*: The service engineering concept lowers the barrier to transfer product-related services from one product to another

Some fields of application of the service engineering concept: FESTO AG & Co., traditionally a production company which produced components for automation, successfully applies the service engineering concept for their customer logistics service (Kanban Service). Lufthansa AG sets up service engineering projects in innovative services for e-procurement. Schenker Germany AG (transport & logistics) worked out customised services with the service engineering concept.

By using the service engineering concept the gap between production and service development has been closed. Product service innovation, service diversification and the development of product supporting services can be developed more professionally even if there are still open questions (see BULLINGER, MEIREN 2001, p. 163).

Service Development Roadmap

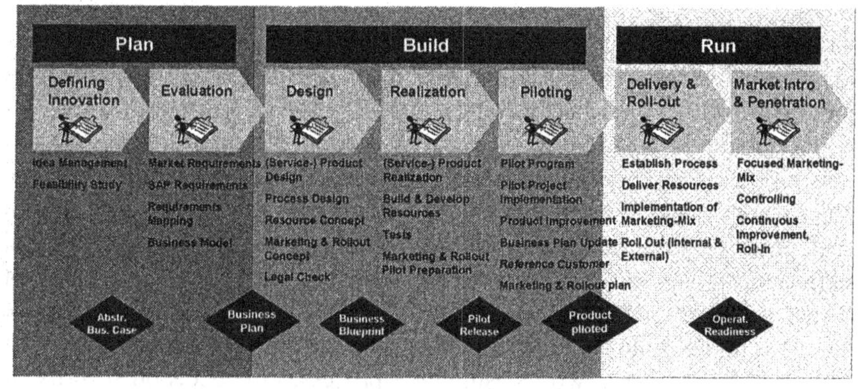

Figure 4. The service engineering concept
(Source: KARAPIDIS 2003, p. 325)

Exactly gap one and gap two are closed. The company does not know what the customers expect. So you need a standardisation and a professional market research. This is the first step en route to more service in production companies.

The second step includes closing gap two - the services do not have the right service quality & standards. Right service quality and standards can be realised with a joint customer engineering.

4. COMPETENCE CARD – MEASUREMENT AND MANAGEMENT OF SERVICE QUALITY AND SERVICE COMPETENCE

Fit for service engineering in production companies also means to start improvement processes for services continuously. A precondition is to install a monitoring system for service quality and processes. Benchmarking is a method production companies are familiar with for a long period of time. Unfortunately, benchmarking concepts from production cannot be transferred one to one to services – in particular, when we need benchmarks for a strategic approach. In the last years, the Balanced Scorecard and the EFQM-model broke the ice to include "soft factors" in the benchmarking topic (KAPLAN, NORTON 1997). The competence card used the strength of these concepts to allow a management and measurement for service quality and service competencies of companies (GANZ, TOMBEIL 2000).

Putting the competence card into practice means to structure strategies and to transfer these into action on production-related and service-related topics.

The work on the contents of a competence card which is specific to a particular field of growth is cascaded from one dimension to another in accordance with a uniform concept (see KARAPIDIS 2002, pp. 17).

Production-related topics will be measured on the control level. Organisational (service) competencies will be measured on the enabler level (see Figure 5) by ranking the status of the competencies on a stage model with qualitative degrees. Measures to develop and/or establish organisational competencies can be derived from the qualitative degrees which have not been reached yet. That means gap three is closed – the companies service standards do not fit to the service performance by their employees.

In combining these two measurement processes the decision security of companies rises (e.g. to improve production and service developments) and a human resource management can be improved, supplied and a demand coordination can be implanted in companies strategy.

Measuring Service Performance

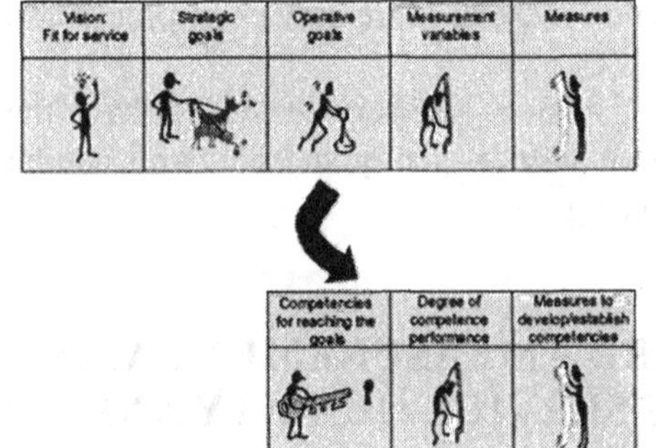

Control Level:
Structuring and balancing strategies from different perspectives and transferring these into action.

Enabler Level:
Structuring and balancing organisational competencies which are critical factors for success for an organisation in handling market-specific requirements.

Figure 5. The Competence Card

5. SUMMARY AND FORECAST

Service management in production companies means a paradigm shift from a customer's point of view: Yesterday, customers bought a product, today they buy a hybrid product to solve problems they have. From a producer's point of view, a production company firstly develops into a service production company which differs in services to the competitors. Its service offers are increasing and the company then yields a growing amount of money with services. Finally, service production companies will develop into "producing service companies" which offer customer-oriented hybrid solutions and integrated products. The service engineering concept and the competence card are the key methods to significantly increase the service quality and the service productivity of production companies as well.

Yesterday a customer bought a product, today they buy a solution that consists of different services around and within products.

REFERENCES

ANTE, S. E.; SAGER, I.:
IBM's New Boss".
In: Business Week,
New York, NY, 11 February 2002, pp. 66-72.

BRÄNNSTRÖM, O.; ELFSTRÖM, B.-O.; THOMPSON, G.:
Functional Products create new Demands on Product Development Organisations.
In: Design methods for performance and sustainability: 13th International Conference on Engineering Design – ICED'2001.
Ed.: CULLEY, S.
Bury St. Edmunds et al.: Professional Engineering Publisher, 2001.

BULLINGER, H.-J.; MEIREN, T.:
Service Engineering – Entwicklung und Gestaltung von Dienstleistungen.
In: Handbuch Dienstleistungsmanagement.
Edts.: BRUHN, M.; MEFFERT, H.
Wiesbaden: Gabler Verlag, 2001, pp. 149-175.

BULLINGER, H.-J.; SCHEER, A.-W. (edts.):
Service Engineering: Entwicklung und Gestaltung innovativer Dienstleistungen.
Berlin, Heidelberg et al.: Springer Verlag, 2003.

EGGERS, T.; WALLMEIER, W.; LAY, G.:
Innovationen in der Produktion 1999. Dokumentation der Umfrage des Fraunhofer-Instituts für Systemtechnik und Innovationsforschung.
Karlsruhe: Fraunhofer ISI, 2000.

FÄHNRICH, K.-P.; MEIREN, T.; BARTH, T.; HERTWECK, A.; BAUMEISTER, M.; DEMUß, L.; GAISER, B.; ZERR, K.:
Service Engineering. Ergebnisse einer empirischen Studie zum Stand der Dienstleistungsentwicklung in Deutschland.
Stuttgart: Fraunhofer IAO, 1999.

GANZ, W.; TOMBEIL, A. S.:
Von der Balanced Scorecard zur Competence Card als strategisches Umsetzungsinstrument für Dienstleistungsbenchmarking.
In: Benchmarking 2000 in der Dienstleistungswirtschaft.
Edts.: MERTINS, K.; KOHL, H.; HEISIG, P.; VORBECK, J.
Berlin: Frauenhofer IPK, 2000.

GROVE S. J.; Fisk R. P.:
Service theater: An analytical framework for service marketing.
In: Service marketing.
Ed.: Lovelock, C.
Englewood Cliffs: Prentice Hall, 2001.
HOLLINS, G.; HOLLINS, B.:
Total Design: managing the design process in the service sector.
London: Pitman Publishing, 1991.
KAPLAN R. S.; NORTON, D. P. (edts.):
Balanced Scorecard: Strategien erfolgreich umsetzen.
Stuttgart: Schäfer-Poeschel, 1997.
KARAPIDIS, A.:
Die Competence Card umsetzen – exemplarische Vorgehensweise aus zwei "Fit for Service"-Clubs.
In: Fit for Service Report 2002, Dienstleistungsbenchmarking (german report) - Service-Benchmarking (english report).
Edts.: GANZ, W.; HOFMANN, J.
CD-Rom.
Stuttgart, 2002.
KARAPIDIS, Alexander:
Fit for Service Engineering in Production Companies.
In: Human Aspects in Production Management.
Eds.: ZÜLCH, Gert; STOWASSER, Sascha; JAGDEV, Harinder S.
Aachen: Shaker Verlag, 2003, pp. 322-329.
(esim – European Series in Industrial Management, Volume 5)
MEIREN, T.; BARTH, T.:
Service Engineering in Unternehmen umsetzen. Leitfaden für die Entwicklung von Dienstleistungen.
Stuttgart: Frauenhofer-IRB-Verlag, 2002.
MILLER, J. G.; VOLLMANN, T. E.:
The hidden factory.
In: Harvard Business Review,
Boston, MA, 63(1985)5, pp. 142-150.
NELSON, P.:
Information and Customer Behaviour.
In: Journal of Political Economy,
Ort, 78(1970)20, pp. 311-329.
NORMANN, R.; RAMIREZ, R.:
Designing Interactive Strategy. From Value Chain to Value Constellation.
Chichester et al.: Wiley, 1994.
RAMASWAMY, R.:
Design and Management of Services Processes: keeping customers for life.
Reading, MA, Wokingham: Addison-Wesley, 1996.
REICH, R.:
Der US-Arbeitsmarkt: Erfolge und Herausforderungen.
In: Die Zukunft der Dienstleistung.
Edt.: MANGOLD, K.
Frankfurt/M.: Frankfurter Allgemeine Zeitung, 1997, pp. 54-58.

ZEITHAML, V.A.; BITNER M. J.:
Service Marketing. Integration customer focus across the firm.
New York, NY: McGraw-Hill, 2003.